# TEGULAE

## Manufacture, typology and use in Roman Britain

Peter Warry

BAR British Series 417
2006

Published in 2016 by
BAR Publishing, Oxford

BAR British Series 417

*TEGULAE*

ISBN 978 1 84171 956 6



BAR Publishing is the trading name of British Archaeological Reports (Oxford) Ltd.
British Archaeological Reports was first incorporated in 1974 to publish the BAR
Series, International and British. In 1992 Hadrian Books Ltd became part of the BAR
group. This volume was originally published by Archaeopress in conjunction with
British Archaeological Reports (Oxford) Ltd / Hadrian Books Ltd, the Series principal
publisher, in 2006. This present volume is published by BAR Publishing, 2016.

Printed in England

BAR
PUBLISHING

BAR titles are available from:

BAR Publishing
122 Banbury Rd, Oxford, OX2 7BP, UK
EMAIL info@barpublishing.com
PHONE +44 (0)1865 310431
FAX +44 (0)1865 316916
www.barpublishing.com

# CONTENTS

# LIST OF FIGURES

# LIST OF PLATES

# ACKNOWLEDGEMENTS

Since its brief flowering in the seventies and eighties, Roman Ceramic Building Material (CBM) has come to be regarded as a worked out area of research with little more to offer. In part this is a result of its focus on fabric analysis which has promised much and delivered relatively little. This book, although largely based on my doctoral thesis at the University of Reading, has its genesis in an undergraduate level assignment at the University of Oxford set by Maureen Mellor who perceived that there were greater possibilities in the study of CBM than appreciated. However I think she would be as surprised as me at how fertile and susceptible to analytical methods this worked out area of research has proved to be.

Perhaps the very abundance of CBM in most excavations has led to the assumption that it can be of little value but, as Pitt-Rivers proposed more than a century ago, the most common archaeological finds may also be the most important. Whatever the reasons, today CBM languishes as a Cinderella artefact unavoidably assigned to the darkest and least accessible parts of a curator's store and, by some immutable law, always with the heaviest boxes on the highest shelves. For almost all the collections I have examined in the course of this research, I have been the only person ever to do so, or at least the first person since Gerald Brodribb twenty five years ago. The curators have been intrigued by my unusual interest in CBM and have been unfailingly welcoming and helpful in locating and extracting these awkward boxes. My thanks must therefore firstly go to all the curators of museums and archaeological units around the country who have assisted in this enterprise: the project could not have been undertaken without them. Their names and organisations are listed on the opposite page. All of these organisations have allowed me to photograph their material and I acknowledge with thanks those that have been used in this book which are also listed opposite.

This book builds upon the pioneering work of Gerald Brodribb who established for the first time that there could be value in the study of ceramic building material. However, it is his work at Beauport Park that has been most influential in this thesis, as it represents the best evidence in this country for how tegulae were made and used. *Roman Inscriptions of Britain* Fascicules 4 and 5 have been a priceless source of data on stamped tiles and saved many months of work.

Chapter 2 on the manufacture of tegulae could not have been written without the assistance of Peter and Tony Minter of the Bulmer Brick & Tile Co Ltd who have tested my experimental moulds and provided invaluable criticism and suggestions on my developing thoughts. Professor Frere provided a detailed criticism of the section on Carpow and my argument has been enhanced by his excellent suggestions.

Many other people kindly helped me amongst whom I should particularly mention Ian Betts, Ian Caruana, Evan Chapman, Professor Barry Cunliffe, Peter Davies, Hella Eckardt, Roy Friendship-Taylor, Kevin Greene, Jill Greenaway, Mark Lewis, John Lucas, John Magilton, Professor John Potter, John Pybus, John Shepherd and Vivien Swan. My thanks are due to all of them and to the many others too numerous to mention individually.

I must especially thank Dr Janet DeLaine and Professor Mike Fulford who were the supervisors for my thesis and whose wise counsel and encouragement have helped this through to fruition. Finally, and most importantly, I must thank my wife who has not only done the proof reading but carried the burden of family, home and garden whilst I have been largely absent in archaeological stores or my study.

| **Organisations** | **Individuals** | **Acknowledgement of plates**<br>All photographs by the author |
|---|---|---|
| Arbeia Roman Fort and Museum | Alex Croom, Paul Bidwell | 2.26, 2.37, 2.38 |
| Aylesbury Museum | Brett Thorn | 2.4, 2.48 |
| Beauport Park Archaeological Trust | John Pybus, Pamela Corbett | 1.1; 2.3, 2.9, 2.11, 2.22, 2.25, 2.49, 2.52, 2.53, 2.56; 6.3, 6.4; 7.1, 7.5 |
| Bignor Roman Villa | Peter | |
| Bulmer Brick & Tile Co Ltd | Peter Minter, Tony Minter | 2.40 - 44, 2.46, 2.50, 2.51 |
| National Roman Legion Museum, Caerleon Photographs by permission of the National Museum of Wales | Mark Lewis | 2.10; 5.1, 5.2, 5.3 |
| Canterbury Archaeological Trust | Louise Harrison | |
| National Museums, Cardiff Photographs by permission of the National Museum of Wales | Evan Chapman | 2.7, 2.20; 4.3; 5.5, 5.6, 5.8, 5.9 |
| Tullie House Museum & Art Gallery, Carlisle | Tim Padley | |
| Chedworth Roman Villa | Mark George | 2.26 |
| Cheltenham Art Gallery & Museum | Julien Parsons | |
| Chester Archaeology | Alison Jones, Gill Dunn | 2.12, 2.19; 5.7, 5.8 |
| Grosvenor Museum, Chester | Dan Robinson | 4.1, 4.2; 5.5, 5.7, 5.8; 6.6 |
| Chichester District Museum | James Kenny, Ian Scrivenor-Lindley, Stuart Milby | |
| Colchester Museums | Paul Sealey | |
| Corbridge Museum | Sarah Lawrence | 2.29, 2.30 |
| Corinium Museum, Cirencester | Paula Gentle | 2.31 |
| Dorset Natural History & Archaeological Society at the Dorchester County Museum | Peter Woodward | 2.27; 6.1a, 6.1b; 7.6a, 7.6b |
| Dover Castle, English Heritage | Kevin Booth | |
| Dover Museum | John Iveson | |
| National Museums of Scotland, Edinburgh | Fraser Hunter, Jim Wilson | 4.5 |
| Essex County Store | Nick Wickendon | 2.21 |
| Exeter Archaeology | | |
| Exeter Museum | John Allan | |
| Fishbourne Roman Palace | David Rudkin, Derek Turner | 2.18; 6.5 |
| Frilford | Gary Lock, Chris Gosden, Sheila Raven | 2.17 |
| Gatehampton | Cynthia Graham Kerr | |
| Gestingthorpe Villa | Harold Cooper, Ashley Cooper | 2.14 |
| Gloucester City Museum & Art Gallery | Sue Byrne, Rachel Atherton | 5.11, 5.12 |
| Guildford Museum | Mary Alexander | 7.1 |
| Hampshire County Museums Store | Kay Ainsworth | 3.1 |
| Tolson Museum, Huddersfield | John Rumsby | 2.13 |
| Lancaster City Museum | Susan Ashworth | |
| Jewry Wall Museum, Leicester | John Lucas | 2.15, 2.23 |
| Leicestershire County Museum Store | Richard Pollard | |
| Liverpool Museum | Robert Philpott | 2.28; 4.4 |
| Museum of London Archaeological Service | John Shepherd, Ian Betts, Susan Pringle | |
| MAT Consultancy, Malton | Paula Ware | |
| Senhouse Museum Trust, Maryport | Ian Caruana | |
| Newark Museum | Glyn Hughs | |
| Newcastle University Museum | Lindsey Allason-Jones | |
| Norwich Castle Museum | John Davies | |
| Oxford Archaeology | Leigh Allen | |
| Oxfordshire County Store | Lauren Gilmore, Francesca Jones | |
| Piddington Museum | Roy Friendship-Taylor | 2.54, 2.55; 5.10; 8.1 |
| Portsmouth City Museum | Jenny Stevens | 2.5 |
| Reading Museum | Jill Greenaway | |
| Rotherham Museum & Art Gallery | Karl Noble | 4.6a, 4.6b; 7 |
| Silchester | Mike Fulford, Amanda Clarke | 2.2, 2.6 |
| Wakefield Art Gallery & Museum | Pam Judkins | 4.6c, 4.8 |
| Winchester City Museum | Geoff Denford, Robin Iles, Helen Rees | 2.1, 2.8, 2.16, 2.24, 2.36; 6.2; 7.3 |
| English Heritage, Wroxeter | Heather Bird | 7.1 |
| Le Yaudet | Barry Cunliffe, Emma Harrison | |
| York Minster | Louise Hampson | |
| York Museum | Elizabeth Hartley, Andrew Morrison | 5.4; 7.1, 7.2 |

Plate 1.1: Beauport Park 167 – tegula with stamp, signature and stepdown

# 1

# INTRODUCTION

## 1.1 Scope

Ceramic building material, particularly roofing material, is one of the most common finds on Romano-British sites, yet despite its abundance, it has been relatively little studied. Whole books have been devoted to relatively minor pottery types, but it is extremely rare for a book to devote as much as a single chapter to ceramic roofing material.[1]

This book is devoted to the study of ceramic roofing material, primarily tegulae. It considers how they were made and develops and dates a typology. It looks at the role of stamps and signatures and how these can inform the study of when and by whom the tegulae were made. It analyses how the tiles were fitted onto pitched roofs, how these roofs were constructed and proposes four stages in their evolution. It suggests that tegulae might also have been used on some vaulted roofs. Finally the logistics, costs and economics of tile manufacture and distribution are addressed.

The book follows a logical sequence considering first how tegulae were manufactured, next their typology and then their dating in order to prepare the ground for the subsequent chapters on stamps and roof construction. The final chapter before the conclusions brings all the evidence together to examine the economic and social data that can be derived from a study of tegulae. The early chapters are essential to the rigorous development of the argument and will hopefully be interesting in their own right to those who are curious as to how tegulae and imbrices were made and how the typology evolved. However for some readers it will be the organisational, constructional and economic information which can be gleaned from tegulae that will be most relevant. Such readers are advised to skim quickly through Chapter 2 (especially Section 2.3) and not to dwell too long on Chapter 3.

The study adopts an empirical approach to the evidence throughout Roman Britain but does not extend to the continent. As will be discussed in Section 1.4, fabric analysis of ceramic building material tends to be descriptive and has not generally yielded results that help with archaeological interpretation. Fabric by its nature is limited to local or regional associations and therefore generally defies any national typology. For these reasons this study primarily focuses on the *form* of tegulae, rather than their fabric, with the intent that the analysis should be *predictive* as well as descriptive. The objective is to cover all aspects of ceramic roof tile material in terms of manufacture, organisation, typology and use, albeit at the expense of depth in some areas. This research does not extend to the design of kilns which have already been well studied,[2] but does cover their organisation.

The results are built around a statistical analysis of the dimensions and features of the tegulae surveyed in this study; the conclusions will therefore be highly dependent upon this data. Whilst the quantity of data is sufficient to produce statistically valid results, the size of the database is inevitably much smaller than, for example, those that have been accumulated in pottery analysis. Many of the results should therefore be regarded as provisional, pending the accumulation of further data.

## 1.2 Data collected

Tiles from 104 separate sites have been examined and these are listed in Appendix 1. Insofar as is possible the sites have been selected to provide a uniform coverage of the whole of Roman Britain. Most of the sites are discrete; however, where isolated pockets of tiles have arisen from several sites within a locality these have been grouped together as a single site. In contrast, where a useful assemblage of tiles has come from an individual site within a town, this has been identified separately from other assemblages within the same town. If these separate assemblages within the same towns are aggregated together then the number of individual sites falls from 104 to 85.

Tiles from the current Silchester excavations have been reported separately from those in the Reading Museum which would have come from the Victorian excavations of the same site. This is because the Victorian excavators

[1] The longest published contribution is Brodribb 1987 which has one chapter on the subject of which just thirteen pages are dedicated to conventional tegulae.

[2] For example McWhirr 1979.

**Figure 1.1: Identification of terms used**

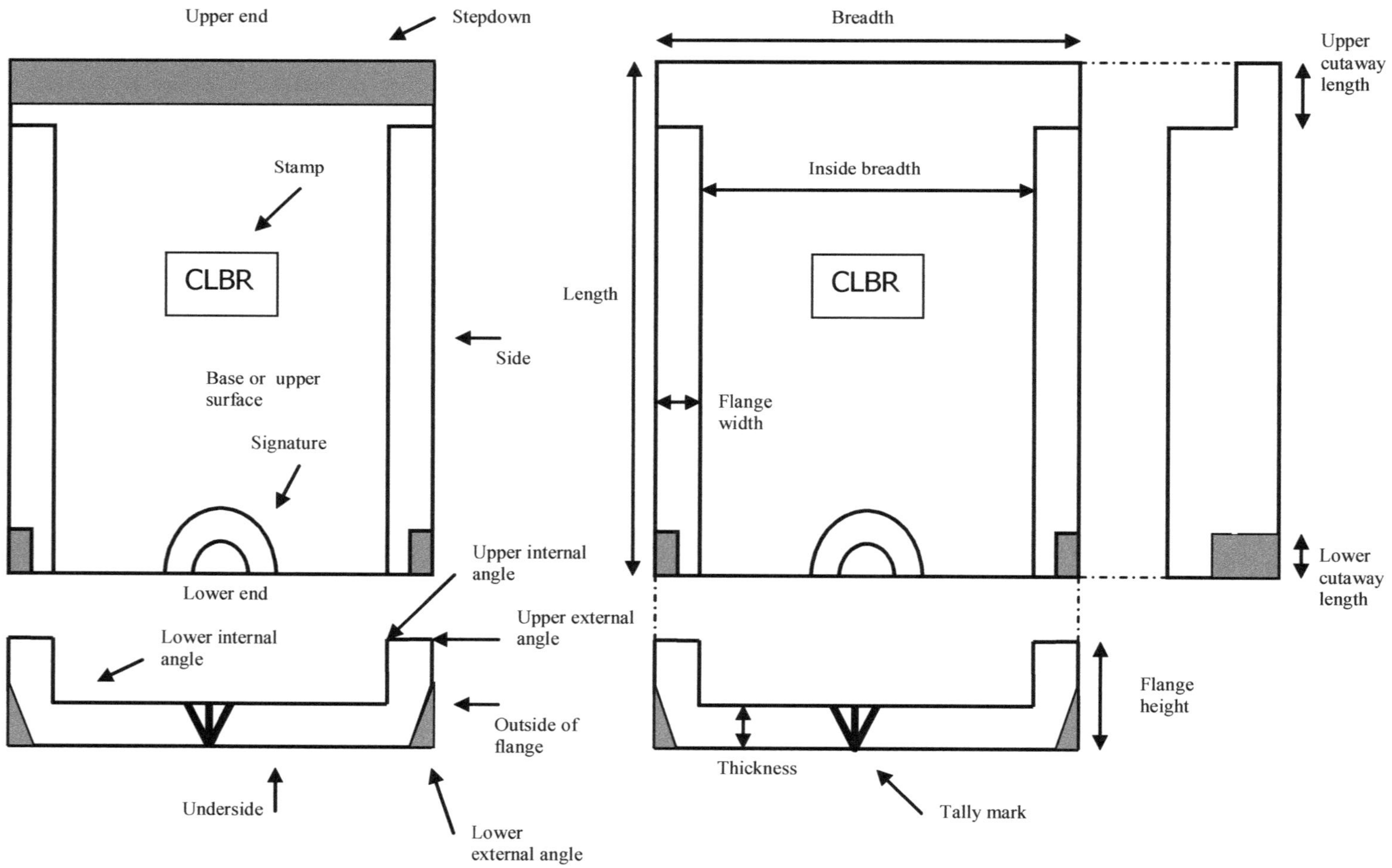

only retained complete tiles which are therefore likely to come from the last phase of building and be rather later in date than the fragmentary tiles emerging from the current exhaustive excavations of *Insula* IX.

Many thousands of tile fragments have been examined during this research: in all 666 tegulae were identified with either a complete length or a complete breadth and of these 427 had both a complete length and breadth. Less complete pieces that met the criterion of having at least a flange and one corner of the tegula present[3] have also been recorded and these bring the total entries in the database to 2317. Each tegula on the database has been assigned a unique numeric reference. A second series of numbers has been used for imbrices. All the tiles, with a couple of minor exceptions which are separately noted, have been personally examined and measured by the author.

## 1.3 Nomenclature and measurements

All tegulae consist of broadly rectangular bases with longitudinal flanges (Plate 1.1). The nomenclature used to describe the various attributes of the tegulae is illustrated in Figure 1.1.

At the upper end of the tile both flanges are cut back flush with the base to form the *upper cutaways*, and at the lower end of the tile notches are cut out of the underside of the flanges to create the *lower cutaways*. These cutaways allow the overlapping tiles to mesh together when they are placed on the roof: the lower cutaways of the higher tile resting within the upper end of the flanges of the tile beneath.

Whilst the profile of the flanges varies from tegula to tegula, which may simply reflect minor manufacturing variations in hand crafting, there are distinctive differences in the shape of the lower cutaways that are manifestly deliberate. Typologically these lower cutaway forms may be divided into four groups based upon the shape and positioning of the notch cut out of the underside of the flange. These are shown generically in Figure 1.2 viewed from the lower end of the tegula looking along the flange.

Broadly cuboid notches emerging from the *side* of the flange have been classed as Group A; basically prism shaped notches emerging from the side of the flange have been classed as Group B; basically cuboid notches emerging from the *top* of the flange have been classed as Group C and basically prism shaped notches emerging from the top of the flange have been classed as Group D.

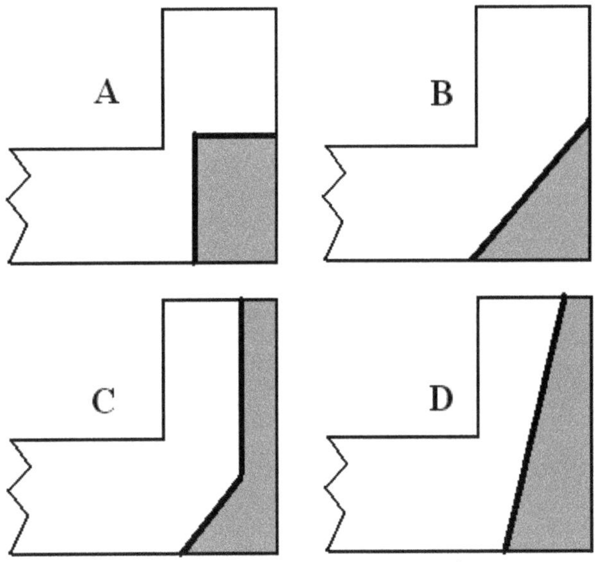

Figure 1.2: Generic lower cutaway forms

Thirteen different cutaway forms appear throughout Roman Britain and a further three appear only in specific regions. Figure 1.3 assigns each of the thirteen cutaway forms found nationally to one of these generic groups (the numbers allocated to the different forms are a result of history and are not intended to be sequential). Some of the cutaway forms do not fit naturally within this classification and their allocation to a particular cutaway group anticipates the results of the analysis in Chapter 3. Moreover, although the cutaway types are all shown as distinct forms, in reality they represent a continuum and the differences, for example between Types 2, 26 27 and 28, can be very subjective. However the allocation of each cutaway form to these generic cutaway groups is always genuinely objective.[4]

Of course it would have been possible to devise alternative typological classifications, for example based upon the number of cuts involved, but as will be shown later, these alternative typologies could not explain the variation in size and other features found on tegulae as well as the chosen typology does. Tegulae with Group A lower cutaways are referred to as Group A tegulae, those with Group B lower cutaways as Group B tegulae and so on. The logic behind this grouping of cutaway forms is discussed further in Chapter 3.

The measurements were recorded in millimetres but this should not imply that accuracy of measurement. Most tegulae have small deformities and most corners are rounded, as a result it is rarely possible to measure to millimetre accuracy[5]. However, as the purpose of most of the measurement was to produce statistical averages

[3] The flange of a tegula is normally tapered reducing in width from the bottom of the tile towards the top. Without the evidence of a corner piece it is impossible to tell where on the tile isolated pieces of flange were situated and such pieces have therefore been excluded.

[4] The only exception is potentially Types 5 and 15 but this is based on measurement as discussed in Chapter 2.

[5] Occasionally it was necessary to re-measure previously recorded tiles and differences of 2mm on individual attributes were common.

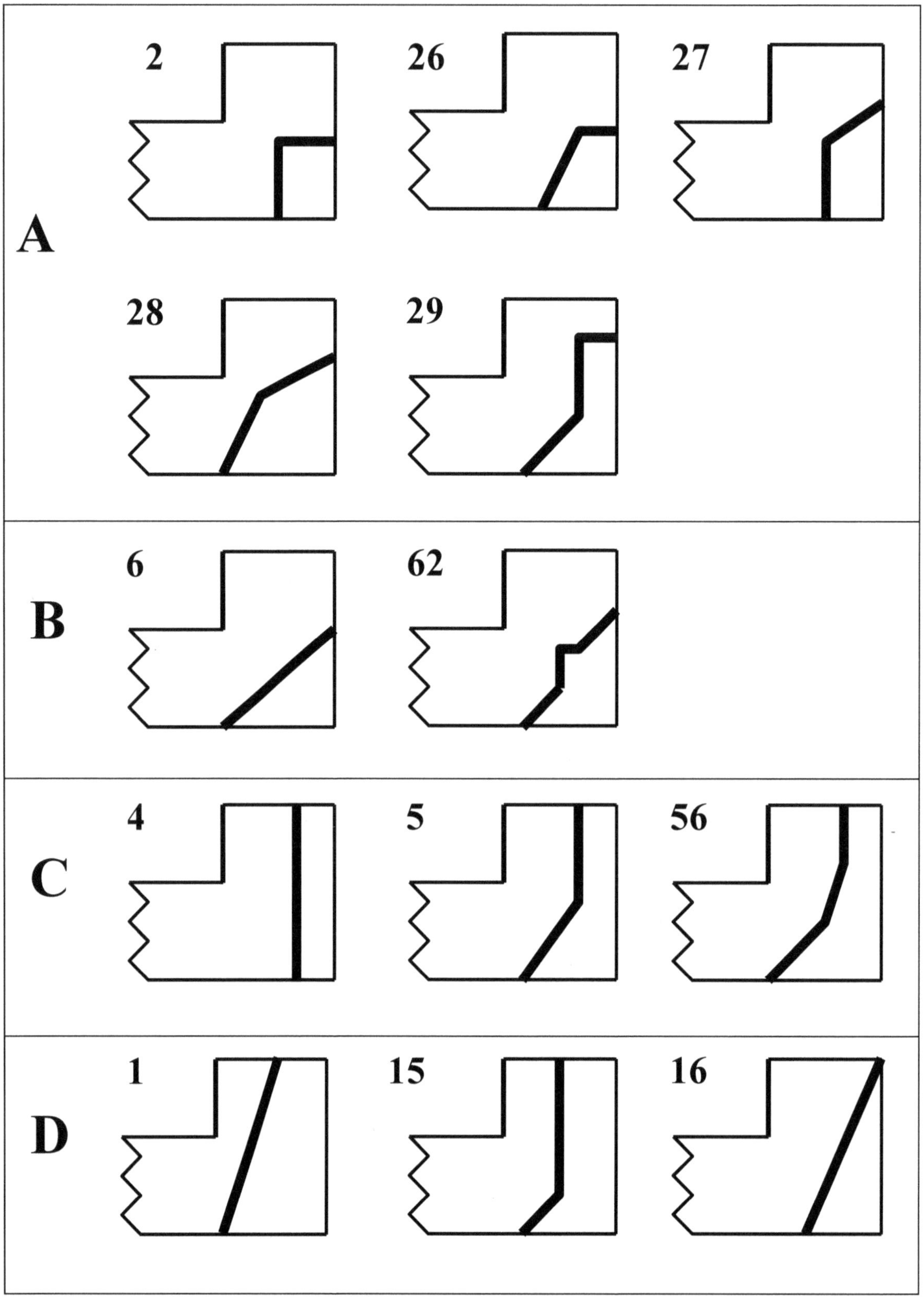


**Figure 1.3: Lower Cutaway Groups**

based upon much larger numbers of tiles, the accuracy actually achieved in these average measurements should be to a millimetre or less. The notch extracted to form the lower cutaway creates a weakness such that tegulae often break at this point which can make it difficult to always accurately identify the cutaway form. Inevitably there will have been occasions when cutaways were wrongly attributed or measurements were misread or mis-transcribed but hopefully these were few.

## 1.4 Previous research

The only previous attempt at a national study has been Gerald Brodribb's pioneering survey of all the ceramic building material in the country for his thesis which was delivered in 1983. This was subsequently published in slightly abridged form in 1987 as his book on Roman brick and tile. His survey recorded a total of 840[6] tegulae comprising 555[7] complete tegulae and 285 fragmentary examples. Brodribb took only five measurements (all in centimetres) and reported on eight other features.

There is a considerable overlap between the tegulae in Brodribb's survey and those in this thesis, particularly those with complete lengths or breadths.[8] Where similar attributes have been identified the two surveys should have produced similar results and in general this is true but there are also some significant differences. For example Brodribb states that 75% of the lower cutaways in his survey were Type 1,[9] whereas the equivalent figure from this analysis is 9.5%.[10] Likewise at Beauport Park, where identification numbers have allowed individual tegula measurements to be compared, there were some fairly material differences in measurement.

The pioneer in any research field is not able to check and compare his results with those achieved later, and in Brodribb's case, the particular results that are the main focus of this study, were only a secondary output of his more all embracing work. Moreover, Brodribb did not have the benefit of an easily usable computer database or digital photography to quickly record relevant details. Fascicules 4 and 5 of Volume II of the *Roman Inscriptions of Britain*[11] (hereinafter referred to as RIBII.4 and RIBII.5) have provided an invaluable data source on stamped tiles for this research but were published after Brodribb had finished his work.

Since Brodribb's work, much of the research into ceramic building material has focused on the analysis of the fabric with the intention of establishing the provenance of the material. However, even when the fabrics can be accurately identified, they do not always provide information from which conclusions can be drawn because several fabrics may be in use contemporaneously, and many of these over long periods. For example, Caruana and Hird[12] identified seven fabrics on a site in Carlisle that had yielded ceramic building material from the first through to the fourth century. Five of these fabrics appeared in most contexts from AD 100 onwards, and although some of these could be residual material, the majority were not. The authors concluded that it was not possible to establish any difference in fabric use at different periods[13] nor was there any correlation between stamp dies and fabrics.[14]

Betts[15] has undertaken a comprehensive scientific analysis of the fabrics used by the tile industry around York, especially tiles stamped by the Sixth and Ninth legions. Neutron activation analysis showed that the same clays were used by both legions but provided little discrimination between them whilst thin section analysis provide a good distinction, albeit with some overlap.[16] However, when the same approach was applied to stamped tiles found outside York, very similar results were achieved. Taken literally this would imply that all the Sixth and Ninth legion tiles were made in York including those found at Binchester and Carpow in Scotland. Whilst this is not impossible, the Binchester and Carpow dies have not been found in York itself and Betts observed[17] that the "similarity may be due to tilemakers from York moving northward, and selecting similar clays for the manufacture of Sixth Legion tiles". Betts has achieved rather more satisfactory results in the London area with some fabrics being datable to a range of fifty years or so, but this is the exception.

McWhirr and Viner[18] attempted to prove a relationship between the Minety kilns and stamped tiles from Cirencester using optical emission spectroscopy but this proved inconclusive. Peacock used a range of techniques to successfully demonstrate that there were only two sources of the Classis Britannica stamped tile but nevertheless described his work as "a disproportionate

[6] This compares with a figure of 1010 stated in his subsequent book (1987, 142) which must include tiles where no measurements were taken.

[7] There are a further thirty tegulae in his listing with no measurements and eight tegulae from Boulogne which are not included in the total used here.

[8] The overlap between the complete tegulae in the two surveys is probably 75% with the largest exception being Colchester where there is now no record of the 31 complete tegulae reported by Brodribb.

[9] Brodribb 1987, 17.

[10] Brodribb used only five different cutaway forms compared to the thirteen in this survey: if Type 6 and 16 cutaways which are similar to Type 1 but not differentiated by Brodribb are also included then the percentage rises to 36%.

[11] Frere & Tomlin 1992 and 1993.

[12] Caruana & Hird, 27.

[13] Caruana & Hird, 30.

[14] Caruana & Hird, 33 although this understates Table 118 which shows that all the stamped tiles appeared in just two of the seven fabric types.

[15] Betts 1985, 234-284.

[16] Fabric identification by visual examination with a hand lens or microscope did demonstrate that some dies were associated with particular fabrics but this was not a general conclusion and even where the relationship was established the same fabric was also used on other dies.

[17] Betts 1985, 257.

[18] McWhirr & Viner 1978, 363.

investment of research time and resources".[19]

## 1.5 Calculations

Where percentages are reported these are always calculated as a proportion only of those tiles where the attribute is definitely present or absent. For example, the percentage of tegulae with signatures is calculated only on complete tiles where the presence or absence of a signature can definitely be detected: tiles with crazed surfaces where no signature can be seen and fragmentary examples where a signature is present are excluded. Inclusion of crazed surfaces where no signature is visible but where one might have originally existed would produce too low a percentage, whilst inclusion of fragments with signatures but excluding fragments without signatures (because it is uncertain whether there was a signature on the rest of the tile) would bias the result in the opposite direction.

## 1.6 Terminology

Within this monograph the word *tiles* normally means *tegulae* and is used where repetition of the same word might become wearisome to the reader. *Tiles* may also be used to refer to *tegulae* and *imbrices* collectively or to ceramic building material more generally where the context permits.

[19] Peacock 1977, 246 although, as will be seen later, his work has proved invaluable in the interpretation of Beauport Park in its wider context.

# 2

# MANUFACTURE

## 2.1 Previous research

In their account of the overall tile making process, McWhirr and Viner suggested that tegulae were made with reasonably firm but still plastic clay. This was thrown into a mould and pressed against the frame with the surplus being removed with a bat or by hand. No detail on the design of the mould was proposed with "the flange being made either in the frame or by hand afterwards".[1]

Rook[2] had experimented with a four-sided mould with an open bottom that was placed on a table (see Figure 2.1 Mould D). The ends of the mould were cut down to match the inside profile of the required tegula and the sides were similarly cut down where the upper cutaway would be formed. Clay was placed into the mould and "batted" in to fill the corners; the excess clay was then removed from the top of the mould with a tensioned wire. The flanges were created by aligning the wire with the side of the tile and running it down the inside of the tile across the centre and up the other side using the ends of the mould as a template. The upper cutaway was formed in a similar manner with the wire now held across the tile. The lower cutaways could be formed by inserts into the mould, but because these inserts would prevent the mould being lifted vertically off the tile he proposed three possible solutions: the tile could be pushed upwards from the frame; the inserts could be attached to the baseboard rather than the side of the mould; or the mould could be made in two parts. He believed his approach was supported by longitudinal markings on the surface of the tile and the bulging form of some flanges, which he equated with the use of an inadequately tensioned wire.

Betts,[3] who also advocated the use of a four-sided mould placed upon a sanded working table, argued against Rook's approach on the grounds that it would have been difficult to push the clay into the corners of the mould; that it would have been awkward to cut the clay with a wire whilst it was still in the mould and that it would have required the removal of large quantities of clay. He therefore assumed that the clay was rolled out into a flat blank which was cut to size prior to being lifted into the mould, but gave no details about how the clay would be lifted and aligned into the mould and then released. The grooves running along the base adjacent to the flanges were created by the tiler pushing the clay into the edges of the mould whilst also smoothing out any irregularities on the flange surface. The upper surface of the tile may have been smoothed with a scraper and a similar instrument may have been used on the underside as well.

He suggested that Type 4 and Type 5 lower cutaways (Figure 1.3) were formed by a block attached to the mould (the additional diagonal cut being made later in the case of the Type 5 cutaway). Other cutaway forms would have prevented the mould being directly lifted off the moulded tegula and he suggested that these were made by blocks attached to the baseboard or that they were also attached to the mould and the mould either split open or the clay was pushed out from underneath the mould. After the tile was removed from the mould excess clay was removed with a knife, the upper cutaways were formed and the lower cutaways finished as necessary.

Based on his own experiments Brodribb[4] believed that flanges were folded in a mould (presumably using clay already rolled flat as suggested by Betts), but he provided no information on the type of mould or how this was done.

Vitruvius provides information on the manufacture of mud bricks[5] but not on fired clay tegulae and none of the surviving texts of the other classical authors cover the subject.

## 2.2 Evidential approach

The best way to resolve these conflicting approaches, and to illuminate the points that they omit, is to consider the surviving evidence in the form of excavated tegulae. However tegulae, whether made in a mould or not, were a hand crafted artefact and thus, even when two tiles were made by the same craftsman using the same mould, each tile would differ slightly. It would therefore not be

[1] McWhirr & Viner 1978, 361.
[2] Rook 1979, 298-301.
[3] Betts 1985, 158-166.
[4] Brodribb 1987, 13.
[5] Vitruvius II.3.1-4.

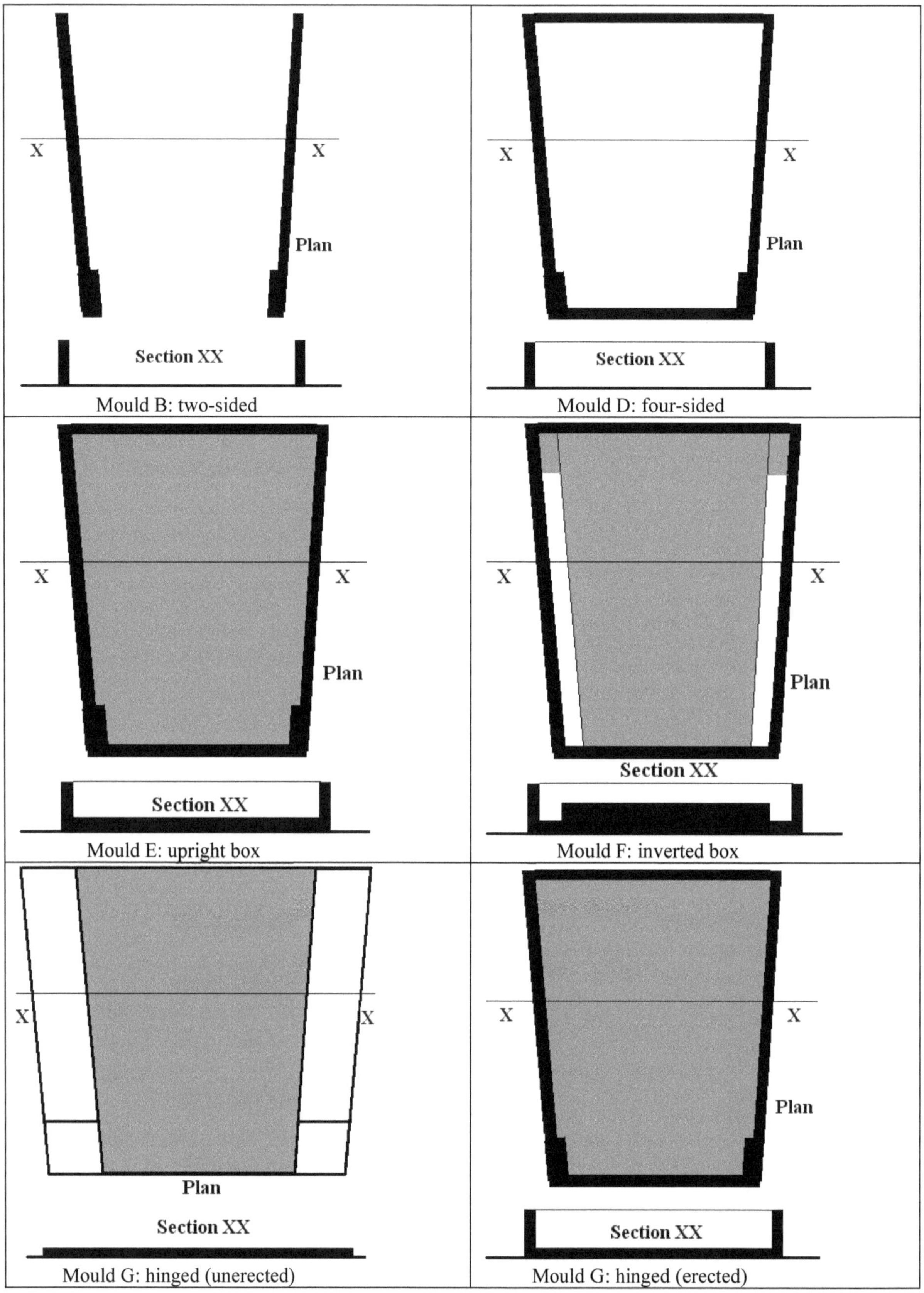


**Figure 2.1: Possible mould configurations**

possible, or sensible, to seek to record every minor variation.

As a general rule only those features that are most frequently observed have been included except where an isolated occurrence nevertheless throws particular light on the underlying manufacturing process. In fact, because some of the manufacturing is slightly blemished, or the tiles eroded or fragmentary, it is normally necessary to see several examples of a feature from a site before one can be confident that the feature has been correctly interpreted and that it is a manufacturing constant. This makes the assessment of features somewhat subjective and it is quite possible that two equally diligent researchers could come to opposing conclusions about the same assemblage of tegula fragments. However, these differences should not be exaggerated because the disagreements will always be small in comparison to the agreements.

Interpretation of the manufacturing details visible on surviving tegulae should allow the mould types in use to be postulated. To provide a structure for this analysis, the most obvious mould forms are listed below and shown in Figure 2.1.[6]

- **A** Hand moulded (not shown)
- **B** Two-sided mould. This would have had to be fixed to the moulding table.
- **C** Three-sided mould: this is a two sided mould integral with a base (not shown)
- **D** Four-sided mould with an open bottom
- **E** Five-sided upright box mould with a closed bottom where the tile was made face up
- **F** Inverted box mould where the tile was made upside down
- **G** Mould with retractable or hinged sides, probably five sided but could be four sided

Each of the approaches listed in A - G above would have produced slightly different features on the resulting tegulae. For example, handmade tiles would be likely to have rounded corners and imperfectly straight flanges whilst tegulae coming from Mould F (the inverted box) would be expected to have sharp corners and straight flanges. Observations of every aspect of tegulae examined in the survey can be systematically compared to what might have been expected from each of the manufacturing methods listed. This should allow some methods to be eliminated and the others to be preferred.

Tegulae were produced for some three hundred years[7] in Roman Britain so it would be surprising if their method of manufacture remained unaltered over this period, nor should it necessarily be expected that a single method of manufacture was uniformly adopted across the country. Thus no single method is expected to satisfy all the observations in the analysis and the results will need to be sifted to determine the most probable combination or combinations of methods adopted.

## 2.3 Analysis

This section utilises the terminology and cutaway forms described in Figures 1.1, 1.2 and 1.3.

### 2.3(a) Overall dimensions and shape

The overall size of tegulae varies widely both between sites and within individual sites. The overall shape of the tiles also varies: tegulae with Group A and B cutaways are normally rectangular whilst Group C and D tegulae are normally trapezoidal with the upper breadth being greater than the lower. Even though the size of the tegulae vary, the ratio of length to breadth is broadly constant.

Most tegulae are reasonably well made but Group D tiles are normally the best made grading down to Group A tiles which are normally the crudest, but within this generalisation there are Group A tegulae that are well made and Group D tegulae that are poorly made. Most tegulae are flat but about 20% of them are longitudinally convex: that is they bend away from the flanges and would have made an arch shape on the roof. No longitudinally concave tiles were observed. Convex tiles are discussed in more depth in Chapter 7.

More rarely tegulae are laterally concave: that is they bend inwards bringing the opposite flanges closer together. This was probably caused by the tegulae being turned onto their sides before they were sufficiently dry and the upper side consequently slumping downwards. No laterally convex tiles were observed. Infrequently tegulae are buckled with bends in more than one direction.

It would be convenient if one could identify tegulae from the same mould to determine the amount of natural variation, but even where tiles have the same stamp and signature, they do not necessarily come from the same mould.[8] There are however three tegulae from Winchester that all have the same blemish in the flange and the same distinctive signature and double finger impression running adjacent to the flanges (Plate 2.1). These are therefore likely to have come from the same mould. The variation in length and breadth between these tegulae is at most 2%. The flange height varies by up to 4mm from the average, which is more than any measurement error and is therefore an indication of the

[6] The majority of tegulae are trapezoidal rather than rectangular in plan thus the moulds have been designed for trapezoidal tegula production

[7] Tegulae were being manufactured soon after the invasion and are plentiful in the Boudiccan destruction horizon at Colchester, whilst a tile kiln at Crookhorn Farm in Hampshire has an archaeomagnetic date of AD 330 +/- 20 (McWhirr 1979, 137). Possible still later tile manufacture is discussed in Chapter 4.

[8] See Chapter 3 for a detailed analysis of stamps and sizes.

amount of natural variation that can be expected. The lower cutaway lengths are virtually identical but the upper cutaways vary from the average by 3mm.

Plate 2.1: Winchester 1895/6/7 flange blemishes

There is a further group of seven tegulae from Beauport Park that also have a common blemish in the flange, but its position varies. This group is considered in Section 2.7.

Plate 2.2: Clay leaking beneath base of mould

## 2.3(b) Outside of flanges

In the table below, and those in subsequent sections, each observation has been given an identification (eg B1) to allow easy reference back. The conclusions given against the observations in the tables are considered at more length in the subsequent discussions.

| | Observation | Conclusion |
|---|---|---|
| **B1** | Always longitudinally straight[9] | Made in a mould |
| **B2** | Outside face of flanges flat with evidence of mould sand | Made in a mould |
| **B3** | Lower external angle of flange square[10] and sharp | Made in a mould |
| **B4** | A significant minority of tiles show a line of clay protruding outwards from the base along the length of the tile that was normally then trimmed with a knife (Plate 2.2) | Only in Mould D would clay be expected to leak beneath the bottom |
| **B5** | Upper external angle of flange normally sharp | Formed against a mould |
| **B6** | Dribble marks over top of flange, rare (Plate 2.3) | Fluid from wet clay forced up inside of mould |
| **B7** | Vertical or horizontal markings rarely observed[11] | Mould sides were clean or dismantled to remove tile |
| **B8** | Stanton Low Group D tiles were stamped on the side of the flange (Plate 2.4) | Side of mould must have been retractable (Mould G) |

The square straight sides (B1) are a convincing demonstration that these tiles were not hand made. The regular observation of clay apparently squeezing under the bottom of the mould (B4) is strongly supportive of the loose four-sided Mould D. However, the Stanton Low tiles with stamped flanges which are in relief two or three millimetres deep (Plate 2.4) could not have been made in Mould D and must have involved some form of retractable side.

[9] Only one exception to this has been observed (Caerleon 618) where one flange was straight and the other apparently hand formed, presumably indicating that the tile was damaged and rebuilt after leaving the mould or that one side of the mould collapsed as the tile was being made.

[10] 2% of tegulae have non square lower corners, extreme examples are Silchester 110 and Chedworth 094.

[11] But see discussion of Beauport Park markings in Section 2.7.

## 2.3(c) Flanges

| | Observation | Conclusion |
|---|---|---|
| **C1** | Fabric structure aligned with base but bent within the flange (Plate 2.6) | Hand formed or clay folded prior to being thrown into mould |
| **C2** | Flanges are wider than base is thick | Suggestive of flanges formed by folding |
| **C3** | Width of flanges normally same on both sides | Suggestive of flanges being cut with a wire rather than folded |
| **C4** | Width of flange tapers by an average of 30% from bottom to top[12] | Suggestive of flanges being cut with a wire guided by a template |
| **C5** | Height of flange tapers by an average 2% from bottom to top | Artefact of the way the flanges were finished after cutting |
| **C6** | Upper internal angle of flange rounded | Artefact of the way the flanges were finished, probably not Mould F |
| **C7** | Lower internal angle of flange normally rounded | Result of flange finishing, not Mould F |
| **C8** | Profile of flange varies within sites and between sites | Result of flange finishing, not Mould F |
| **C9** | Normally one or more channels running adjacent to flanges | Result of flange finishing, not Mould F |
| **C 10** | Some Group C and D flanges and all Type 7 Regional tiles have sharp corners and a rectangular cross-section (Plate 2.5) | Probably made in an inverted box Mould F |
| **C 11** | Some tegulae from Leucarum and Carlisle have triangular flanges (Plate 2.7) | Made in an inverted box Mould F |
| **C 12** | Some Group C tegulae from Sparsholt have a rectangular insert cutting into both flanges (Plate 2.8) | Possibly to house a device for stacking tiles whilst drying |
| **C 13** | Flanges show no distortion above lower cutaway insert | Wet clay punched into mould not folded |

[12] Roughly 10% of tegulae have no or minimal flange width taper, but only eight tiles (1.5%) have a reverse width taper and these are all isolated cases except for Corbridge where three of the four probable Group A tegulae had a reverse taper: Corbridge 403, 404 and 411.

Plate 2.3: Beauport Park 206 – Excess fluid in mould

Plate 2.4: Stanton Low 2238 – Stamp on outside of flange

Plate 2.5: Crookhorn 2187 –Rectangular cross-section

The fold in the fabric structure (C1) either follows the profile of the flange (Plate 2.6 lower) or follows the profile up and then comes down to the base as if the fabric were buckled (Plate 2.6 upper). These folds would be consistent with a flat piece of clay having the sides bent upwards to form a flange or, in the case of the buckled profile, with a flange being formed by pushing the side of a flat piece of clay to force it to buckle into the shape of a flange. Alternatively, if the clay was too wet to be rolled, it could have been cut very roughly to shape, formed into a crude ball by folding the clay upwards and then thrown into the mould. Once in the mould the clay could partially unwrap but the bits resting against the sides of the mould would not be able to do so and would either remain bent upwards or get pinched against the sides forming a buckle in the fabric. This is the modern craft tile making practice and it must be possible that the Roman craftsman adopted the same method with similar results.

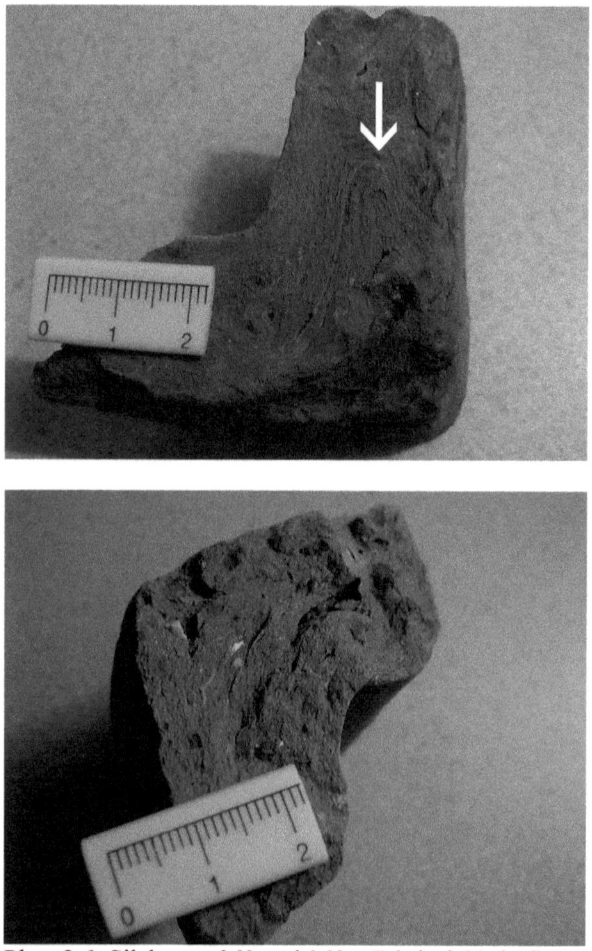

Plate 2.6: Silchester 060 and 068 – Fabric folds in flange

The fact that flanges are normally wider than the base is thick (C2) could be consistent with the flanges being folded upwards from a flat piece of clay which would result in a bulge of surplus clay (Figure 2.2) which would then have been displaced into the width of the flange as part of the hand finishing process. Alternatively, if the tile were made with wet clay, the flange width would simply have been a parameter of the design produced by running a wire across the template provided by the ends of the mould as described by Rook.[13]

The fact that there is no distortion of the flange over the lower cutaway (C13), even when an insert has been used, is persuasive of wet clay being punched into the mould and around the insert. If a flat sheet of stiffer clay were inserted into the mould and folded to form the flanges, then it seems highly probable that there would be some distortion between the parts of the flange where there was no obstruction beneath and those bits lying over the lower cutaway inserts.

Plate 2.7: Leucarum 1692 – Triangular cross-section

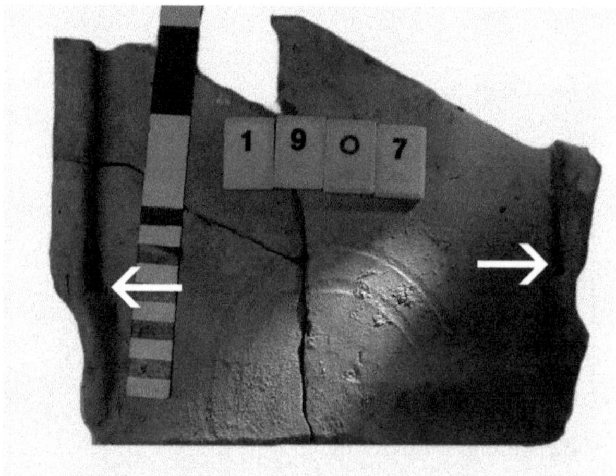

Plate 2.8: Sparsholt 1907 – Rectangular insert

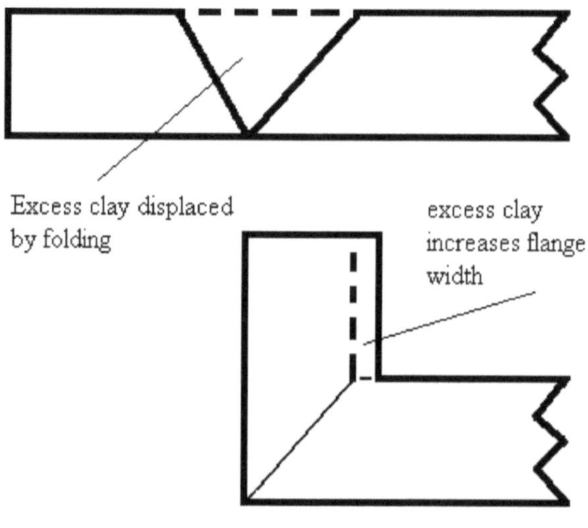


Figure 2.2: Flange folding displaces surplus clay

[13] Rook 1979, 298-301.

The flange taper (C4) is a useful engineering feature to help overlapping tiles mesh when they are placed on the roof (see further discussion in Chapter 6): the taper is normally similar on both flanges as is the width (C3). Although there are minor differences between the opposite flanges on most tegulae, these differences are small compared with the similarity of width, height and flange taper that are generally observed. This would be expected if any of the Moulds D, E, F or G were used with wet clay, but if the clay was sufficiently dry to be rolled flat and cut to shape prior to insertion into a four-sided mould (as suggested by Betts), then a tegula with symmetrical flanges would only be produced if the clay was precisely aligned with the mould. If the clay were placed too far to one side then one flange would be wider and the other thinner than planned and the difference between them therefore exaggerated. If the clay were put in short of one end of the mould then the clay would buckle at the further end and possibly fail to align with cutaways at the nearer end. Perhaps the most likely misalignment is that the clay would be put in slightly askew which would not only result in thicker and thinner flanges but also one flange would be extremely tapered and the other not tapered at all. It therefore seems unlikely that the clay was rolled out prior to insertion into the mould.

The observations in C6 to C9 are all indicative of smoothing the flanges by running the hands along them which would not be possible[14] if the inverted box Mould F were used. Depending upon how the hands were held, different flange profiles would be produced and none, one or two finger[15] channels would be left on the base running parallel to the flange. The evidence of a small height taper (C5) suggests that tegulae were smoothed by running the hands from the top to the bottom of the tile.

Tiles with square flanges (C10 – Plate 2.5) could have been made in the same way as other tiles and simply smoothed using a square block of wood rather than a hand,[16] or as in the case of tiles from Gestingthorpe[17] a sharp corner and flat top to the flange could have been achieved by pinching the flange between thumb and forefinger as they were run down its length. Nevertheless it seems likely that some of these tiles were made in an inverted box Mould F which would produce similar results. This finding is reinforced by observations of the undersides (discussed in Section 2.3(f)). The triangular flanges from Leucarum and Carlisle (C11 – Plate 2.7) could only have been produced in Mould F because the exceptional regularity of the flanges makes it most unlikely that these could have been the product of wire cutting and subsequent smoothing.[18]

All the six Sparsholt Group C tegulae that have been retained from the excavation have the insert (C12 – Plate 2.8). This is difficult to explain because it has not been produced by an object crushing the flange, as this would have left some evidence of displaced clay which is not observed. Thus it seems that the shape was created when the tile was formed even though the remainder of the flange has the normal rounded form left by running the hands down the flange. Moreover whilst it is suggested that it was used to house a device for stacking tiles during the drying process when space would have been at a premium, it is by no means clear how such a system could have worked.

The bulging flange profile (ie wider in the middle than at the upper or lower ends) cited by Rook[19] in support of his manufacturing proposal is extremely rare and does not justify inclusion in this analysis.

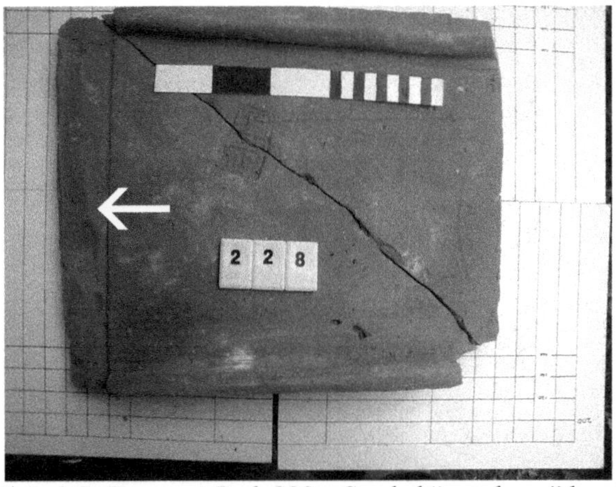


Plate 2.9: Beauport Park 228 – Sanded "stepdown" band

Plate 2.10: Caerleon 560 – Linear "striker" marks

[14] If Mould F were used then smoothing could have been undertaken after the tile was removed from the mould.

[15] There are rare examples of three finger channels eg Batten Hanger 2170, Silchester 065 and Holt 1113.

[16] Frilford 015.

[17] Gestingthorpe 2512.

[18] An isolated example of a tegula with a triangular flange has also been observed at Caerwent which could possibly have been imported from Leucarum.

[19] Rook 1979, 301.

## 2.3(d) Upper surface

| | Observation | Conclusion |
|---|---|---|
| **D1** | Upper surface always smooth | Smoothed with striker |
| **D2** | 13% of tegulae have a sanded band or step reduction across the width at upper end (Plate 2.9) | Part of production process or for stacking tegulae |
| **D3** | 80% of tiles have signatures drawn with a finger[20] which are almost always on the lower part of the tile[21] | Tradesman's mark |
| **D4** | A minority of tiles are stamped | Military unit or manufacturer's mark |
| **D5** | Where a signature clashes with a stamp, the signature almost always underlies the stamp | Stamps added later |
| **D6** | About 10% of military tiles have boot marks but these are rare on civilian production | Military method of testing tile hardness |
| **D7** | Animal impressions often observed on upper surface but never beneath | Always dried face up |
| **D8** | 25% of tiles have preformed nail holes which are normally at the top of the tile but occasionally adjacent to flange (Plate 2.14) | For securing the bottom row of tegulae or the edge of the roof |
| **D9** | Some Group D tiles have larger preformed holes (Plate 2.16) | Pegs rather than nails were used |
| **D10** | Chipped holes are rare | Created when no tile with preformed hole available |
| **D11** | Some tiles appear to have been coloured | Dipped or painted with slip |

Although almost all upper surfaces are smooth they occasionally have longitudinal lines of slightly raised clay which would be consistent with two passes of a striker to smooth the surface failing to overlap and thus leaving a linear raised island between them (Plate 2.10). Striation marks caused by a wire as suggested by Rook[22] have not been observed in this survey.

The sanded band (Plate 2.9), referred to as a *stepdown,* is not part of the process for forming the upper cutaway because it is not always aligned with the cutaway (Plate 2.11) and occasionally it would appear to have pushed against the end of the flanges after the cutaway was formed causing the flanges to turn back (Plate 2.12).[23] The band is always sanded which indicates that it was placed against the tile whilst it was still wet (and therefore probably still in the mould) otherwise sanding would have been unnecessary. This is reinforced by the observation that when the flanges are turned back as in Plate 2.12, they are invariably turned inwards, suggesting that the sides of the mould were still in position when this occurred. It therefore seems likely that the stepdown was a device to facilitate transport whilst still in the mould or to assist in the process of removing the tile from the mould.[24]

Plate 2.11: Beauport Park 1195 – unaligned stepdown

The stepdown is primarily a feature of military production: 57% of Chester/Holt/Tarbock, 36% of Caerleon and 44% of Beauport Park tegulae have stepdowns, perhaps suggesting that it represents a more sophisticated manufacturing process for dealing with larger volumes of production. The only non-military locations to have produced more than an isolated stepdown are Gloucester (9) and Silchester/Reading (7).

Some 80% of tiles have a signature mark[25] (D3) which was almost certainly drawn with a finger at the same time

[20] On some Group D tiles the signature is drawn with a comb: eg Sparsholt 1911, Bignor 2265.

[21] Whilst some signatures may be in the centre of the tile there is only one example of a signature at the upper end: Corbridge 401. Betts 1985, 196 notes that some tegulae from York had signatures at the upper end.

[22] Rook 1979, 301.

[23] Brodribb 1987, 17 asserts that the Beauport Park stepdowns were made with a knife and were how the upper cutaways were formed but this seems to be contrary to the evidence especially where the stepdown fails to align with the upper cutaway. There is, however, an isolated example from Caerwent (1142) where exactly this approach was adopted leaving an unsanded stepdown.

[24] An alternative proposal is that they are *skink* marks which result from resting one tile over the end of another during the drying process in order to minimise the amount of space required, but this does not explain why the band always appears sanded. Also see Section 2.7.

[25] Brodribb 1987, 101 gives a figure of 60% for the proportion of tiles with signatures; the figure in this survey has been calculated by only including complete or virtually complete tiles where there is no chance that a signature has not been counted because it was on the missing fragment. Betts 1985, 197 comments that the incidence of signatures in York was much higher than Brodribb's 60%.

as the tile was made, although signatures made with a comb are frequently observed on later tegulae from the south coast area.[26] With just one exception,[27] these signatures are always on the lower part of the tile and would therefore have been visible when the tiles were placed on the roof.[28] Either the signatures were designed to be visible[29] or their placement was simply a consequence of the lower end of the tile being more accessible in the manufacturing process: in other words tiles were made with the upper end furthest away from the tradesmen.[30]

Plate 2.12: Chester 926 – Flange bent against stepdown

Semi-circular signatures (including arcs and three-quarter circles) account for some two-thirds of all signatures that are found on tiles. If the signatures were intended to distinguish the work of individual tradesmen, then it is curious that so few different forms of signature are observed; indeed at Silchester where there is a large assemblage of tiles, almost all marks are semi-circular. One possible conclusion is that there were relatively few tradesmen producing tiles in each tile manufacturing unit and thus a range from one to four semi-circular signatures was sufficient to distinguish each of the tradesmen's work.[31] Military units tend to have a slightly wider range of signatures and this could indicate that a wider range of people were detailed to participate in tile making than was the case in civilian tileries.[32]

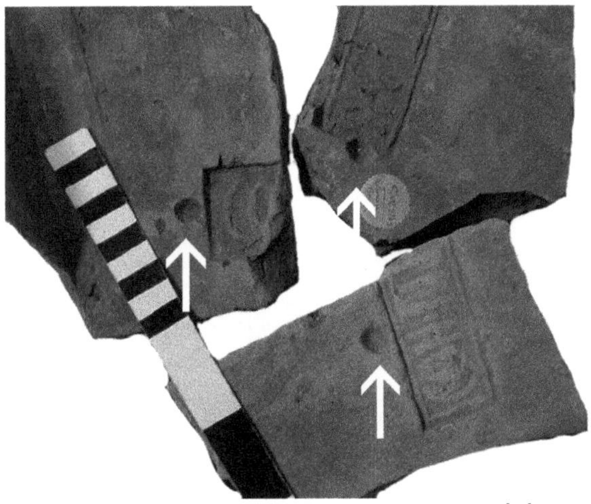
Plate 2.13: *Cohors IIII Breucorum* stamp retaining nails

A minority of tiles were stamped (D4) and these are the subject of more detailed discussion in Chapter 5. The dies were normally carved in wood to produce a stamp in relief.[33] Wood was subject to wear and dies show occasional signs of re-cutting and, more rarely, splitting.[34] Some dies were made of metal which typically produced incuse stamps.[35] Stamps made of fired clay are also known.[36] The dies would have been fixed to a handle normally with nails which often leave impressions on the tiles.[37] There are a group of three dies from *Legio II* (RIB 2459.43-5) which, as discussed in Section 5.3, would appear to have been consecutive replacements for the first die, each secured by the same overlarge nail. A more unusual situation existed with *Cohors IIII Breucorum* at Grimescar where the impression of the retaining nail appears in at least six different positions (Plate 2.13). It is difficult to find an explanation for this. Perhaps the most plausible is that the die was secured to the handle, not by a permanent nail but by a clip which made a similar impression, such that the die could be removed after use: each day the die was re-clipped to the handle in a random position.

Most sites where tiles have been stamped produce examples where the signatures and stamps overlap on the same tile and where this happens the signature almost always underlies the stamp.[38] The most likely explanation is that, on most sites, stamping the tiles was a secondary

[26] Tegulae with signatures made with a comb have been observed at Batten Hanger, Bignor, Crookhorn, Exeter, Grateley and Sparsholt. They also occur on the Isle of Wight (Tomalin 1987, 110).
[27] Corbridge 401.
[28] Whereas if the signature had been placed at the top of the tile then it would have been wholly or partially covered by the overlapping roof tile.
[29] It has been suggested that the signature was simply a method of testing the dryness of the tile but some tiles, such as Silchester 065, have very sharp edges to the signature that could only have been produced when the tile was visibly wet. It is also inconsistent with the combed signatures above.
[30] Corbridge is the only site with systematic reverse flange width taper and it is the only site to have produced a signature at the upper end (although this was the only one of the four Group A tiles that had a small normal taper) which suggests that these Corbridge tiles were manufactured with the upper end closest to the tilemaker. None of the five tiles from other sites with a reverse flange width taper had a visible signature and four definitely had no signature.
[31] The size of the workforce is considered in Chapter 8.
[32] Birley 2002, 90 quotes from one of the Vindolanda tablets that 343 men were "in the workshops" including *inter alia* some "at the kilns" although the number is missing.
[33] Brodribb 1987, 119 notes that the wood grain is sometimes visible on stamp impressions. Betts 1985, 207 makes a similar observation.
[34] For example Betts 1985, 207 on the *Legio* IX die RIB 2462.9.
[35] For example Benwell 1997 with a die of *Ala I Asturum*.
[36] Hassall & Tomlin 1984, 342 No 17 and Plate XXVI shows a ceramic die albeit this was not necessarily for use on tegulae.
[37] For example RIB 2481.7A and 7B.
[38] York 464, Chester.A and Tarbock.A all have signatures added after the tegula was stamped.

process undertaken by an overseer who would have been unconcerned with the signature marks so that it would have been purely chance whether the stamp was superimposed on the signature or not.

Beauport Park differs from the general rule because almost every tile was stamped; yet despite this, only once does the stamp clash with the signature. This could suggest that, at the *Classis Britannica* tile-works, the stamping was undertaken by the tradesman when the tile was formed which resulted in virtually every tile being stamped. The tradesman would, of course, have had more regard for his signature and therefore would have attempted to place the stamp such that it did not clash, which is precisely as observed.[39] There is virtually no overlap between the stamp types used on tegulae and those on imbrices at Beauport Park.[40] As manufacture of tegulae and imbrices were likely to have been separate activities, this tends to reinforce the notion that these tiles were stamped by the tradesman rather than subsequently by an overseer.

Brodribb has identified ten instances at Beauport Park where a tile has been stamped twice and notes that double stamping also occurred on *Classis Britannica* tiles at Lymme and Dover.[41] Double stamping is extremely rare elsewhere and only four other instances have been observed in the survey.[42] Whilst one would not expect the tilemaker to frequently accidentally double stamp his tiles, it would be more surprising if the overseer did so regularly. This, therefore, probably supports the notion that in the *Classis Britannica* the tilemaker did his own stamping.

Some 10% of military tiles have boot marks (D6), but such impressions are very rare in civilian production.[43] This could mean that military personnel were careless about how they walked in the tile drying area or that civilians could not afford boots. However a more likely explanation is that the impressions were made by the officer in charge of the tile making works testing the hardness of the tiles to see if they were sufficiently dry to be turned for stacking onto their sides[44] (a similar method is sometimes adopted today). Civilian tile works either tested the hardness in a different way or were closer to the work and did not need to test, whereas the military officer might only have been making an occasional visit to supervise production and hence was less sure of the state of his tiles.

Impressions of animal feet (D7) are a regular occurrence on tegulae but they only appear on the upper surface and never on the underside. These marks were obviously made whilst the tiles were drying,[45] but if they were dried in the open where they were accessible to animals then one would have expected that they would be covered with straw or other material to control the rate of drying which would have prevented most animal impressions. If they were left uncovered not only would there be a danger of cracking because the sun baked the surface faster than the core of the tile, but also a danger of damage from rain. Cracked and rain damaged tiles[46] are extremely rare so it would seem that either the animal impressions occurred in the hour or two before the tiles were covered (which gives information about the human/animal interface) or they were dried under cover and the animals accessed them there. Modern tileries use *hacks* (Plate 2.50) which are long low structures with open sides that animals can easily access.[47] Hartfield[48] may represent just such an arrangement which appears to be an open sided shed 3m by 5m adjacent to the kiln. At some sites animal prints are a relatively common occurrence whereas at others they are rare, for example Beauport Park only has one animal print[49] on the forty complete or near complete tegulae, which suggests that drying procedures varied between sites.

Plate 2.14: Gestingthorpe 2512 –Holes adjacent to flange

[39] Stamping the experimental tegulae (discussed in Section 2.4) just after they were made produced satisfactory results although this was not apparently the experience of Brodribb 1983, 232 who suggested they needed to be leather hard.

[40] Brodribb 1980, 189.

[41] Brodribb 1980, 187.

[42] Chester (two), Caerleon and Dorchester, Dorset.

[43] Most of the hobnail patterns created by boots on tegulae are narrow and sharply pointed at the toe end of the shoe: this is at variance with the rounded end of the traditional legionary *caliga* unless the hobnails did not follow the outline of the leather.

[44] There is also an apparent correlation on military sites between tegulae that were stamped and those that had boot impressions suggesting that these tiles were the most easily accessible, but as stamped tiles have been retained disproportionately to unstamped tiles in collections, this correlation may be misleading.

[45] York 512 has a dog print on top of a boot mark.

[46] Piddington 2145 is an example of a rain damaged tile that was subsequently fired even though the lower cutaways had not been formed. Presumably it was used as packing for the kiln.

[47] Further support for this comes from the process of stamping tiles because if tiles were stamped after they had been moved to the drying area then this would be inconvenient if they had already been covered with straw.

[48] Rudling 1986, 200.

[49] Beauport Park 178.

Roughly one tegula in four has a nail hole (D8) that was formed during the manufacturing process. Most of these holes are at the top centre of the tile but occasionally they are adjacent to the flange (Plate 2.14). The functioning of nail holes is discussed in more depth in Chapter 6, but in terms of manufacture it should be noted that a minority of nail holes are blind: that is there is a good hole on the upper surface but it does not fully penetrate the depth of the tile. Moreover many complete holes exhibit chipping around their undersides which would be consistent with blind nail holes being knocked through.

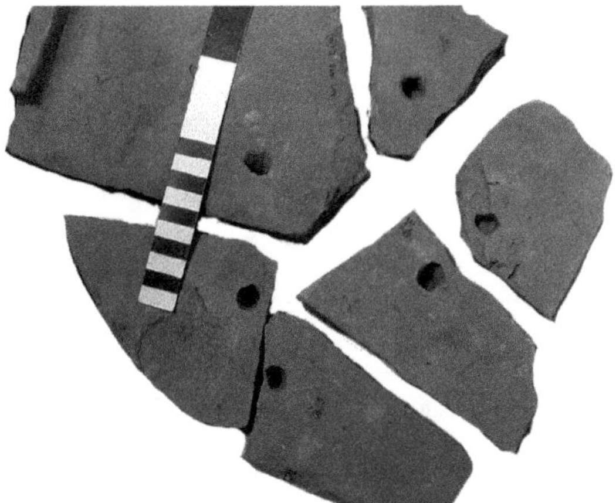

Plate 2.15: Leicester – Blind nail holes

All of the mould designs present a difficulty in making a complete nail hole because this would require the gouge used for forming the hole to penetrate either the palette or table on which the clay was resting (although this would not have been a problem if the tiles were made directly on the ground). So it is possible that many of these holes were either deliberately made blind or were an unavoidable imperfection of the manufacturing process. At Norfolk Street, Leicester there appears to have been a period of production when all the holes were deliberately made blind (Plate 2.15), perhaps with the intention that they would only be broken through as required.

Plate 2.16: Winchester 1895 – Peg hole

Although most holes were shaped to accommodate nails (indeed nails are sometimes still found in situ[50]) some Group D tiles have much larger perfectly cylindrical holes[51] which were presumably designed to accommodate wooden pegs (Plate 2.16). These holes appear in the same place on all the tegulae and so it is possible that the mould contained a protruding peg such that the holes were automatically formed when the clay was placed into the mould.[52] This would have limited the ways in which the tile could have been removed from the mould and this approach is therefore only likely to have been used with the inverted box mould F.

Some tiles exhibit chipped holes (D10) which must have been made after the tile was fired; these were presumably necessary when a roofer ran out of tegulae with pre-formed holes. However, as the tiles with nail holes were the first to be placed on the roof (see Chapter 6), this would suggest some disorganisation in the supply of tiles to the site.

Plate 2.17: Slip on tegulae from Frilford

Some tegulae appear to have been coloured by an application of slip (Plate 2.17). The colours appear to have been limited to red, white/cream and dark blue, perhaps in imitation of colours that can be achieved on fired clay to produce a pattern on the roof. In most cases both sides of the tile are coloured indicating that it was dipped in the slip, but on others only one surface is coated, suggesting that the slip was painted on. Slip would be applied whilst the tile was still "green" such that it was absorbed into the tile and became integrated into the outer part of the fabric.[53] Slip tends to be associated with later sites[54] but Drury has noted white slip at the earlier sites of Chelmsford and Stonea.[55] Colouring

[50] Reading 342.
[51] Two Group C tegulae from Winchester also have peg-holes.
[52] But note Plate 2.1 showing three Winchester tegulae which are presumed to come from the same mould where two have a peg hole and one does not.
[53] Pers Comm: Peter Minter of Bulmer Brick & Tile Co Ltd.
[54] Examples have been noted, *inter alia*, at Frilford, Stanton Low and Gestingthorpe.
[55] Drury 1988, 79.

on tiles has also been noted at Piddington[56] but this is not slip and similar observations have not been made elsewhere.

## 2.3(e) Ends

| | Observation | Conclusion |
|---|---|---|
| **E1** | Ends straight and square often with a sanded band running along centre of the end face, end is rarely knife cut (Plate 2.18) | Not Mould B or C |
| **E2** | Often trimmed above and below sanded band (Plate 2.19) | Not Moulds E or F |
| **E3** | Sanded band normally covers flange (Plate 2.20) | Acts as template for wiring out centre (Mould D) |
| **E4** | Tally marks on ends of some military tiles but rarely on civilian tiles (Plate 2.22) | Identification number of detachment? |

Plate 2.18: Fishbourne 2158 – Sanded band along end

The ends of most tegulae appear to have been made in a mould with closed ends because it is rare to see a tile where the ends have been cut with a knife as would be required for an open-ended mould.[57] Frequently, the end of the mould appears to take the form of a narrow bar marginally thinner than the base of the tile (Plate 2.18). As a result clay was often squeezed above and beneath the narrow end of the mould and was subsequently trimmed with a knife (Plate 2.19). This suggests that the open ended mould forms B and C were not employed, nor were the box moulds E and F because these would not have resulted in a sanded band and there should have been no gap between the end of the mould and its base for the clay to squeeze through to be subsequently trimmed.[58]

Plate 2.19: Chester 901 - Clay trimmed around end band

The sanding normally extends to cover the end of the flange as well (E3). This suggests that the side of the mould around the flange was used as a template to guide the wire when the centre of the clay blank was removed to form the flanges (Plate 2.20). If this was the case, then it is unlikely that the end of the mould was made of wood because this would have been quickly frayed away by the action of the wire being rubbed against it. Metal mould ends are therefore likely to have been used. If such metal ends extended around the corners of the mould these would have had the additional advantage of providing structural rigidity. This would have been particularly helpful in the case of trapezoidally shaped moulds where it would have been difficult to maintain the integrity of a mould where the corners were not square.

Plate 2.20: Holt 1112 – Sanded band over flange ends

Plate 2.21: Witham 2517 – Sanded band not over flange

[56] Piddington 2105.
[57] Brodribb 1987, 101 notes that the lower part of some signatures had been foreshortened because the end of the tile had been trimmed.
[58] It is slightly surprising that there is any leakage over the top of the bar because the wire should have been run along it as a template. This might be a consequence of the elasticity of the clay which could be compressed by the passage of the wire or the result of clay displaced in subsequently striking the tile.

Occasionally the end of the mould does not appear to have fully covered the flange (Plate 2.21) which is difficult to understand if this was used as a template for cutting out the centre of the tile. It is just possible that a further "profile" piece was attached to the metal band to allow the combined assembly to act as a template but this does not seem very satisfactory.

Plate 2.22: Beauport Park 172 – Tally mark

A small number of tiles from military sites have markings (tally marks) incised into the ends of the tiles (Plate 2.22). These markings look like numerals but, based on the sample from Beauport Park (which has more of these than any other site), they do not appear to constitute a counting system because the range of numbers is far too limited.[59] An alternative explanation could be that they were the identification number of the military detachment assigned to making that batch of tiles. This explanation would also help explain the extreme rarity of such marks on civilian tiles.[60]

Plate 2.23: Leicester 710 – Linear markings on underside

## 2.3(f) Undersides

| | Observation | Conclusion |
|---|---|---|
| **F1** | Underside normally rough | Tile formed facing upwards, not Mould F |
| **F2** | Rare raised linear markings on underside (Plate 2.23) | Gaps in palette planking, Moulds C, D or G |
| **F3** | Some Group D and regional Type 7 tegulae have smooth undersides as if finished with a striker (Plate 2.24) | Mould F |
| **F4** | Wire marks sometimes complete length of underside (Plate 2.25), other tiles with knife marks on part of underside | Releasing tile from palette, not Moulds E or F |
| **F5** | Rare plant impressions but never animal prints | Always dried face up |

Most tegulae have rough undersides (F1) consistent with the clay being thrown into the mould. As a generality, Group A and Group B tegulae tend to have rougher undersides than Group C or Group D. Some of the rougher tiles could have been made in a mould placed on the ground rather on a table or palette. If this were so then these would have used the open bottomed four-sided mould D and the tiles would then have been left in situ to dry prior to being turned onto their sides. It is arguable that if the clay was rolled flat prior to insertion into the mould then the quality of the underside would have been superior to that observed (at least on Group A and B tegulae) and that there would have been some evidence for a spade or other lifting device being inserted beneath the flat sheet of clay to transport it to the mould. Although the knife marks noted in F4 could just as easily be spade marks, they always appear to be polished indicating that they were made when the clay was leather-hard and are therefore inconsistent with transporting the clay to the mould.

Very occasionally undersides are noted with very clear raised sections running the length of the tile (F2): a group of such tiles have been noted at Leicester (Plate 2.23). These markings could be related to some system for lifting the tegulae when they were wet, but more likely they reflect a deformity in the palette on which the tile was made, most probably a gap in the planking caused by shrinkage.

Some Group D, and all of the regional Type 7 tegulae, have smooth undersides (F3) consistent with the surface having been finished with a striker (Plate 2.24). These tiles must have been made in an inverted box Mould F. The tegulae with triangular flanges from Leucarum and Carlisle also have smooth undersides which further

[59] See discussion in Chapter 5.

[60] Leicester 727 is the only civilian tegula in the survey on which a tally mark was observed, however 25 tally marks were recorded from the Silchester forum/basilica (Timby 2000, 116) and Brodribb 1987,132 reports that a further seven (civilian) sites have produced single examples of tally marks but these were not necessarily on tegulae.

confirms the expectation that these were also the product of an inverted box Mould F.

Plate 2.24: Sparsholt 1903 – Smoothed underside

Some tegulae have longitudinal striations running the full length of the underside (Plate 2.25) which are consistent with a wire being run between the palette and the base of the tile to release it when it has become stuck. However, they would also be consistent with the initial state of a tegula made in an inverted box mould prior to striking to smooth the surface, which in these cases may have been omitted. Knife marks covering part of the underside are also observed and are consistent with a knife being slid under the tile to release it from the palette.

### 2.3(g) Upper cutaways

| | Observation | Conclusion |
|---|---|---|
| G1 | Upper cutaways normally knife formed but not normally polished (plate 2.26) | Cut at the same time as the tile was formed |
| G2 | Cutaways are normally of equal length but not always cut flush with base of tile[61] | Side of mould provided a template for cutting |
| G3 | Some Group C and D tiles and all regional Type 7 tiles have surfaces that do not appear cut and occasionally have lips of residual clay around the cutaway form (Plate 2.27) | Cutaway formed by a mould insert, most likely Mould F |
| G4 | Some military tiles have the ends of the cutaway cut diagonally across the flange rather than at right angles to it (plate 2.28) | Military practice |
| G5 | Type 9 regional tiles have a unique form of upper cutaway (Plate 2.29) | To fit with the form of lower cutaway |

[61] Leicester 714.

Plate 2.25: Beauport Park 164 – Wire marks on underside

With the exception of observation G3, moulding sand is never observed on the upper cutaways which mostly appear to have been formed with a knife (G1). Only a minority of these knife cuts are polished indicating in most cases that the upper cutaway was formed at the same time as the tile, rather than when the tile was leather-hard and out of the mould.

Cutaways are almost always of equal length (G2) which would be unlikely to occur if they were produced freehand and it therefore seems probable that some sort of template was employed. The obvious template is a cut out section in the side of the mould which would allow a knife to be inserted to remove the upper part of the flange (Plate 2.26). This also reinforces the view that most of the upper cutaways were formed whilst still in the mould. Such an arrangement would also explain why a proportion of upper cutaways are not flush with the base of the tile but are typically two to three millimetres above it.[62]

Plate 2.26: Chedworth 094 – Knife cut upper cutaway

Some tegulae have upper cutaways that have the appearance of being created by an insert into the mould (Plate 2.27) often with residual lips of clay either still upstanding or subsequently crushed showing where the clay wrapped itself around the mould insert. Forty-two of

[62] Leicester 714.

these upper cutaway inserts have been observed, and 29 of these tegulae have both an upper cutaway insert and lower cutaway present: all of the lower cutaways were Group C and D or regional Type 7.[63] A mould insert of this sort would have been used with the inverted box Mould F.

Plate 2.27: Dorchester 2450 – Upper cutaway insert

Some upper cutaways are cut on a slope thereby producing a gradual reduction in the height of the flange rather than a sharp step. Rarer are cutaways where the cut is made diagonally across the flange (Plate 2.28) rather than being square. With the exception of South Shields regional Type 8 which is discussed in the next section, there does not appear to be any pattern to the sloping cutaway. However, the diagonal cutaways are found exclusively on military sites[64] with the exception of two ostensibly civilian sites in East Anglia.[65] Only a small minority of tegulae on these sites had the diagonal cuts. There is no obvious functional reason for these unless they were required for the row next to the roof ridge and interacted in some manner with the ridge tiles.

Regional Type 9 cutaways from Corbridge (Plate 2.29) have a unique form designed to mesh with the equally unique lower cutaways from the same site (Plate 2.30). This would have produced a single overlap of the tiles rather than the normal double overlap.[66] This approach, which has not been observed on any other site, would have been extremely time-consuming and difficult to achieve.

[63] However there was one Group B tegula from London (Dominant House 381) with a possible upper cutaway insert.

[64] Caerleon, Carlisle, Chester, Exeter, Fishbourne, Holt, South Shields, Tarbock and York. The Fishbourne tegulae with diagonal upper cutaways come from one of the abortive foundation trenches for the palace and must therefore date to circa AD 65 or earlier. In terms of size and design they are very similar to tegulae from the fortress baths at Exeter. It therefore seems likely that were produced by *Legio II Augusta*.

[65] Gestingthorpe, Suffolk and an observation by Brodribb (1983, 34) at Fiddlers Hamlet, Essex.

[66] A single overlap is where the lower cutaway of the tile higher up the roof rests on the upper cutaway of the tile beneath, and a double overlap is where the lower cutaway rests on the top part of the flange of the tile beneath: in round terms this increases the overlap from 50mm to 100mm. This topic is discussed in more depth in Chapter 6.

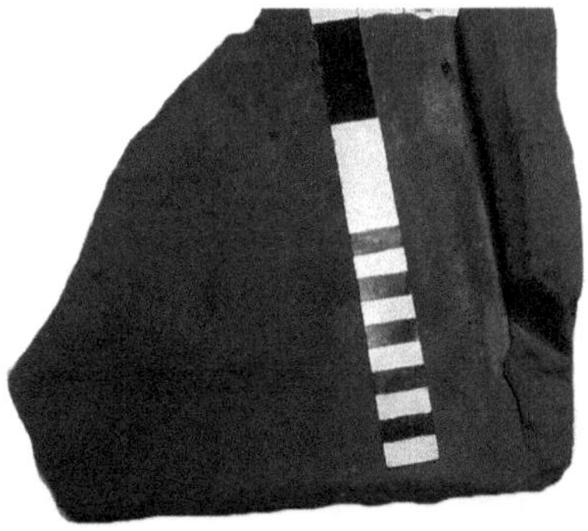

Plate 2.28: Tarbock 279 – Diagonally cut upper cutaway

There is a final class of upper cutaway which appears to have been formed by simply pressing down upon the flange with a thumb. Whether this was a deliberate policy or a response to a failure to form the cutaway whilst it was still in the mould is difficult to say, however four such cutaways have been observed at Caerwent,[67] which might tend to suggest that it was a deliberate act.

Plate 2.29: Corbridge 407 – Regional upper cutaway form

Plate 2.30: Corbridge 410 – Regional lower cutaway form

Cirencester (Plate 2.31) has an unusual upper cutaway which has resulted in a splayed base to the tegula: this could not have occurred whilst it was in the mould and

[67] Caerwent 1164.

may have resulted from the cutaway being hand formed by pressing down on the flange, but it is much cleaner than the Caerwent examples.[68]

Plate 2.31: Cirencester 130 – splayed upper cutaway base

## 2.3(h) Lower cutaways

There are thirteen different variants of the lower cutaway form which have been divided into four basic national groups A – D as shown in Figure 1.3. The Type Numbers allocated to each of the forms are a result of history and are not intended to constitute a numeric sequence. The methodology behind the allocation into the four groups A – D was discussed in Section 1.3 and the implications of this choice are discussed in Chapter 3. Figure 2.3 shows the regional distribution of the cutaway groups. With the exception of the regional (Group R) forms which are also discussed in Chapter 3, the analysis shows that all the groups are present in each region although Groups A and B tend to predominate in the northern regions and Groups C and D in the southern regions. However the paucity of data when broken down by both cutaway group and region means that caution should be used when drawing conclusions from this data.

Group A cutaways either comprise three orthogonal cuts to create a cuboid notch beneath the flange emerging out of its side or a cuboid mould insert which is then finished with a diagonal stroke normally to the base.[69] Plates 2.32 a-e show examples of each of the Group A cutaway forms which are also illustrated schematically in Figure 1.3. These are presented as discrete types but, with the exception of Type 29, they really represent a continuum of forms. Although the majority of sites tend to have cutaways that have either been knife cut or have mould inserts, there are some sites[70] where both inserts and knife cuts were used.

Plate 2.32a: Type 2 lower cutaway

Plate 2.32b: Type 26 lower cutaway

Plate 2.32c: Type 27 lower cutaway

Plate 2.32d: Type 28 lower cutaway

---

[68] Gloucester 680 has a similar bulge on the base but this seems to have resulted from pressure by the stepdown but it should be noted that Cirencester 130 also has a stepdown.

[69] However even those variants with sloping cuts, where logically a cuboid insert should have been used, frequently have the appearance of being entirely knife cut: for example Castleford 2327.

[70] For example Caerleon, Caerwent, Chelmsford and Colchester.

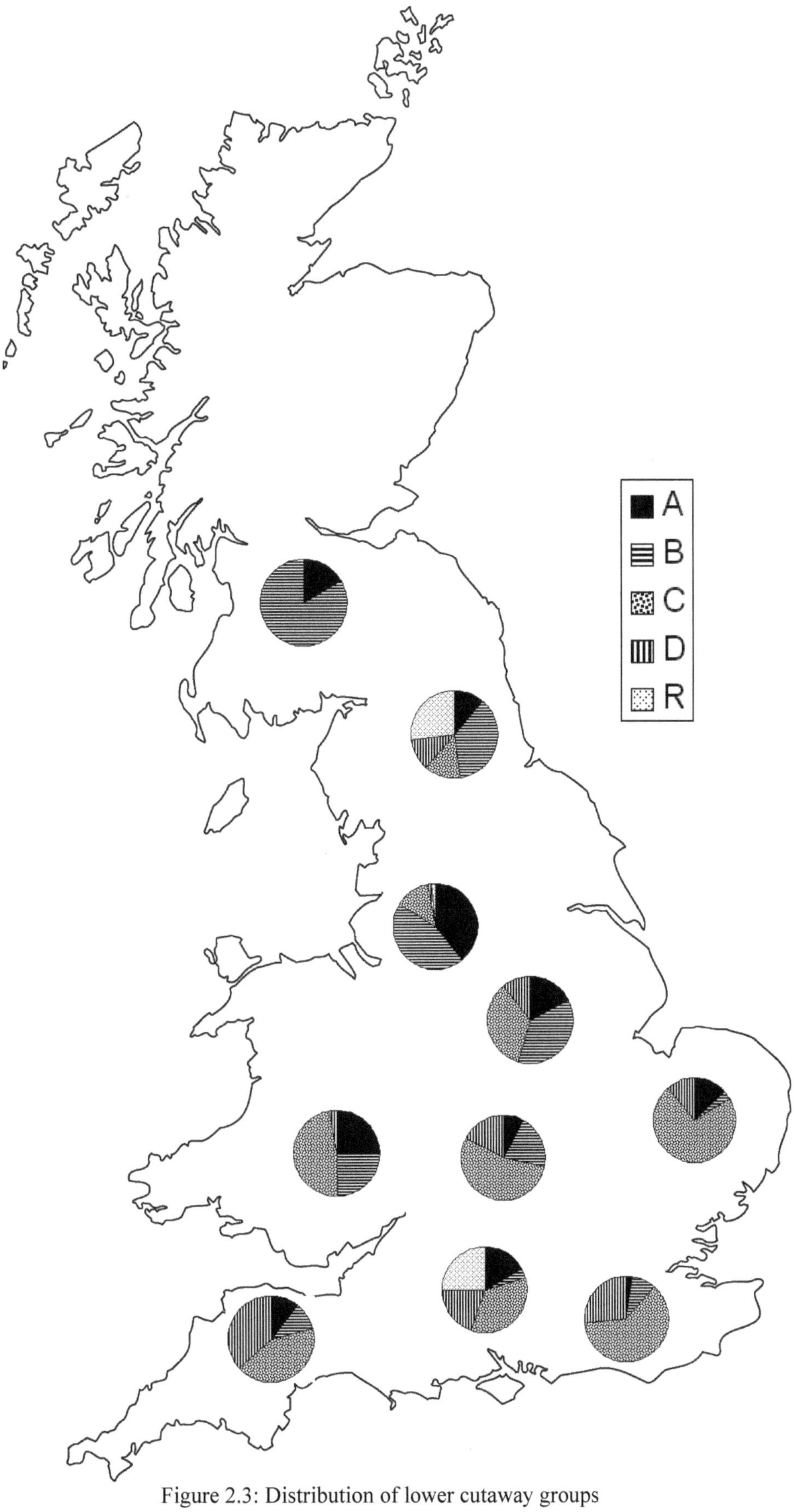


Figure 2.3: Distribution of lower cutaway groups

Plate 2.32e: Type 29 lower cutaway

The inserts were most probably attached to the palette rather than the side of the mould which would then provide both a location for the mould onto the palette, and also allow the mould to be removed vertically from the tegula which would not be possible if the inserts were attached to the mould.

Plate 2.33a: Type 6 lower cutaway

Plate 2.33b: Type 62 lower cutaway

Group B cutaways are made with two knife cuts: one a diagonal along the line of the flange and the other at right angles to it to release the notch of clay (Plate 2.33a/b and Figure 1.3). The Type 62 has a cuboid mould insert followed by the same cuts whereas the Type 6 shows no evidence of an insert. It is possible that all Group B tegulae had a cuboid insert but that on the Type 6 tiles the diagonal cut removed all trace of it whereas the Type 62 diagonal cut only removed the outer edges. Sometimes the Type 62 cutaways look as if a thumb has been placed on a Type 6 cutaway in order to grip the tile to lift it, giving it the appearance of a Type 62. However, as there are no matching finger marks (or other marks) on the upper surface and the marks in the cutaway all look to have mould sand adhering to them, it seems likely that these are Type 62 cutaways formed with an insert.

In contrast to Groups A and B, Group C cutaways have a mould insert that runs to the full height of the flange and emerges through the top (Plate 2.34a-c and Figure 1.3). Type 5 tegulae are finished with a diagonal knife cut which normally appears to be made when the tile was leather-hard. Type 4 tegulae have no cut and their relative rarity (only one Type 4 to every 25 Type 5) could suggest that they were manufacturing errors where the subsequent diagonal cut had been overlooked. This conclusion is supported by the three tegulae from Winchester (discussed in 2.3(a)) that were made in the same mould, where two of them have Type 4 cutaways and the third a Type 5.

Plate 2.34a: Type 4 lower cutaway

Plate 2.34b: Type 5 lower cutaway

Plate 2.34c: Type 56 lower cutaway

Although most Group C tegulae show signs of mould sand from the insert, there are some where the whole cutaway has been formed by knife cuts. These knife cuts could be over-zealous fettling but it seems more likely that these tegulae were actually made without an insert:

this would entail four separate knife cuts to create the cutaway and could therefore be evidence of a powerful requirement for this manufacture to conform to the prevailing style.

Type 56 cutaways have been placed in Group C but are really a hybrid between Group B and Group C because they do not necessarily incorporate an insert but consist of a series of diagonal cuts. This cutaway form is rare.

Plate 2.35a: Type 1 lower cutaway

Plate 2.35b: Type 15 lower cutaway

Plate 2.35c: Type 16 lower cutaway

Group D cutaways have the diagonal stroke either emerging through the top corner of the flange or through the top of the flange itself (Plate 2.35a-c and Figure 1.3). Some show clear evidence of an insert whilst others give the appearance of being knife cut. The diagonal stroke on the Type 1 can sometimes be vertical but can be distinguished from a Type 4 by the amount of flange removed: if the indent is less than 15mm then it is classed as a Type 4 unless it comes from a site where there are many similar tegulae that are clearly Type 1. Fettling of the lower part of the diagonal insert is common and occasionally this takes the form of a further small diagonal cut (Type 15). This can make it awkward to distinguish from a Type 5 but again if the size of the indent is greater than 15mm and the "fettling" diagonal is less than 20mm then it has been classed as a Type 15.

The most abundant regional cutaway form is Type 7 which is found in Hampshire and the Isle of Wight but only on villa sites (Plate 2.36). Examples have been examined from Sparsholt, Crookhorn, Grateley and Houghton Down and there are descriptions from Combley, Brading and Rock villas on the Isle of Wight[71] which match these. The cutaway is not present at Fishbourne, Bignor, Batten Hanger or Chilgrove to the east or in Dorset to the west, whilst Grateley is the most northerly observation. The Sparsholt and Crookhorn fabrics appear similar and they share a similar set of unusual signatures which also appear on the Isle of Wight examples. It therefore seems possible that at least some of the Type 7 tegulae came from the same manufacturing source.

Plate 2.36: Type 7 lower cutaway

The Type 7 is unique in that the whole of the flange is removed at the lower end of the tile which simplifies manufacture and meshing but which might make them less watertight (see discussion in Chapter 6). All of the examples have smooth undersides and upper cutaway inserts which makes it highly probable that these tiles were made in an inverted box Mould F.

The Type 8 regional cutaway form has been principally found at South Shields where one of the examples also carries the stamp of *Cohors V Gallorum* which was the third/fourth century garrison of the fort. The cutaway takes the form of a diagonal cut into the side of the flange running from the top of the lower end of the flange down to the base, a second vertical slice is then made to remove the notch of material (Plate 2.37). This lower cutaway form is presumably designed to mesh with the unusual

[71] Tomalin 1987, 110.

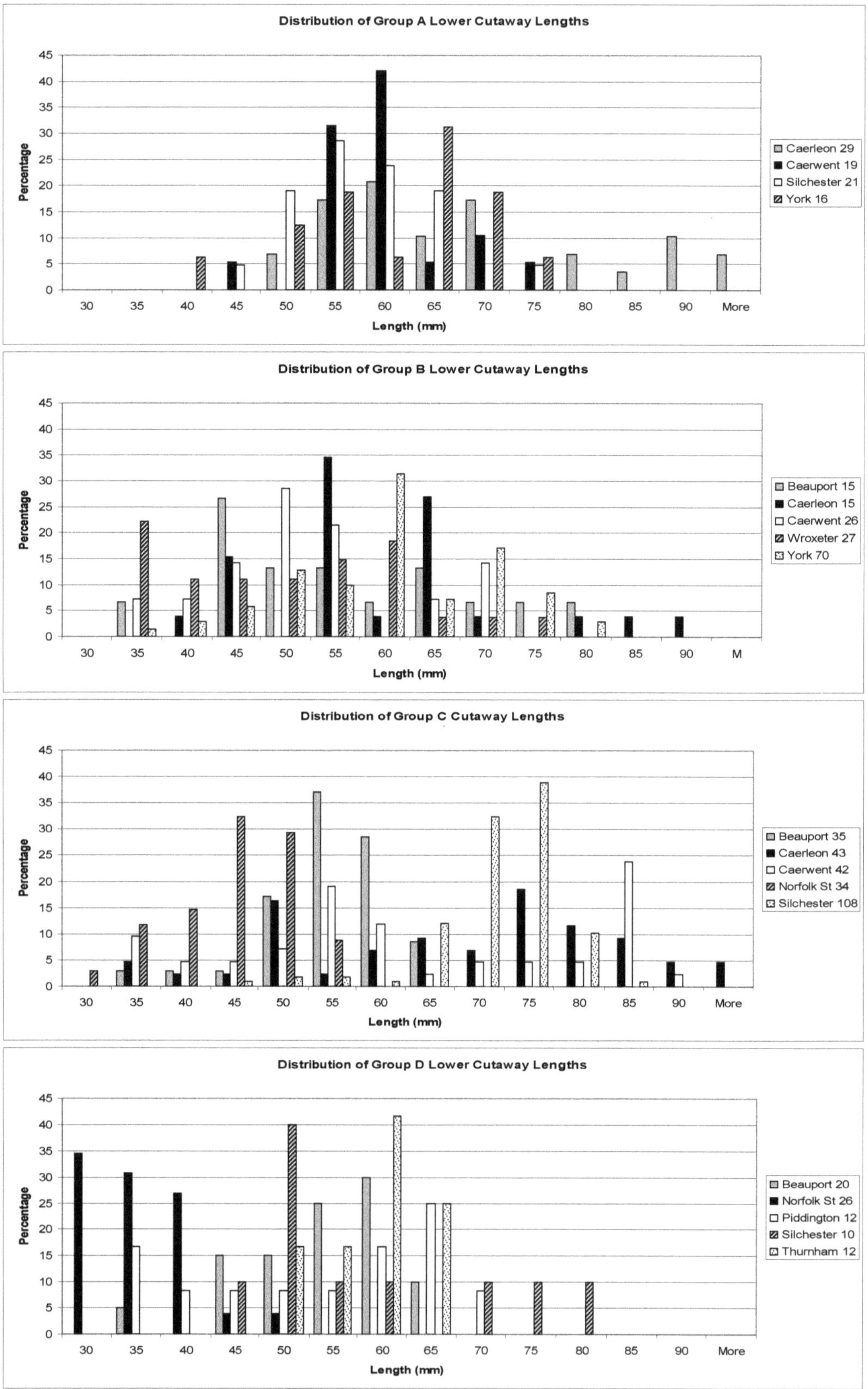


Figure 2.4: Distribution of lower cutaway lengths

sloping upper cutaway which also appears on this tegula (Plate 2.38). This would have produced a single overlap, rather than the normal double and may not have been watertight.

Plate 2.37: Type 8 regional lower cutaway

Plate 2.38: Type 8 upper and lower cutaways

Curiously the Type 8 cutaway also appears with a conventional upper cutaway (Plate 2.39) which would have produced a very unsatisfactory mesh. The only other site where similar lower cutaways have been observed is Quernmore, near Lancaster,[72] but no sloping upper cutaways were discovered in the assemblage. There is no evidence for *Cohors V Gallorum* being stationed at Lancaster.

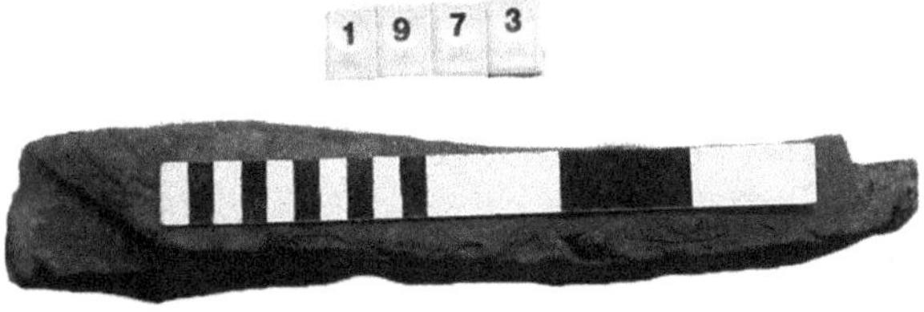

Plate 2.39: Type 8 lower and conventional upper cutaway

There is one tile from Grimescar in Yorkshire that is superficially similar to a Type 8 but this has been made with three distinct cuts and so is more likely to be a Group A tegula where the normal horizontal cut has been made at an angle along the flange.[73]

The final regional form, Type 9, only occurs at Corbridge. The lower cutaway is created by cutting away the clay when leather-hard from across the whole lower end of the tile (Plate 2.30); this would have allowed it to mesh with the upper cutaway which had just the inside of the flange removed with a knife (Plate 2.29). These cutaways would have been very time consuming to make, and based on the separate examples with either upper or lower breadths intact, would have produced a single rather than a double overlap on the roof.

Having summarised the different lower cutaway forms it is possible to explore empirically the extent to which inserts were used in their manufacture. If lower cutaways were entirely knife cut then one would expect some variation between the lengths of the cutaways on opposing flanges because it would not always be possible to judge this by eye, especially as the general evidence of tegula manufacture suggests that the craftsmen were not overly concerned with precision. If, on the other hand, they were either wholly produced by mould inserts or the insert was used as a guide (for example in the case of the Type 62) then it would be surprising to find significant differences between the mould inserts used on the same mould or attached to the same palette. In fact, of the 480 tiles with a complete lower breadth, and therefore both lower cutaways present, only 18 have a difference of one centimetre or more. This may slightly understate the incidence because it is always possible that some mismatches have been overlooked, nevertheless this correlation seems to be a strong endorsement of the use of inserts in almost all cutaways.

The question of whether the insert was attached to the palette or the mould can also be explored empirically. If a separate palette[74] was used for each tegula to dry on (where it may have remained for as much as a week), but only a single mould was used for the day's production, then there would be far more palettes than moulds. Thus, in this scenario, the variety of cutaway lengths would be expected to be greater if the insert was attached to the palette rather than the mould. Only Group C cutaways are likely to have been attached to the mould because the other forms prevent the mould being lifted off, so only Group C cutaway lengths might show such a distinction.

Figure 2.4 shows the distribution of cutaway lengths for each of the four lower cutaway groups. Only those sites with the largest data sets have been included and the number of entries for each site is recorded against the site name in the shade key box. It can be seen that only

[72] Lancaster 2068.

[73] Grimescar 2034.

[74] But see further discussion in Section 2.6.

Silchester, and to a lesser extent Beauport Park, of the Group C cutaways have a tight length distribution. All the other sites and cutaway groups show a wide spread of cutaway lengths. It is therefore reasonable to conclude that Group C tegulae at Silchester, and possibly Beauport Park, were made with the cutaway insert attached to the mould. All other sites and cutaway groups, including the Group C tegulae from Caerleon and Caerwent, probably had the insert attached to the palette.

The design of the inverted box Mould F would have required any lower cutaway inserts to be attached to its side. Loose inserts could also have been used but this would have been more cumbersome. This means Mould F was unlikely to have been used to manufacture Group A or B tegulae or for Type 16 from Group D because the tile could not be removed from the mould. However, it is possible that some of the more vertical forms of Types 1 and 15 from Group D could have squeezed out past a cutaway insert attached to the side.[75]

## 2.4 Experimental results

Modern craft (rather than mass production) brick and tile making practice[76] is to cut a lump of clay larger than the size of the intended mould from the mound of prepared clay. Next this lump is cut very approximately to the shape of the mould, sprinkled liberally with sand, and the clay is then folded into a rough ball such that it can be more readily picked up. The mould is prepared by being wetted and sprinkled with sand and the ball of clay is literally thrown into it. The clay is punched into the corners of the mould and, in the event that there is not a surplus of clay in the mould, then more clay is added and punched into place (Plate 2.40). Next the surplus clay is removed by dragging a tensioned wire (referred to as a *bow* because it is traditionally tensioned by tying it between the ends of a bent hazel stick) across the top of the mould and the surplus clay can then be peeled off (Plate 2.41). This leaves a rough surface (Plate 2.42) which is smoothed by running a straight-edged piece of wood across the surface in a process known as *striking*. Striking also increases the surface tension so producing a harder (and therefore more weatherproof) surface.

After the forming process is completed, the mould is shaken to release the clay from the sides and corners before being turned out of the mould onto a palette or former to dry. When the tile is leather-hard, which might take up to a week depending upon temperature and humidity, the tile is turned off its palette. Rough edges are fettled off with a knife and any additional features required are cut. The tile is then typically rested on its side to complete the drying process. To optimise space utilisation the tiles may be stacked in a herringbone fashion which also provides mutual structural support to the tiles to ensure that they do not slump. They are then left to dry for up to a further week before being loaded into the kiln.

Plate 2.40: Loading clay into three-sided Mould C

Plate 2.41: Peeling clay off mould after wiring

Plate 2.42: Surface left after wiring

[75] This suggestion is reinforced by the observation that most Type 15 cutaways appear to have originally been Type 1 forms that have been fettled at the base of the cutaway which could have resulted from the need to clean up the area where the clay lapped over an insert attached to the side of the mould.

[76] This account is based upon the practice adopted by the Bulmer Brick and Tile Co Ltd in Suffolk.

Four different moulds were tested at the Bulmer Brick and Tile works, each with a Group C cutaway insert attached to the side of mould. The clay used was the same clay as was in regular use for their commercial tile making.[77] This clay was too moist and adhesive to be rolled out and was therefore cut and folded as described above before being thrown into each of the moulds in turn.

## 2.4(a) Two and three-sided Mould B and C

The three-sided Mould C proved difficult to fill because the clay tended to spill out of the open ends when it was punched into the sides (Plate 2.40) thus requiring more clay than would have been necessary for four-sided and box moulds. It also caused the mould to break apart because the sides had no bracing and were entirely reliant on the nails in the baseboard to hold them in position. The experimental mould was rather crude, but this problem might still be anticipated in a better made production version.

Removing the surplus clay off the top of the mould with a wire presented no problems, but running the wire into the centre of the mould to create the flanges proved extremely difficult because of the absence of a template to press the wire against. Although this was ultimately achieved the flanges were not symmetrical and it seems unlikely, even with more practice, that symmetrical flanges could be regularly achieved without the aid of a template. Striking the surface of the tile and running the hands along the flanges to smooth them worked well. Once the tile was finished the mould was lifted and tipped and the tile slipped out without difficulty.[78]

A two-sided Mould B fixed to the table should give identical results to the three-sided mould with the exception that the tegula would have to be lifted out of the mould which might be a challenge.

## 2.4(b) Four-sided Mould D

The four-sided Mould D (Plate 2.43) worked well. The clay was thrown in, punched into the corners and the surplus clay wired off the top. The wiring out of the centre of the mould using the mould ends as a template proved satisfactory, although a slight radius to all the corners of the template would have made it easier for the wire to run. The smoothing of the flanges with the hands was entirely satisfactory. The surface of the tile was struck with a piece of wood which yielded a smooth finish and the mould was easily lifted off the tile which was finished by making a finger mark signature and stamping the tile, both of which looked authentic (Plate 2.44).

Plate 2.43: Four-sided Mould D

Plate 2.44: Tegula made in four-sided Mould D

A test was also done to see if the flange taper could be produced purely by squeezing the flange with the hands rather than cutting the required taper. This proved very difficult to do, and even with practice, it seems unlikely that the taper could be produced by hand alone.

Vertical marks on the outside the flanges are very seldom seen which is somewhat surprising if the mould is removed vertically from the tile. To test this issue a nail-head was inserted into the side of the mould protruding out by some 2mm. After the tile was made and the mould removed, a dimple the size of the nail head was visible in the outside face of the flange, but there was no vertical scratch caused as the mould was removed. The absence of marking was presumably because the clay is somewhat elastic and therefore shrinks back from the sides of the mould despite being initially punched against them.

[77] Peter Minter of Bulmer Brick and Tile Co Ltd believes (pers comm) that, although his clay was too wet to be rolled, it was still stiffer than that used by the Roman tilemakers where the water content would have been around 24%.

[78] However if the Roman tilemakers had used a wetter clay mix then the suction from this mix might have prevented the tile from slipping easily off the palette.

## 2.4(c) Upright box Mould E

The procedure for using the upright box Mould E (Plate 2.45) was exactly the same as for the four-sided mould until the point where the tile was to be removed from the mould. The one difference in the experiment was that the four-sided mould omitted a cut down section in its sides to act as a template for forming the upper cutaways whereas the upright box mould employed this feature. Removing the upper cutaways with the wire proved a little awkward (in part because of the square corners) and it would probably have been quicker and easier to have used a knife.

Plate 2.45: Upright box Mould E

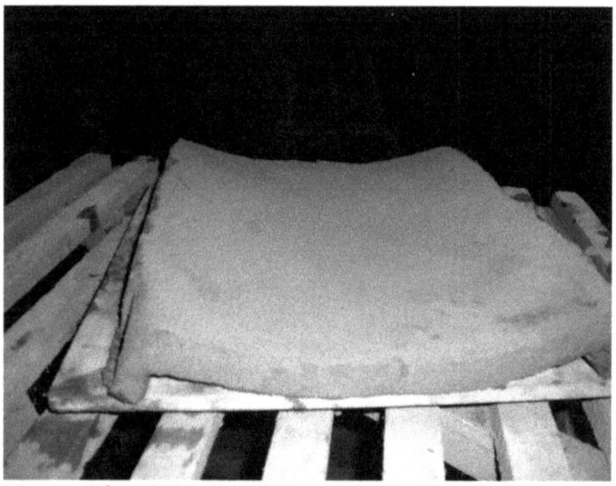

Plate 2.46: Tegula rotated out of Mould E

Removing the tile from the mould was done by placing a board over the top of the mould and then rotating the two pieces together such that the tile fell out of the mould onto the board. Although this procedure worked satisfactorily, the clay was insufficiently rigid and the unsupported centre of the tile collapsed (Plate 2.46). This approach could only work if the clay were left in the mould for a sufficient period for it to gain some structural rigidity (which would therefore have necessitated a multiplicity of moulds) or if it was turned out onto a baseboard with a raised centre section to provide support for this part of the tile. There is no evidence for this.[79] An alternative procedure would be to turn the tile out onto a flat board and then immediately rotate it again onto another board so as to restore it to the correct way up where the tile would be properly supported but this would be difficult to achieve without damage and would take more time.

## 2.4(d) Hinged Mould G

The final mould examined was one with hinged sides and end pieces[80] (Plate 2.47). The concept was based upon the observation that trapezoidal tegulae tend to have flanges which are tapered in the opposite direction to the tapering of the overall form such that if the flanges could be laid down flat then they would counterbalance the trapezoidal shape of the base to create a rectangular slab of clay. In fact at Beauport Park, allowing for shrinkage, the flat slab of clay so created would be a square with a side equal to one and a half Roman feet: thereby vindicating the appellation *tegula sesquipedalis* which is the full Latin name for the roof tile.

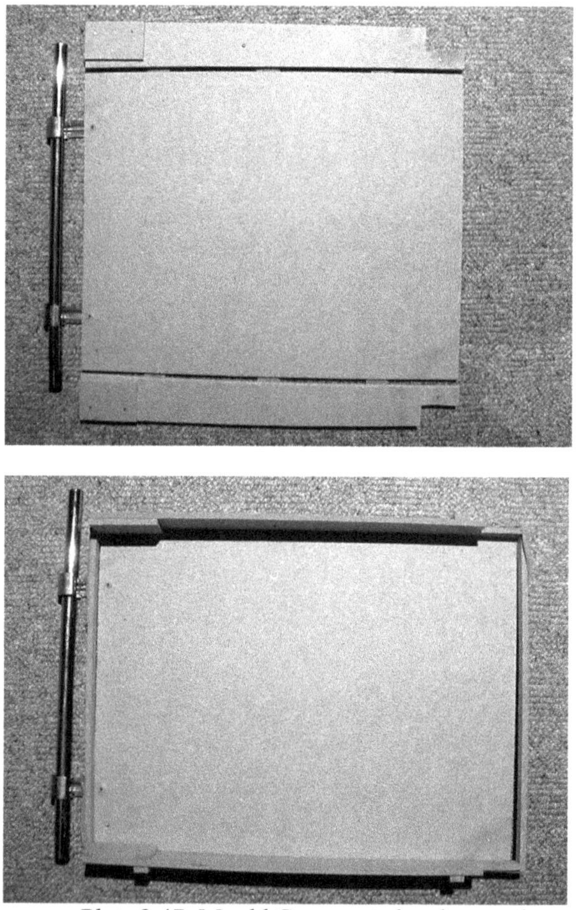

Plate 2.47: Mould G unerected and erected

[79] There is also no evidence to support Rook's contention (1979,301) that striation marks might be left from cutting the centre of the tile out with a wire: the surface left by the wire would have to be smoothed before firing because it is far too rough and pitted to be weather resistant.

[80] The inverted box mould F was not experimentally tested

In this approach the intention was that a lump of clay would be placed on the flat mould with all its hinged sides down, rolled flat and then trimmed to the outside profile of the flat mould. The sides of the mould would then be raised with the clay inside being bent up to form the flanges. The ends of the mould would be swung over into position locking the sides in a vertical position. The tile could then be finished in the same way as for the other moulds. Removing the tile from the mould would be done by rotating the mould about the lower end (in the model tested there was also a hinge to facilitate this at the lower end) where it would have been placed on a transitional baseboard with a raised centre to provide support whilst the lower cutaways were formed before being rotated backwards onto a flat baseboard to be taken away to dry. The rotation of the tile onto the transitional baseboard would have required a retaining strip to be inserted at the upper end of the tile to ensure the tile did not prematurely fall out of the mould in the process. Such a retaining strip would have been sanded and would have created the stepdown that is observed on many tiles.

However the underlying presumption that the clay could be rolled flat proved false. So whilst it was possible to make tiles in this way by placing clay on top of the mould and then producing the flat surface by wiring off the surplus clay, the whole process proved time-consuming and would not have been adopted unless the consistency of the clay was such that it could be rolled.

## 2.5 Summary of evidence

The best arguments for and against the different mould types can now be summarised below. To avoid repetition further arguments have not been added once a persuasive case has been made. Where certain characteristics produce conflicting evidence these have been juxtaposed to emphasise the alternatives. The codes in brackets refer back to the observation numbers in Section 2.3.

| **TEGULAE WERE HAND FORMED** | |
|---|---|
| FOR | AGAINST |
| | Flanges longitudinally straight (B1) |
| | Flanges flat with evidence of mould sand (B2) |
| | Leakage from bottom of mould should not be observed (B4) |
| | Ends rarely knife cut (E1) |

| **CLAY WORKED WITH WIRE, NOT ROLLED** | |
|---|---|
| FOR | AGAINST |
| Profile of fabric in flange consistent with rolled ball thrown into mould (C1) | Profile of fabric in flange consistent with bending (C1) |
| Difficulty of lifting rolled blank into mould | |
| Absence of asymmetric flanges (C3) | |
| Flanges show no distortion above lower cutaway insert (C13) | |
| Lower external angle sharp (B3) | |
| Leakage from bottom of mould inconsistent with bending (B4) | |
| Rare evidence of very wet clay (B6) | |

| **TEGULAE MADE IN TWO-SIDED MOULD B OR THREE-SIDED MOULD C** | |
|---|---|
| FOR | AGAINST |
| | Leakage from bottom of mould should not be observed (B4) |
| | Width of flanges normally equal: would have required a template (C3) |
| | Ends rarely knife cut (E1) |
| | Experiment showed that this mould was difficult to fill and the flanges problematic to cut symmetrically |

| **TEGULAE MADE IN FOUR-SIDED MOULD D** | |
|---|---|
| FOR | AGAINST |
| Sharp external but rounded internal angles on flanges (B5, C6 and C7) | Sharp corners on some Group C/D and Type 7 tegulae (C10) |
| Width of flanges normally equal: would have required a template (C3) | |
| Underside normally rough (F1) | Some Group D and Type 7 tegulae have smooth undersides (F3) |
| Most upper cutaways formed with a knife (G1) and against a template (G2) | Some Group D and Type 7 tegulae have upper cutaways made with mould inserts (G3) |

| TEGULAE MADE FACE UP IN OPEN BOX MOULD E | |
|---|---|
| FOR | AGAINST |
| | Leakage from bottom of mould should not be observed (B4) |
| | End trimmed above and below sanded band (E2) |
| Underside normally rough (F1) | Some Group D and Type 7 tegulae have smooth undersides (F3) |
| | Some Group D and Type 7 tegulae have upper cutaways made with mould inserts (G3) |
| Lower cutaway inserts straight forward | Knife cut lower cutaways not expected |
| | Experiment showed that turning the tile out of the mould was problematic |

| TEGULAE MADE FACE DOWN IN INVERTED BOX MOULD F | |
|---|---|
| FOR | AGAINST |
| | Leakage from bottom of tile should not be observed (B4) |
| Sharp corners on some Group C/D and Type 7 tegulae (C10) | Rounded internal angles on flanges (C6 and C7) not expected |
| | Finger channels adjacent to flanges (C9) |
| Triangular flanges from Leucarum and Carlisle (C11) | |
| | End trimmed above and below sanded band (E2) |
| Some Group D and Type 7 tegulae and those with triangular flanges have smooth undersides (F3) | Underside normally rough (F1) |
| Some Group D and Type 7 tegulae have upper cutaways made with mould inserts (G3) | Most upper cutaways formed with a knife (G1) |
| | Group A and B tegulae could not be satisfactorily made with fixed lower cutaway inserts |

| TEGULAE MADE IN MOULD G WITH RETRACTABLE SIDES | |
|---|---|
| FOR | AGAINST |
| Stanton Low stamps on outside of flanges (B8) | Experimental version was cumbersome and any mould that has to be dismantled will add extra work |

The virtual absence of asymmetric flanges makes a compelling case for the use of wet clay that was cut rather than rolled. *Inter alia*, the straightness of the flanges rules out hand forming and the impracticality of the two/three sided moulds, combined with few knife cut ends and regular leakage from the bottom of the mould, make these methods of manufacture unlikely. It is possible that the upright open box Mould E was used but it was somewhat impractical and the existence of end banding and leakage from the base of the tile make it unlikely to have been in general use.

The four-sided Mould D was an efficient production method and the evidence points to its use for a majority of tegulae.[81]

The inverted box Mould F would have been an efficient method for slightly higher quality tiles and must have been used for the manufacture of the Type 7 and some Group D tiles as well as those with triangular flanges. In particular the Type 7 cutaway seems to have been developed with inverted box mould manufacture in mind because the unconventional lower cutaway means that it can be formed and removed from the mould with minimum effort.

Plate 2.48: S.L. 2248 – Evidence of lower cutaway insert

Mould G with retractable sides is not an efficient manufacturing method if the clay is wet; nevertheless it seems some version of this mould must have been used for the Stanton Low tegulae which were stamped on the outside of the flange (Plate 2.4). This view is reinforced by the observation of a lip of clay beneath the Type 1 lower cutaway inserts on these tiles (Plate 2.48) which is inconsistent with the use of either Mould D or F.

[81] Although almost all tegulae have a smooth upper surface and symmetrical flanges, the thickness of the tile is less consistent which would suggest that some might have been made with a four-sided Mould D that was placed on (uneven) ground rather than on a table.

## 2.6 Probable manufacturing method

Based on the foregoing evidence and analysis it is now possible to postulate the manufacturing process, first of all based on the four-sided Mould D. Manifestly not all tegulae made in a four-sided mould will have been made in exactly this manner as small manufacturing differences have already been noted in Section 2.3; however, the proposed procedure could be modified in almost all cases to accommodate the alternative observations. The procedure could have worked as follows:

### 2.6(a) Manufacture using the four-sided Mould D

*1. The mould was wetted, sprinkled with sand and placed on a palette that had been similarly sanded*
The palette was made of sawn planks joined together,[82] cut slightly larger than the outside of the open four-sided mould; the overall size of the palette would be minimised to optimise the space required for drying. The mould was positioned with its upper end furthest away from the tilemaker because the signatures are always at the lower end.[83]

Many tiles were made with lower cutaway inserts that must have been fixed to the palette: if these were coupled with an end stop positioned just outside the lower end of the mould, then the mould would have been securely locked in position to prevent both longitudinal and lateral movement. However, some tegulae do not appear to have had lower cutaway inserts and it is also possible that Group C tegulae inserts were attached to the side of the mould (although, as discussed earlier, the evidence from the majority of sites would suggest that the Group C inserts were also attached to the palette).[84] These moulds could have been located onto the palette using end stops outside all four sides.[85]

*2. Cut clay from mound, rough to shape, fold into ball, throw into mould, and punch into corners*

If the lower cutaway inserts were loose then care would need to be taken not to dislodge these, however there is no evidence of dislodged cutaways so presumably they were fixed as suggested above.

*3. Run wire over top of mould and peel off surplus clay*

*4. Wire out centre of mould using the ends as a template to form outline of the flanges*
As discussed earlier, the mould ends were thin metal straps which widened out adjacent to the flanges to create the required profile and taper for the wire to follow. They would also have provided the strength to hold the mould in the (normal) trapezoidal shape. The upper end strap would have needed to narrow again once the corner was formed so as not to impede the cutting of the upper cutaway.

*5. Run hands along flanges to smooth into finished form*
The hands were run simultaneously from the upper to the lower end as tiles show a small (2%) increase in flange height between the upper and lower ends. This process creates the finger channels running down the sides of the flanges and, according to how each craftsman positions his hands, the different flange profiles.

*6. Cut upper cutaway*
With a knife using the side of the mould as a template

*7. Strike the base to smooth and remove surplus clay*
A short piece of wood with a straight side was run down the tile from top to bottom in a series of passes to smooth the internal surface. It is possible that some tilemakers did this prior to smoothing the flanges.

*8. Mark with signature*
A finger was used to trace a (typically semi-circular) mark at the lower end of the tile.

*9. Sponge tile down*
This produced the final smooth surface and removed any sharp lips created by the signature and flange smoothing.[86] Sponging also helps produce a harder surface to the tegula.[87]

*10. Shift tile on its baseboard to drying area*
To avoid congestion tiles would need to be moved immediately after manufacture and it would be sensible if this was directly to their proposed drying position to avoid double handling.

*11. Lift mould vertically off tile*
Leaving the mould in position until the tile had been placed in the drying position would provide some

[82] Joined planks are indicated by the linear features on the underside of some tegulae eg Leicester 710 (Plate 2.23).
[83] The height of the flanges and the thickness of the base of the tile increase from upper to lower end, as there is no engineering requirement for this, this presumably results from the natural actions of smoothing the flanges and striking the tile towards the tile-maker. The thickness of the base could also be a consequence of holding the bow at right angles to the body to cut out the centre of the tile.
[84] Group C inserts were too thin to be nailed to the palette without splitting so the only way they could have been successfully fixed to palette is if they were much wider and fitted into a rebated slot in the sides of the mould. This would have required much tighter tolerances than would have been expected but would have secured the mould without the need for an end stop.
[85] Modern moulds tend to adopt either a rebated fixing or use dowels. A rebated fixing could have been achieved by nailing a flat piece of wood shaped to the inside of the profile to each palette, but this seems unlikely. Dowels require rather more precision over several hundred palettes than seems likely to have been achievable.
[86] This was not always done last because some tiles show sharp edges eg Silchester 065.
[87] Pers comm: Peter Minter, Bulmer Brick and Tile Co Ltd.

stability for the wet tile in transit. It is also suggested by the interpretation of the marks on the side of some of the Beauport Park and York flanges discussed in the next section. This makes it likely that it was the tilemaker who carried the tile and returned with his mould because otherwise double the number of moulds would have been required if work was not to be interrupted.

It is possible that instead of leaving the tile on the palette it was slid off onto a drying board. This may seem an unnecessary step, but if output per tile-maker of 220 tegulae per day is assumed (see later discussion in Section 8.2) then as many as 1000 palettes would be required to support each tilemaker in the absence of secondary drying boards. Each of these palettes would require lower cutaway inserts fixed in precisely the correct position if the manufacture was to be successful. The use of a simpler drying board could therefore make sense and slipping the tile off the palette should be straight forward as was demonstrated with the experimental three-sided mould.

However, the wide variation in most lower cutaway lengths, as shown in Figure 2.3, argues in favour of a multitude of cutaway inserts and therefore a multitude of palettes. Thus, although the approach suggested above would have been the most economic, it may only have been adopted on a minority of sites.

Even drying boards would represent a significant investment so it is not surprising that some tegulae look as if they were made on the ground (where neither palette nor drying board would be required). One might anticipate that this would be particularly prevalent in the early years of the occupation and in rural areas where the demand would not have supported this scale of investment.

*12. Overseer stamps a proportion of the day's output*
This may have been performed at the work bench or later.

*13. Up to seven days later (depending upon the weather) rotate tile off palette, fettle as necessary, finish lower cutaway then stack on side to complete drying*
When the tile was deemed to be robust enough, it would have been carefully rotated onto a spare board. Some tegulae may have stuck to the palette and these would have been eased off by sliding a knife between the tile and the palette. The tegulae would have been fettled as necessary, in particular any lip of clay beneath the sides of the mould and over the top of the ends would have been removed (this may have been done prior to rotating it off the original baseboard). The lower cutaway would have been completed with a knife (or formed in its entirety if no insert had been used). The tile would then have been moved to the stacking area where it would have been placed on its side, presumably supported by other tiles set out in a meshing arrangement.

*14. Up to a further seven days later (depending upon the weather) the tile would have been sent to the kiln*

## 2.6(b) Manufacture using the inverted box Mould F

The first two steps of the procedure for preparing and filling the mould would have been the same as for the four-sided mould, as would be steps 13 onwards once the tile has been placed on the drying board. The intervening steps are as below:

*1. Run wire across top of mould, peel off surplus clay and use striker to smooth*
Wire cut surfaces are very rough and some smoothing would be necessary. However not all tiles showing the features of the inverted box mould (square flanges and upper cutaway inserts) also have smooth undersides, so perhaps not all tiles made in this mould had their undersides smoothed.

With this mould any lower cutaway inserts would have to be attached to the side of the mould. As discussed in Section 2.3(h), this means that they would not have been suitable for use with Group A, B and some D tegulae unless the inserts were pushed loose into the mould after it had been filled with clay but prior to the surplus being wires off.[88]

*2. Place drying board over top of mould, rotate the mould and board, lift mould off vertically*

*3. Smooth flange top and upper surface of base*
With a striker or wet cloth. Sometimes the flanges were smoothed in a similar manner to Mould D because finger channels were produced.

*4. Mark with signature*

*5. Move to drying area*

*6. Complete as for four-sided mould*

## 2.7 Further consideration of stepdowns[89]

Some 40% of the Beauport Park tegulae in this study had stepdowns all of which were on tiles stamped with RIB 2481.102. There were four tegulae with this stamp that did not have a stepdown[90] and of these two had a tally

[88] Leucarum 1692 (Plate 2.7) which had a triangular flange and smooth underside, and which was therefore almost certainly made in an inverted box mould, has a Type 62 cutaway made with an insert. This must have been plugged loose into the mould as suggested above. Leucarum 1694 has an upper cutaway that could have been formed by a mould insert.
[89] See Section 2.3(d) for the initial discussion of stepdowns
[90] One of these (Beauport Park 161) had a partial line on its surface and a slight turning back of the flanges suggestive of the first stage of forming a stepdown.

mark; however, none of the tiles with stepdowns also had tally marks. Overall, of the 39 tiles at Beauport Park where the lower edge was intact, 13 had incised tally marks; there were also 38 tiles (comprising the majority of these 39 tiles but also including some tiles where the lower edge has been lost) where the upper edge was also intact and 16 of these had stepdowns at the upper end. However, no tile had both a stepdown and tally mark implying that these were two exclusive groups: tiles could either have a stepdown or a tally mark but never both.[91] The chances of this being simply a statistical aberration are extremely remote.[92]

Plate 2.49: Examples of Beauport Park flange feature

This correlation suggests that whatever the purpose of the stepdown it either impeded access to the other end of the tegula such that a tally mark could not be cut, or the process that required the stepdown negated the use of a tally mark.

Seven[93] Beauport Park tiles also share another curious feature in the side of their flange which takes the form of a vertical line emerging from the base of the tile followed by a horizontal line (travelling towards the lower end of the tile) followed by another horizontal line (Plate 2.49). The feature is of similar size on each tile although there is no pattern to where along the flange it occurs; it appears on either flange but never on both. The marks that are on the left hand flange are the mirror image of those that occur on the right hand flange. A similar, but more rounded feature, appears on three tiles from York.[94]

There is no obvious explanation, but one possibility is that a nail or piece of dried clay was projecting from the inner surface of these moulds. When the mould was lifted upwards the first vertical line would have been created, the mould was then moved horizontally for a short distance before being lifted vertically again off the tile. If the projection were big enough then this movement could make the observed marks with a mirror image on the opposite flange.[95]

The output of a single tile-maker could have been over one thousand tegulae a week and ways of optimising the significant amount of space required to dry them would have been required. It is possible that the stepdowns and flange features are both connected with the storage of the tiles for drying. The movements that caused the marks in the flange could have related to slotting the palette into a drying rack and then removing the mould where there was insufficient headroom to take the mould off cleanly in a single movement. Instead a small upward movement was required to lift the mould clear of the base at the upper end, and then a horizontal movement (perhaps the length of the upper cutaway) before the mould could be released vertically. It is significant that the seven tegulae with these marks were all stamped with RIB 2481.102 which was the only die with stepdowns.

If the drying racks were incorporated into low standing drying hacks (Plate 2.50) this might be another cause of restricted headroom. Such drying racks could also have led to pressure on the ends of the upper cutaways that caused stepdowns.[96] Tiles in such drying racks would not

[91] Beauport Park 227 had a scratched as opposed to incised tally mark and this tile did in fact have a stepdown; however this does not necessarily contradict the hypothesis.

[92] For example if one assumed that stepdowns and tally marks were independent variables and the probability of a tally mark was 13 in 39 or 1 in 3, then the probability of none of the sixteen tegulae with stepdowns also having tally marks would be $(2/3)^{16} = 0.0015$. In other words, if stepdowns and tally marks were unrelated then the observation above should only occur one time in 650.

[93] Beauport Park tegulae 161, 178, 191, 205, 214, 1195 and 1208.

[94] York tegulae 474, 478 and 1434.

[95] Given that the 2mm nail-head tested with the experimental mould failed to leave a vertical mark when the mould was removed, the nail or lump of clay would have needed to project at least 5mm.

[96] The stepdowns could have been caused by a different stacking method where tiles were laid out in facing rows but with sufficient distance between them to allow someone to walk along to place further tiles. Later this space was backfilled by placing further boards at right angles resting on the upper cutaways of the two facing rows beneath.

have been accessible for tally marks. Other tiles may have been dried in a more accessible manner and therefore received a tally mark but not the stepdown.[97]

Plate 2.50: Drying hacks at Bulmer Brick and Tile Co

## 2.8 Manufacture of imbrices

One hundred and twenty-one complete imbrices were recorded in the survey. They ranged in length from 315mm to 540mm and all of them were trapezoidal with one end being wider and taller than the other (which facilitates overlapping of the imbrices on the roof). The widest imbrex was 225mm at the wider end and the narrowest 95mm at the narrower end. The cross-sectional profile of imbrices varied from completely semi-circular to two straight edges set at approximately right angles joined to each other by a relatively sharp arc.

Plate 2.51: Imbrex type products on modern formers

Imbrices were probably formed by throwing clay into a flat trapezoidally shaped mould, running a wire across the top and peeling away the surplus. The clay would then have been turned out into a quasi semi-cylindrical former which could either have been inverted such that the clay was laid inside it, or upright such that the clay was laid over it. In either case the process would have been to hold the former in one hand and the filled mould in the other, bring them together and deftly flip the clay either into or over the former as appropriate. Plate 2.51 shows imbrex-like products placed on formers in a modern craft tile works where exactly this approach was adopted.[98]

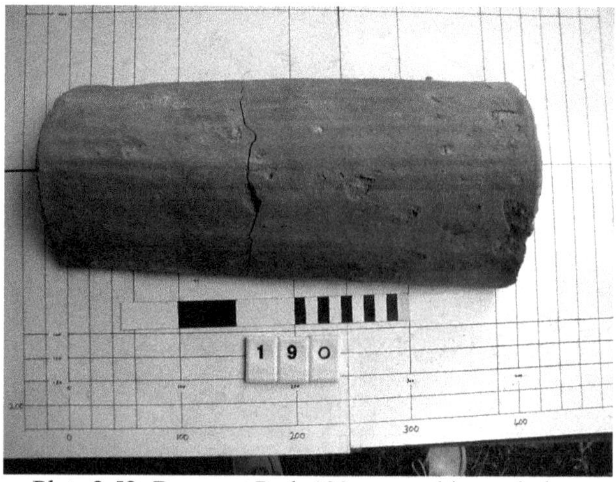


Plate 2.52: Beauport Park 190 - smoothing striations

Where the clay was placed over an upright former it would have been the underside of the clay in the mould that was then exposed as the outside of the imbrex. This would be rough and covered with sand and probably unsuitable to be placed on a roof in that state. It would therefore be smoothed over with a wet cloth or hand which would cause the longitudinal striations frequently observed on the outside of imbrices (Plate 2.52). This process would have increased the surface tension and therefore hardened and weatherproofed the outer surface.[99] The inside of the imbrex would correspond with the upper surface in the mould and this could be left in its crude "wired" state (Plate 2.53).

Some clays have a low natural adhesion and would be unsuitable for draping over a former: consequently these clays would need to use the inverted former. Plate 2.54 shows an imbrex with a deer's hoof on the inside of the imbrex which could only have been made whilst the clay was still soft and, as the impression is not distorted by any subsequent folding of the clay, it is reasonable to conclude that the tile was upside down in a former when the imprint was made. Imbrices made in this manner would have the upper surface of the clay in the mould as

Access to any of the tiles to cut the tally marks would have been difficult in this arrangement so it does not provide a possible explanation of the correlation.

[97] However Beauport Park 161, which has a tentative stepdown, has both the flange feature and a tally mark, albeit only a single diagonal stroke which could possibly be accidental.

[98] Bulmer Brick and Tile Co Ltd.

[99] An imbrex was observed at Bignor (photograph not available) that had two lines about 30mm apart running down the full length of the inside centre of the tile. Such markings would be consistent with the clay being draped across a pole rather than a former. However, whilst this could be a simpler manufacturing approach it is unclear how the outer surface could be smoothed if the clay was not resting against a firm surface such as would be provided by a former.

the underside on the former and thus the outside on the roof. Thus, provided the upper surface of the clay in the mould was cleaned with a striker after it had been wired flat, there would be no need for any further treatment on the former.

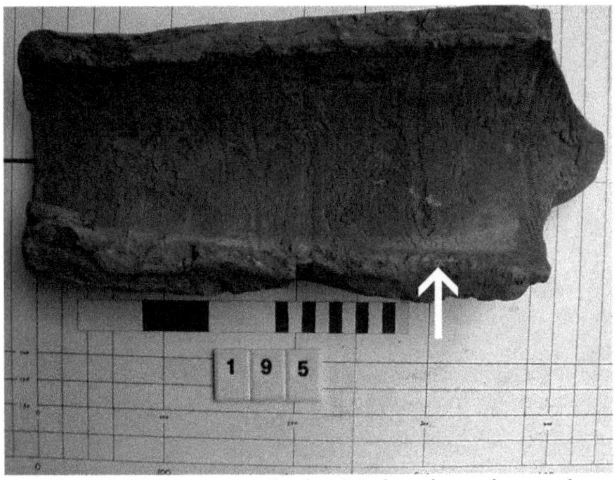


Plate 2.53: Beauport Park 195 showing wire marks

Some imbrices from later Roman sites[100] have patterns on the outside (Plate 2.55). Whilst it would be possible to etch these onto the tiles whilst they were draped over the formers, it would be a lot easier if it were done whilst the clay was still in the flat mould. This would only be possible if the inverted former was used because this approach had the outside face of the imbrex uppermost when the clay was in the mould. This method is consistent with Piddington imbrex 2114 which has a pattern on the outside and the deer's hoof on the inside (Plates 2.55 and 2.54), and is consistent with the appearance of other patterned imbrices. It is significant that many of the sites that have produced patterned imbrices made in an inverted former have also produced tegulae made in an inverted box mould.[101]

Plate 2.54: Piddington 2114 – Imbrex with deer's hoof

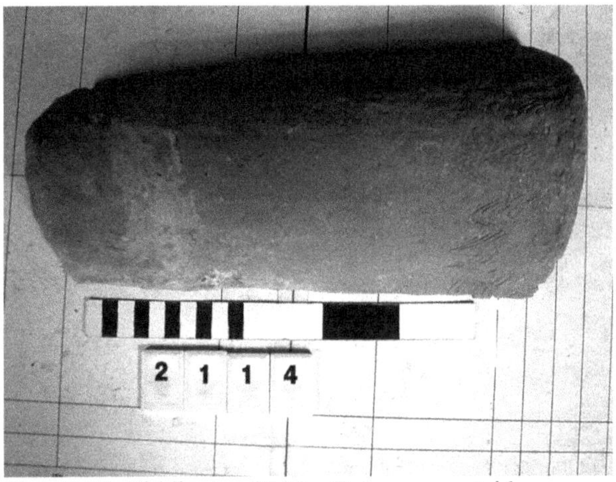


Plate 2.55: Piddington 2114 – Pattern on outside

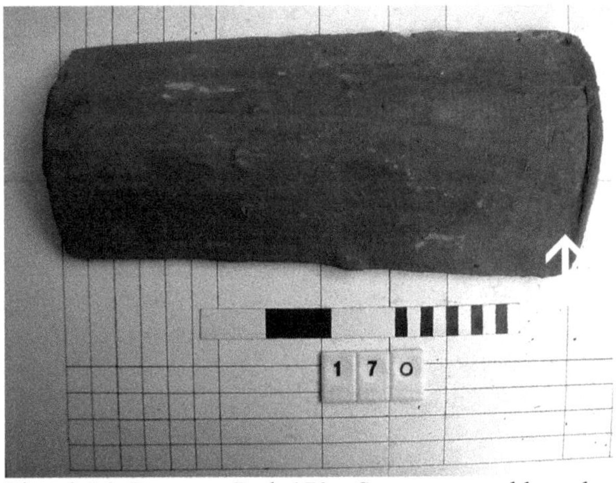


Plate 2.56: Beauport Park 170 – Stamp over gable end

Some imbrices are stamped, and whilst most stamps are on the flank of the tile, some are across the gable end (Plate 2.56). As with the patterns, stamping would be most easily undertaken in the mould rather than on the former, especially if it was to be done across the gable end. However, most of the stamped imbrices have evidence of longitudinal striations consistent with manufacture on the upright former, and some of the stamped imbrices from York[102] have dog prints on the outer surface which could only have occurred when the upright former was used. Stamped imbrices from York and Beauport Park[103] are likely to be second or early third century so it is possible that upright formers were used for these and were replaced by inverted formers later in the third century.

[100] Inter alia Crookhorn, Dorchester (Dorset), Grateley, Norton Disney, Piddington, Sparsholt and Stanton Low.

[101] Crookhorn, Dorchester (Dorset), Sparsholt, and arguably Stanton Low

[102] York imbrices 473 and 511.

[103] See discussion of dating in Chapter 5.

# 3

# TYPOLOGY

## 3.1 Introduction

Previous attempts to establish a tegula typology have mainly focussed on the profile of the flange.[1] This varies significantly and different sites often have distinctly dissimilar flange profiles. These undoubtedly reflect the idiosyncrasies of individual tilemakers and the way they ran their hands down the flanges in the smoothing process. However, whilst the profiles may give good differentiation between sites and within sites, because they were simply the result of the characteristics of individual tilemakers, they are extremely unlikely to form the basis of a *national* typology. The only exception is flanges with a square profile which result from the use of an inverted box mould and not variations in craft technique.

This chapter considers an alternative approach using tile stamps and tegula dimensions to establish a typology based on the evolution of the lower cutaway form. Tegula dimensions vary significantly: for example the longest tegula in the survey was 590mm and the shortest 305mm,[2] whilst the greatest flange height at the lower end was 82mm and the smallest 28mm. The most significant variations arise in comparisons between one site and another, but there are also large variations within individual sites. Chapter 2 has demonstrated that tegulae were made in moulds so these variations can be attributed to three possible causes: firstly, differential shrinkage whereby two tiles that were identical when manufactured shrank by different amounts in the drying and firing process; secondly, inadvertent disparities in the manufacturing process whereby two tiles that were intended to be the same size were hand finished in a slightly different way; and thirdly, the use of different sized moulds either deliberately or unintentionally.

Provided the impact of inadvertent discrepancies and differential shrinkage is not too great, then any remaining difference should be due to the difference in the size of the moulds and thus could form the basis of a typology. Ideally this typology would be based upon a comparison of lengths and breadths but in reality relatively few complete tegulae are found in excavations, so if a typology is to be of any practical use, it needs to be capable of use with tile fragments as well.

Flat fragments are unlikely to convey much information; indeed one cannot always be sure that they are tegulae. Fragments with some flange in addition to the base offer more hope. However, as has already been noted in Section 2.3(c), the width of the flange is tapered, so unless the location of a flange fragment is known, its use in a typology will be diminished. The fragments that are likely to be most useful in a typology are therefore those with a piece of flange and either the upper or lower cutaway, as this will fix the position of the flange and also provide data on the cutaway.

In order to ascertain the extent to which size variation is deliberate it is first necessary to identify how much of the variation can be due to differential shrinkage.

## 3.2 Differential shrinkage

A figure of 10% for tile shrinkage has been widely quoted but it has sometimes been used in relation to the *absolute* shrinkage of the tile from wet clay to the finished product and at other times has referred to the *differential* shrinkage between tiles that were originally the same size in the mould. Brodribb[3] states that his experimental tegulae shrank by 10% which is clearly an absolute rather than a differential figure. However, it is not necessarily a good indicator of the absolute shrinkage of Roman tegulae because this will be strongly influenced by the initial moisture content of the clay used which may not have been matched in Brodribb's experiments. Betts, Black & Gower[4] give a figure of 10-12% for the variations in size of box flue tiles implying that this is differential shrinkage, but they provide no evidence to support the statement. Boon[5] stated that impressions of the same die at Caerleon were often identical in size, but could vary up to 7%. This is manifestly a measurement of differential

[1] For example Chauffin 1956, 81-88.

[2] Adam 2005, 213 lists some tegulae sizes from Italy, the largest at Ostia was 72x48cm, at Rome 66x49cm and at Pompeii 69x47.5cm.

[3] Brodribb 1987, 4.

[4] Betts, Black & Gower 1997, 13.

[5] Boon 1984, 20.

shrinkage but it is unclear whether 7% from the mean is intended or a difference of 7% between the smallest and largest examples.[6] Moreover, from a typological standpoint, it will be important to know not just that a maximum difference of 7% was observed, but whether this was normal or unusual: in other words what the distribution of differential shrinkage is about the mean.

Given the difficulty of demonstrating that two tegulae come from the same mould, the best way of assessing the relative shrinkage is through the examination of tile stamps. If two tiles bearing the same tile stamp are examined and the dimensions of the tile stamps are measured then the difference in size of the two stamp impressions should be wholly due to the relative difference in shrinkage of the two tiles on which they had been impressed. Most of the dies were made of wood and so it is possible that some variation was due to the shrinkage of the wood; for example, a die may be dry in the morning when it is first used but will absorb water during the day, but this variation is likely to be small.

There will however be some variation due to the manner in which the die is applied to the wet clay. In an ideal world the die would always be applied vertically onto the tile to a uniform depth, but in practice some impressions were made with the die sliding in at an angle to the vertical and being removed often at the opposite angle to the vertical. Although this does not necessarily affect the quality of the stamp it means the measurements based upon the overall size and shape of the impression made by the die are likely to be greater than those that would have been made had the stamp always been applied vertically onto the tile.[7] Any calculation of shrinkage based upon the size of similar tile stamps must therefore be based on the dimensions of the die itself and not the overall dimensions of the entry and exit hole of the stamp tool.

Another factor to be considered (alluded to by Boon[8]) is the point in the manufacturing process at which the tile is stamped because shrinkage occurs both through air drying and in firing. Thus if some tiles were stamped in the mould and others immediately prior to firing then their size should automatically differ by the proportion of shrinkage that takes place during the air drying stage. The absolute air-drying shrinkage is believed to be small relative to the absolute firing shrinkage[9] but nevertheless it is a potentially significant factor in determining the relative shrinkage.

Traditionally tile stamps have been recorded either by making a squeeze[10] or a rubbing, both of which are relatively time consuming processes. Digital photography offers a quicker and possibly more accurate approach. In this research all the stamps were recorded by taking digital photographs vertically above the stamp using a raking light to bring out the contours of the stamp with a 14 by 10 cm frame placed around each stamp to provide a scale. The resultant photographs were then digitally normalised to bring them to the same scale using the reference frame. The frame was retained on the photographs such that the impact of any optical distortions could be monitored. The photographs were then digitally enlarged to several times life size such that measurements could be most accurately taken.

The raking light will have accentuated the leading edge of the lettering at the expense of the overall profile of the letters although this does not appear to have been a major problem. Of greater concern was the eroded state of many of the stamps and the fragmentary nature of some of the tiles that were recorded. On balance these inaccuracies are likely to have been the same or less than those adopted using more traditional methods.

Section 2.3(d) discussed the stamping process and concluded that although most tegulae were stamped at some point after they had been removed from the mould, tiles from Beauport Park were probably stamped by the tilemaker whilst the stamps were still in the mould. These stamps should therefore avoid any distortion due to being stamped at different stages in the drying process.

There were twenty one examples of RIB 2481.102 from Beauport Park recorded in the survey and these were selected for measurement.[11] Not all of these stamps were complete and some of them were eroded or inadequately impressed. To mitigate this four different parameters[12] were identified for measurement on each stamp and as many of these measurements as possible were recorded for each of the examples. For each parameter a mean was calculated from all the examples where that measurement could be recorded and then the difference from this mean calculated for each example. For each example the differences from the means for each of the four parameters were then averaged to produce an overall average difference for that example. The averages for all twenty-one examples are shown in Figure 3.1. This shows that differential shrinkage from the mean lies broadly within the range +/-2%. The standard deviation is 1.3%.

[6] The measurements discussed later in this section show that Boon was most probably referring to the difference between the largest and smallest examples, implying a 3.5% variation around the mean.
[7] Brodribb 1980, 187 makes precisely this point from his own experiments with stamps.
[8] Boon 1984, 20.
[9] Boon 1984, 20 and pers comm. Peter Minter of Bulmer Brick & Tile Co Ltd but a scientific analysis by Rosenthal 1949, 68 showed that with china clays the air-drying and firing shrinkages were broadly equivalent.

[10] A squeeze utilises wet blotting paper to create a papier-mâché negative image of the die which has then to be manually traced to produce an orthograde version of the stamp.
[11] Beauport Park 164 was excluded from this analysis as it was clearly inserted at a significant angle to the vertical.
[12] The two diagonals and two horizontal lengths, one taken at the top of the stamp and the other at the bottom.

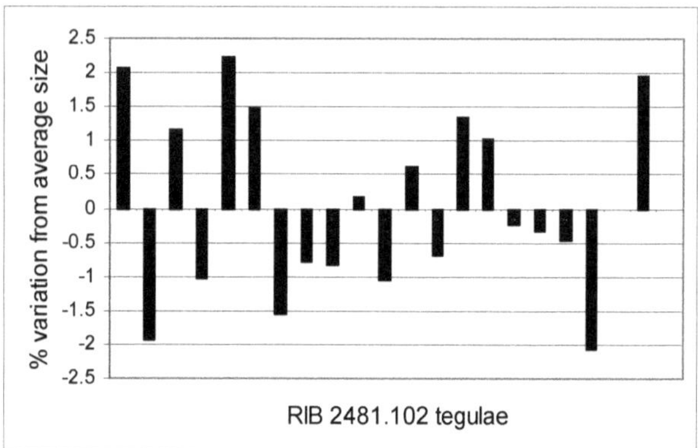


Figure 3.1: Variation in sizes of die RIB 2481.102

This approach has been repeated with the nine next most popular stamp types in the survey. These comprised two dies each from *Legio II*, *Legio VI* and *Legio XX*, one die from *Legio IX*, a further die from the *Classis Britannica* and one from *Cohors IIII Breucorum*. On some examples it was only possible to collect data on one of the parameters but for the majority of examples it was possible to record two or more parameters. Figure 3.2 shows the number of tegulae recorded, the number of measurements per tile that could be taken and the resulting estimate of the differential shrinkage derived from the standard deviation of the measurements.

| Stamp | Number of Tiles | Measurements per tile | Differential shrinkage |
|---|---|---|---|
| 2459.36 | 11 | 1.7 | 2.5 |
| 2459.56 | 8 | 2.5 | 0.9 |
| 2460.12 | 7 | 1.9 | 1.5 |
| 2460.53 | 7 | 2.7 | 1.9 |
| 2462.9 | 15 | 2.7 | 2.8 |
| 2463.29 | 6 | 2.7 | 2.4 |
| 2463.36 | 11 | 2.9 | 1.5 |
| 2470.2 | 11 | 1.7 | 1.6 |
| 2481.7 | 8 | 2.1 | 2.2 |
| 2481.102 | 21 | 3.6 | 1.3 |
| TOTAL | 105 | | 1.9 |

Figure 3.2: Estimates of differential shrinkage

It can be seen that all the dies have a standard deviation of less than 3%.[13] The *Legio* IX stamps had the largest standard deviation and it is possible that these were most affected by air-drying shrinkage caused by disparate timing of the stamping.

The results can be put together to produce an overall distribution as shown in Figure 3.3. This has a standard deviation of 1.9% which means that, in principle, 95% of all examples will lie within two standard deviations from the mean: that is within less than 4% of the mean. The largest differential shrinkage on a single tile in the 105 examples was 5.4%. This suggests that the relative shrinkage is somewhat less than has become accepted. The result is consistent with the maximum 2% difference measured on the three Winchester tegulae (discussed in Section 2.3(a)) which appear to have come from the same mould.

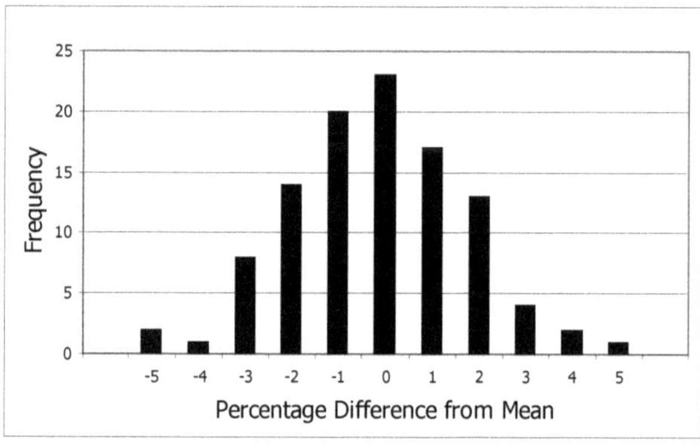


Figure 3.3: Distribution of differential shrinkage

This methodology is likely to overstate the actual degree of relative shrinkage because, although the shrinkage of the stamps may be affected by the timing of their striking, the actual tegulae will be unaffected because they all have to go through the complete air-drying and firing process.

If a mould was transferred to another site where different clays were used then the differential shrinkage would be greater than estimated here. However there is no evidence for this and, to the extent that the stamps were transferred as well (for which there is also no military evidence), this would already be reflected in the above analysis.

## 3.3 Comparison of stamped tegula sizes

It is clear that differential shrinkage can only be responsible for a very small proportion of the size variation that is observed. Stamped tiles may give a way of ascertaining how much of the remaining variation is due to random manufacturing quirks and how much to deliberate intent. It is reasonable to assume that stamp dies had a limited life and that not all dies were in use at the same time: as a consequence tegulae bearing different dies may well represent different periods of tile production. If the size variation is simply due to accidental variations in manufacture then there should be no correlation between size and stamp types; on the other hand if the size variation is linked to stamp types, then this may well be indicative of some sort of temporal shift in tegula sizes. Some size overlap should be expected because it is clear that at least a proportion of stamps were used contemporaneously; for example as is demonstrated on Holt 1110 which was stamped by two different dies

[13] Reassurance on the accuracy of the measurements can also be gained from the difference between the means of the different parameters on each example: this was never more than 2% and normally 1% or less which indicates that each of the parameters were independently producing similar estimates of the differential shrinkage on each example.

(Plate 5.6). Moreover it is perfectly possible that, even if sizes of tiles changed through time, more than one size could have been deliberately produced at a single point in time.[14]

The three principal legionary fortresses at York, Caerleon and Chester, together with the *Classis Britannica* bathhouse at Beauport Park, have produced the most stamped tegulae and the findings from these sites are considered below. Figure 3.4 shows a comparison of the length versus lower breadth for the main groups of stamped tiles from York. Clearly sizes are not a random feature because there are two distinct discrete groups of stamped tiles. One group is focused around a breadth of circa 360 mm and the other around a breadth of 395 mm, which is 10% larger, and therefore much greater than could have been produced by differential shrinkage.

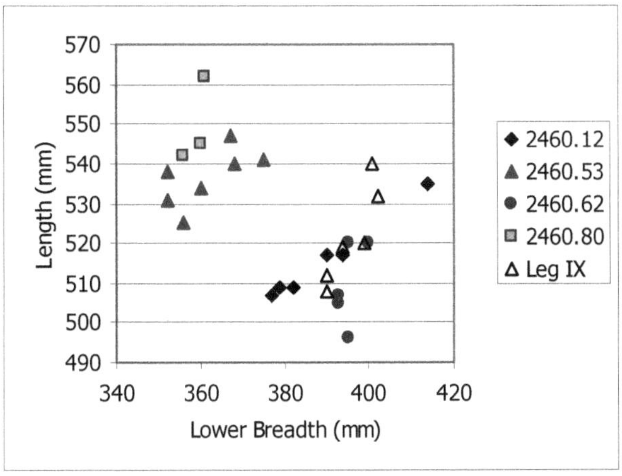


Figure 3.4: York length vs breadth

Within these two groups most of the stamps form discrete clusters but with a range of sizes which in the main run diagonally up the graph: that is with lengths and breadths increasing proportionately. In principle this could reflect differential shrinkage but in fact the differences are still too large to make this a likely explanation. For example for RIB 2460.53 the difference between the smallest and largest breadths is 7% and for RIB 2460.12 it is 10%. It therefore seems probable that these size ranges represent different mould sizes within individual dies which are nevertheless visibly separate from the size clusters of neighbouring dies.

At Beauport Park there were four principal dies used on tegulae and Figure 3.5 shows the size distribution of complete tiles. The discrimination between different dies is, if anything slightly better than York although the RIB 2481.102 tegulae overlie the separate groups of RIB 2481.7 and 2481.75 tiles. Again the clusters have a linear dimension which, in contrast to York, is purely along the breadth axis.

[14] See discussion of size ranges in Chapter 6.

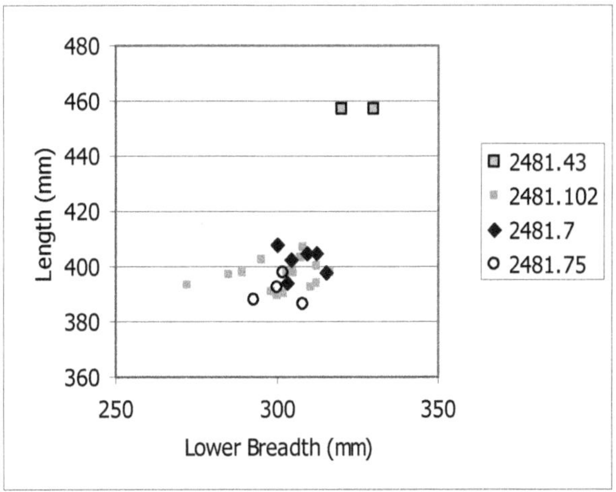


Figure 3.5: Beauport Park length vs breadth

Caerleon has rather fewer complete tegulae but Figure 3.6 shows that this limited data would appear to cluster with the sizes arranged broadly along a diagonal. Although Chester has produced a number of complete tegulae there is inadequate duplication of dies to produce a meaningful graph of their size distribution.

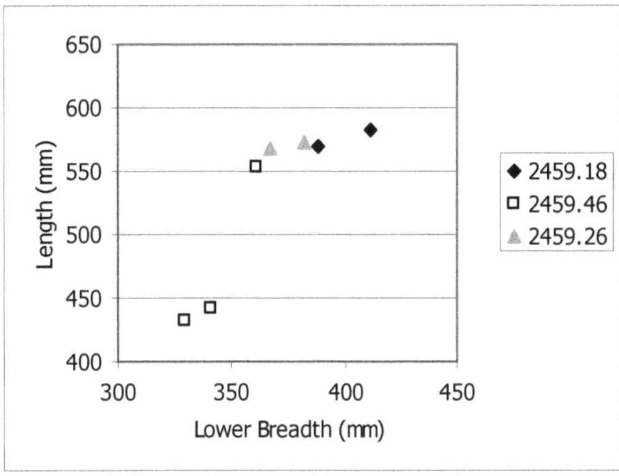


Figure 3.6: Caerleon length vs breadth

This analysis demonstrates that there is a correlation between dies and the length/breadth of tegulae and therefore that the variation in sizes is not due to accidental manufacturing differences. Moreover, because most of the dies are represented by several different moulds which nevertheless form discrete size clusters, the basic dimensions of the moulds in each cluster must have been a deliberate decision. This is an important conclusion; however, there are insufficient tegulae to apply this approach more widely unless a similar correlation also applies to tile fragments.

Chapter 2 demonstrated that most tiles were made either in a four-sided mould or an inverted box mould; in either case the height and width of the flange would be determined by the mould, subject to some small finishing variations caused by the tilemaker smoothing the flange. Tile thickness should also be reasonably consistent if the tiles were made on a flat palette but will be very inconsistent if they were placed directly on the ground. It appears that upper cutaways were formed whilst still in the mould so they should be consistent but the lower cutaway length may be more variable if the insert was attached to the palette rather than the mould. Flange height, width and upper cutaway length[15] should therefore have the strongest relationship with the mould in which they were made and so offer the greatest chance of being correlated with the mould sizes.

The largest group of stamped fragments came from Beauport Park and their lower flange height versus lower flange width is compared in Figure 3.7. It can be seen, in contrast to the data from the Beauport Park complete tegulae, that the RIB 2481.102 stamps, which overlay the other dies on the overall dimensions, are now entirely separated from the other stamps. Conversely RIB 2481.7 and 2481.75 now overlap. This would appear to demonstrate that at Beauport Park, although several moulds were in use for each die (perhaps reflecting the different overall sizes), generally tegulae of each die had the same flange dimensions and were intended to be dimensionally different to tegulae of the other dies.

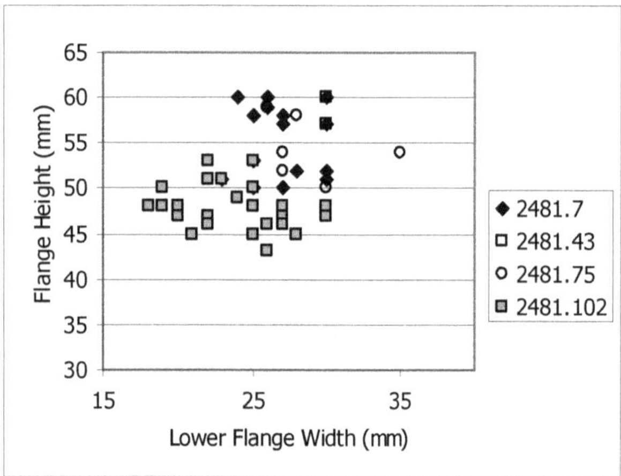


Figure 3.7: Beauport Park lower flange height vs width

Data for Caerleon are shown in Figures 3.8 and 3.9 which show good clusters for each of the eight most popular dies from the site. However, the size variation is again greater than should have arisen from a single mould and thus reinforces the Beauport Park finding that several moulds were in use but all designed around roughly the same flange parameters which were distinct from the parameters used for the moulds of the other tightly clustered stamps. Note that this graph is based on flange height versus cutaway length which gives better discrimination than flange height versus flange width.[16] This is perhaps significant because it will be remembered that Figure 2.4 showed that Caerleon lower cutaway lengths varied far more than those, for example, from Silchester.

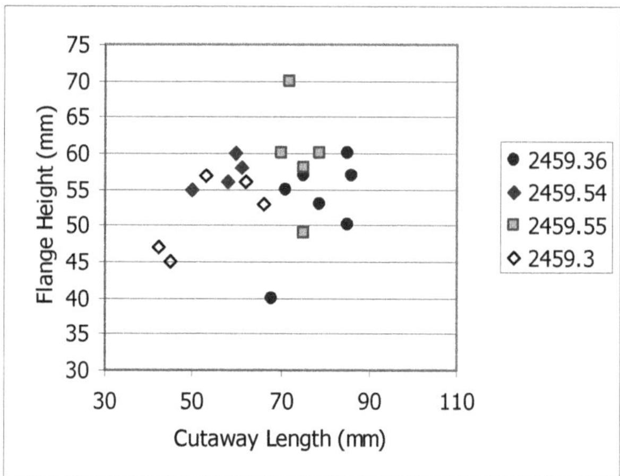


Figure 3.8: Caerleon flange height vs cutaway length(1)

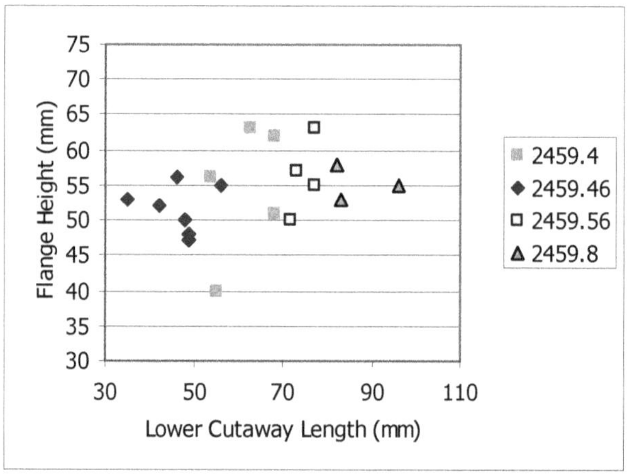


Figure 3.9: Caerleon flange height vs cutaway length (2)

Figure 3.10 shows measurements for the more popular York stamps. Three of the stamps: RIB 2460.8, 53 and 62 show reasonably tight groupings but the other two stamps show a wide range of flange sizes.[17] At Chester there is some clustering as shown in Figure 3.11 but, in part because of the low numbers attributable to any individual stamp type, the result is less impressive than for the other sites.

[15] However upper cutaway lengths can only be correlated with lower cutaway forms on complete tegulae which therefore reduces the usefulness of this parameter (as will be demonstrated later).

[16] A number of the Caerleon fragments were eroded and their cutaways broken leading to some doubt about the identification. As a result four tiles have been omitted from this and subsequent analyses where there was doubt about the identification and also the tiles themselves were distinct outliers from the cluster of the stamp to which they had been attributed (Caerleon 630, 632, 638, and 1725).

[17] Apart from the two extremes, all the Legio IX tegulae have RIB 2462.9 dies and its derivatives.

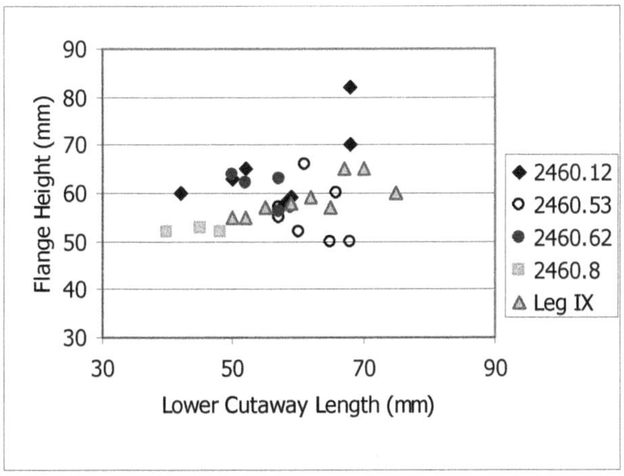


Figure 3.10: York flange height vs cutaway length

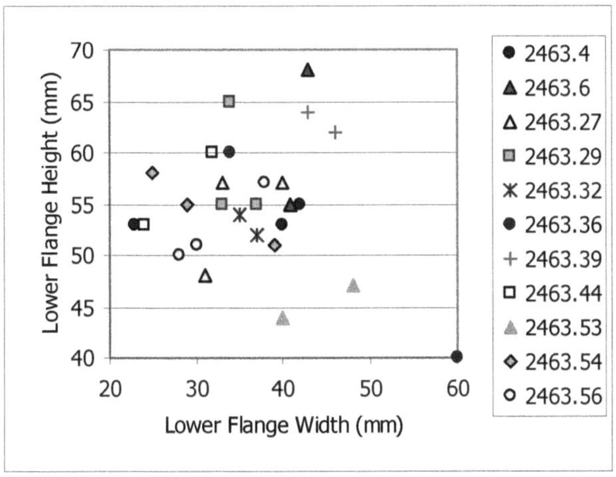


Figure 3.11: Chester flange height vs cutaway length

The foregoing figures show that at all four sites there was a practice of different tegula dimensions attaching to different dies although the practice appears uneven: some of the sites showing correlations most visibly on the overall dimensions and others on the flange dimensions. Discrete groups of tegula moulds must have been deployed for individual stamp types because several different overall mould sizes are implied by the cluster of dimensions attached to each stamp type. Each mould would then have had flange dimensions attributable to that stamp type.

Although some stamps were in contemporaneous use with others, it is likely that many, perhaps the majority, of different stamps represented different periods of production. It may therefore be concluded that differently sized groups of tegulae were made at different times. However, whether these size changes were purely arbitrary or represented some developing typology through time could only be ascertained if it were possible to date the individual stamp types. Whilst some of these stamps can be dated, for example those from Caerleon bearing the *Antoniniana* cognomen, the bulk of the stamps are not directly datable.

## 3.4 Evolution of cutaway forms

Having established that the tegulae size ranges vary through time it is necessary to look for another mechanism to establish whether these variations are arbitrary or form part of a typology. The dimensional changes themselves are probably stylistic rather than functional and therefore offer no obvious clue to the sequence in which they may have occurred. The one aspect of tegula design that had an important functional role and also appears in several different forms, is the shape of the lower cutaway. The purpose of this cutaway was to facilitate the meshing of tiles when they were placed one upon another going up the roof. It is unlikely that the choice of cutaway form was arbitrary because the cutaway would have needed to match the flange on which it was to sit such that the clearance between the flange and cutaway was sufficient to allow both sides of the overlapping tile to sit within the flanges of the tile beneath. (This issue is discussed further in Chapter 6). The cutaway form could therefore be the key to unlocking the typology if one exists.

Section 2.3(h) identified thirteen different cutaway forms that had a national distribution and placed these into the four Groups A to D (see Figure 1.3). The rationale behind this grouping reflected their morphology and the different approaches to their manufacture. Group A cutaways all have a basically cuboid notch removed whereas Group B cutaways have a basically prism shaped notch removed and both of these cutaway Groups emerge through the *side* of the flange. Similarly Group C cutaways all have basically cuboid notches and Group D prism shaped notches but both of these cutaway Groups emerge through the *top* of the flange.

As discussed in Section 2.3(h), inserts into the mould were used to assist in forming the lower cutaways, but only in the case of Groups C and D do the majority of tegulae show signs of the use of such an insert. Recognising the regularity of insert use, Group A cutaways were probably the most time consuming to form requiring three large orthogonal cuts (or two if an insert was used); Group B was the next most time consuming requiring two large cuts whether or not a cuboid insert was used; Group C almost always had a cuboid insert and therefore only required two rather smaller cuts to complete whilst Group D only required minor trimming when an insert was used. Thus in terms of the manufacturing effort required, the cutaway groups can be ranked in the order A to D with Group A being the most time consuming.

The greater the clearance that the cutaway form allows for the tegulae to mesh when placed on the roof then the more successful the roof is likely to be (see Chapter 6). The Group A cutaway only works if the notch out of the underside of the flange of one tile is larger than the flange it has to fit over on the next, but the size of this notch is

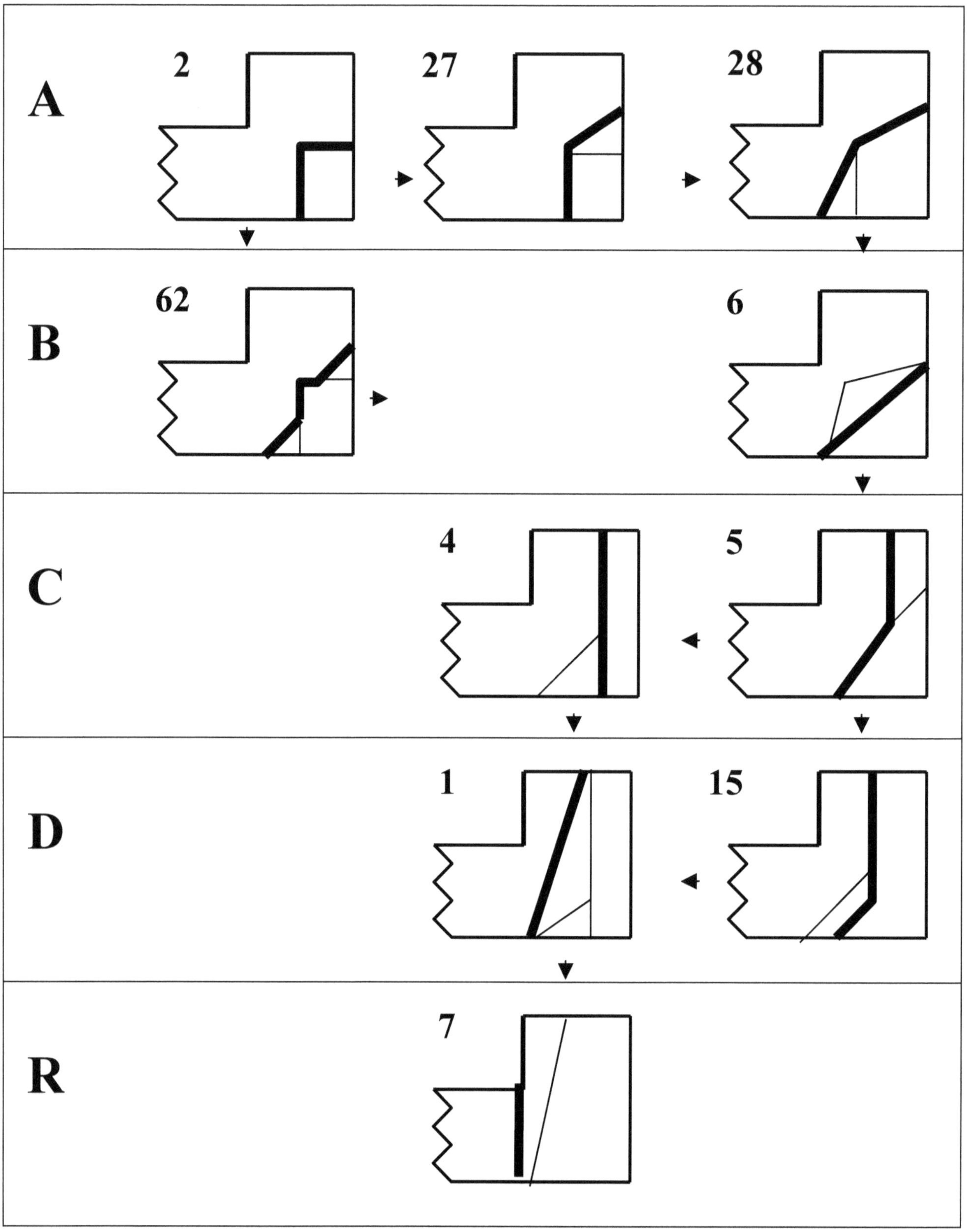


Figure 3.13: Possible Cutaway Evolutionary Pathways

limited if it is to avoid undermining and indeed breaking through the flange from which it is excavated. As result Group A tegulae can be regarded as the least effective form from an engineering standpoint. This is shown schematically in Figure 3.12 where the potential interference is coloured black. (Plates 6.1-4 demonstrate meshing of tegulae in practice).

Group B cutaways represent an improvement on Group A because generally the diagonal cut emerges higher up the flange than the horizontal cut in Group A and thus is able to bridge over the flange of the tile beneath more successfully. Group C cutaways are better again because the vertical insert reduces the whole effective width of the tile and therefore is able to more easily fit within the flanges of the tile beneath. Group D is the most effective from an engineering point of view because on average these had the greatest reduction in effective overall width of the tile at the lower end.

Thus in terms of both engineering effectiveness and ease of manufacturing, it would appear that the cutaway forms should be ranked from A to D. This should therefore be the sequence in which they appear if they represent technological improvement through time. The suggested evolutionary pathways for the main cutaway types are shown in Figure 3.13 which supports this proposed sequence. Although it is possible to imagine other evolutionary routes, they are likely to be broadly similar to the proposed scheme because, for example, it is extremely difficult to imagine any direct link between the Group A and Group D forms.

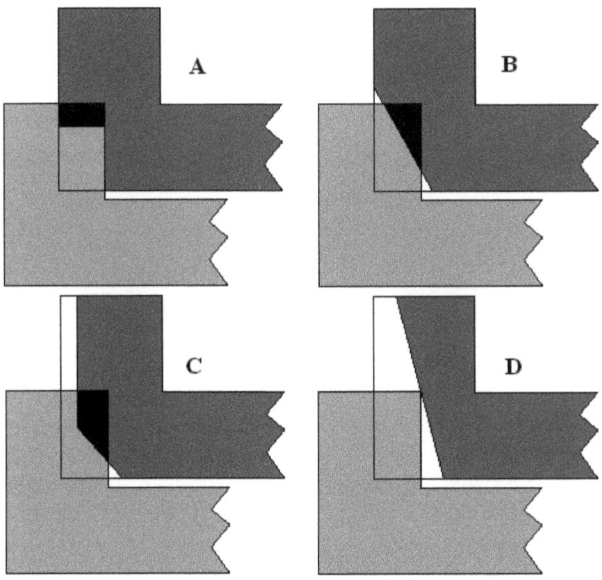


Fig 3.12: Interference in meshing of cutaway groups

Although almost all tegulae have similar cutaways on both flanges, there are some tegulae that have a different form on the left hand side to the right-hand side, but both taken from the same cutaway group: for example Type 2 and Type 26. This is supportive of the typology because if they belonged to different groups it might imply that they should have been produced at different periods. There are, however, seven tegulae in the survey where the cutaways are from different groups: three (all from Beauport Park[18]) where one side is Group B and the other Group C and four from other sites[19] where one side is Group C and the other Group D.[20] These errors are between adjacent cutaway groups so they probably represent transitional errors when production was shifting from one form to the next.

## 3.5 Size comparison based upon cutaway forms

The proposition that cutaway forms represent a chronological sequence can now be tested. Chester produced the least satisfactory correlation between stamps and dimensions so it is a suitable point to start assessing whether there is any correlation between cutaway forms and dimensions. Figures 3.14 and 3.15 show the lower flange height versus width, firstly undifferentiated, and then identified by cutaway forms. It can be seen that the cutaway groups create distinct dimensional clusters, even though each cluster has a range of sizes that are considerably greater than can be attributed to differential shrinkage. This is therefore powerful evidence that cutaway types are correlated with flange sizes which in turn are correlated with stamp types and therefore probably represent different periods of production.

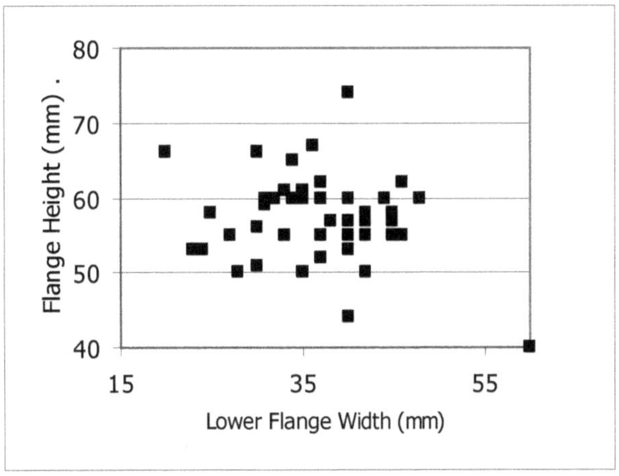


Figure 3.14: Chester lower flange height vs width

[18] Beauport Park 165, 198 and 228.

[19] Beauport Park 173, Bignor 2261, Buckinghamshire (unprovenanced) 2247 and Lime Street, London 252.

[20] Brodribb 1983, 45 reports on a set of tiles from Havant that have a Type 1 cutaway on one side and a Type 4 on the other. He suggests that this may have been caused by a difficulty such as left-handedness of the tilemaker but it could simply have been due to an ill-fitting or ill-formed cutaway insert or a transitional mistake as discussed above.

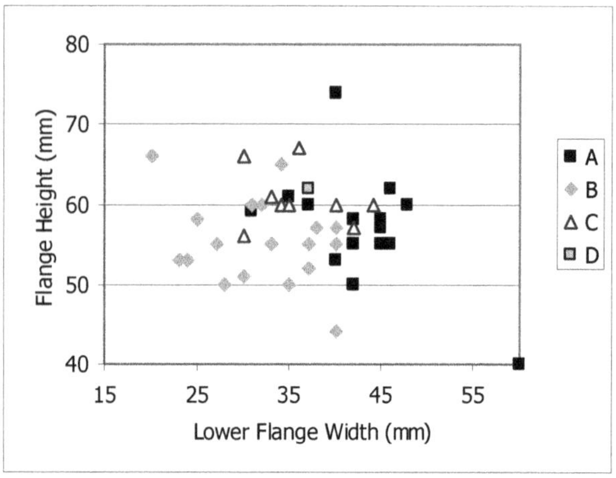


Figure 3.15: Chester flange height vs width by cutaway

Holt and Tarbock are also believed to have provided tiles to Chester, so these may be added to the data to produce the distribution shown in Figure 3.16. Although the graph is now very busy and the boundaries now somewhat blurred, the cutaway clusters still remain.

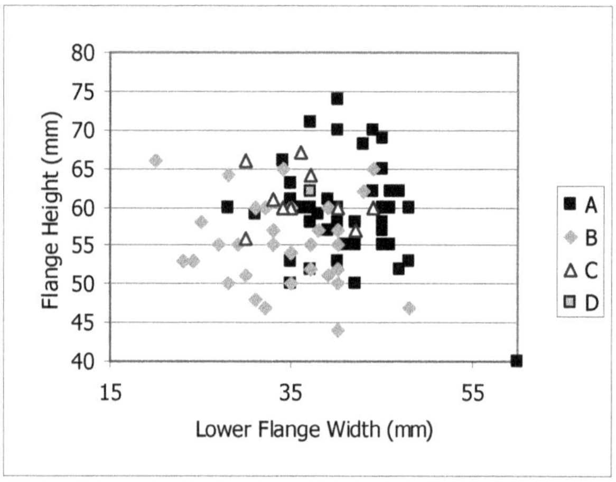


Figure 3.16: Chester/Holt/Tarbock flange height vs width

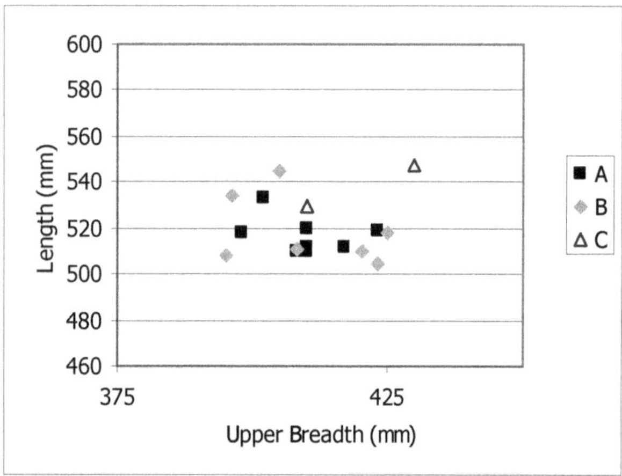


Figure 3.17: Chester/Holt length vs breadth

The correlation does not appear to carry through to the overall length and breadth of these tegulae as shown in Figure 3.17. However, as the analysis based upon stamps has already established that each stamp appears in a number of different overall sizes, it may simply be that there is insufficient data to show a proper differentiation.

Figures 3.18 and 3.19 show the dimensional spread for York flanges and overall measurements respectively. The cutaway form gives a complete distinction between Group C and the other forms but there is no differentiation in either graph between Group A and B (there were no Group D tegulae in the survey from York).

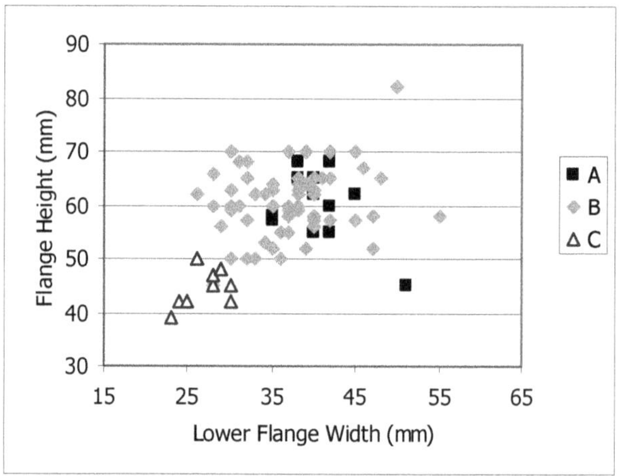


Figure 3.18: York lower flange height vs width

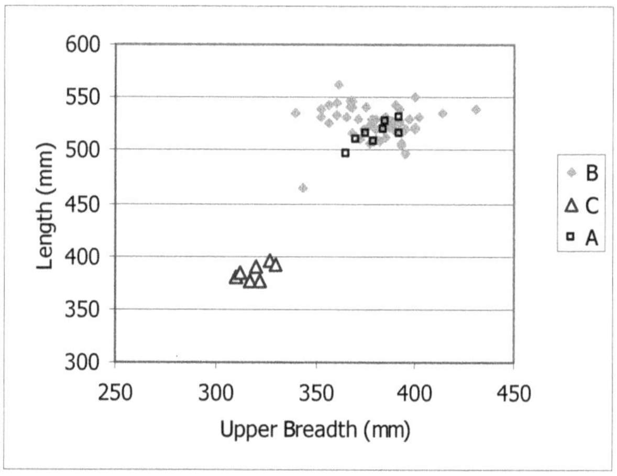


Figure 3.19: York length vs breadth

Caerleon showed comparatively good dimensional differentiation using stamps but Figure 3.20 shows no satisfactory discrimination between cutaway forms based upon flange measurements although Figure 3.21 gives a better picture using the overall dimensions.

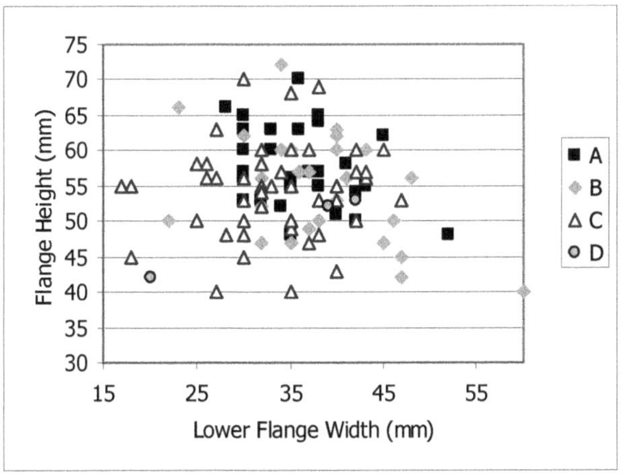


Figure 3.20: Caerleon lower flange height vs width

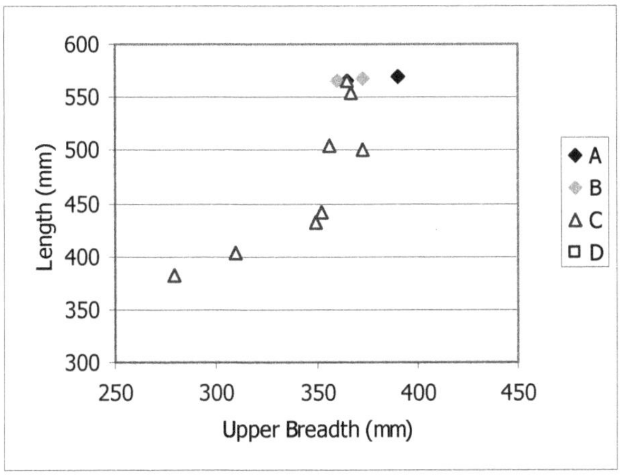


Figure 3.21: Caerleon length vs breadth

Beauport Park flanges show a reasonable distinction between Group C and D tegulae (Figure 3.22) but not with the Group B tegulae which have been omitted from the graph for clarity. There is no apparent clustering on Beauport Park overall dimensions.

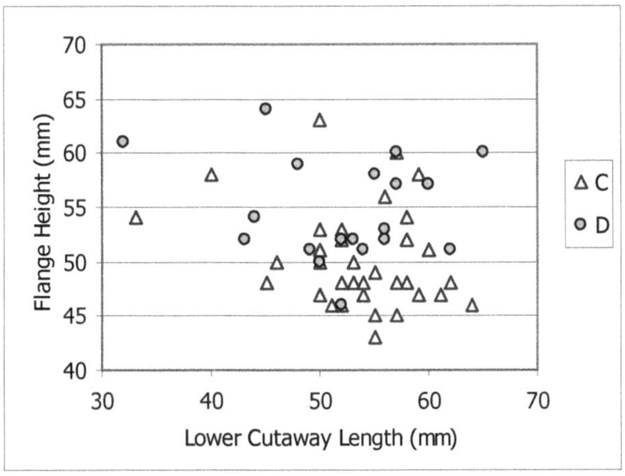


Figure 3.22: Beauport Park height vs cutaway length

So far only military sites with tile stamps have been considered and it is just possible that the lower cutaway forms could have simply been determined by the tile stamps without any temporal component: it is therefore necessary to examine civilian manufacture. Figures 3.23 and 3.24 show the flange data for Silchester undifferentiated and then identified by cutaway type. It can be seen that with the exception of Group B, which largely matches Group C, the cutaway forms give an excellent explanation of the different dimensions. Note that this graph uses flange height versus cutaway length and that, as discussed in Section 2.3(h), Silchester appears to have used a cutaway insert that was attached to the mould for Group C tegulae which should therefore give more consistent cutaway lengths.

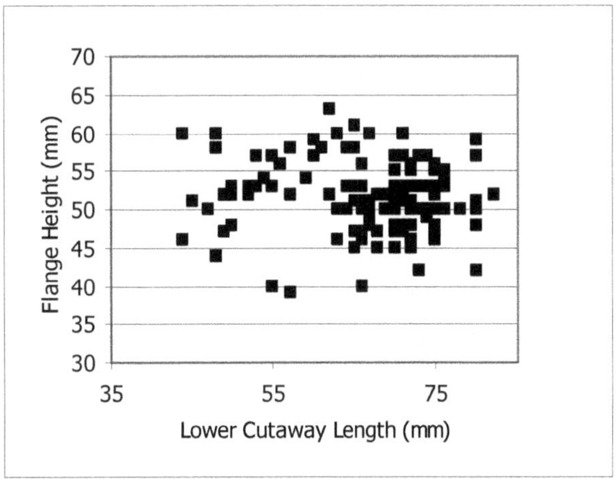


Figure 3.23: Silchester flange height vs cutaway length

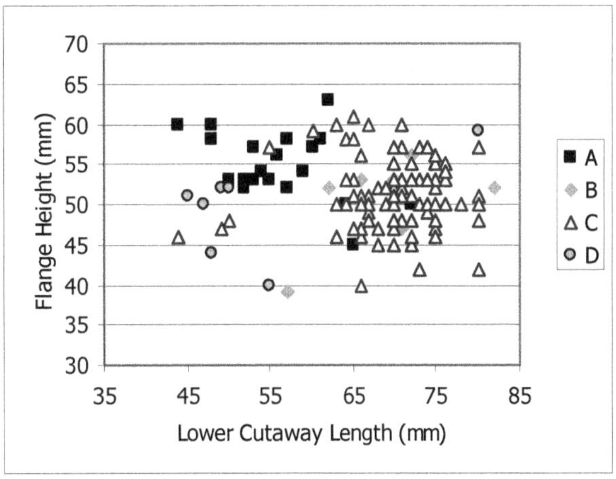


Figure 3.24: Silchester height vs cutaway length by group

Other civilian sites follow the same pattern as military ones with some showing excellent dimensional differentiation: for example Norfolk Street, Leicester (Figure 3.25) and Frilford (Figure 3.26); whilst others such as Caerwent (Figure 3.27) appear in several discrete clusters.

# Figure 3.28: Site Seriation Evidence

| | A | B | C | D | | A | B | C | D | | A | B | C | D |
|---|---|---|---|---|---|---|---|---|---|---|---|---|---|---|
| Alchester | | 2 | 6 | | Dorchester (Dorset) | | | 6 | 5 | Maiden Castle Temple | | | | 3 |
| Ashtead | 2 | | 9 | | Dorch'r Greyhound | | | 11 | 2 | Maiden Hatch | | | | 3 |
| Batten Hanger | | | | 2 | Dorchester Hospital | | 1 | | 4 | Maryport | | 1 | | |
| Beauport Park | 1 | 15 | 35 | 21 | Dorch'r Wollaston | | 1 | 1 | 4 | Medbourne | | 2 | 3 | 6 |
| Benwell | | 1 | | | Dorchester (Oxon) | | | | 6 | Mumrills | | 1 | | |
| Bignor | | | 10 | 6 | Dover | | 1 | 4 | | Narborough | | 2 | 27 | |
| Buckinghamshire | | | 4 | | Drayton | | | 3 | 12 | Newsteads | | 2 | | |
| Caerleon | 31 | 28 | 49 | 3 | Eccles | | | 4 | | Norton Disney | | | 9 | |
| Caerwent | 22 | 16 | 65 | 4 | Exeter | 6 | 4 | 6 | | Oxfordshire | | | 3 | |
| Caerwent House | | 1 | 1 | | Fishbourne | 14 | | | 3 | Piddington | | 2 | 32 | 13 |
| Caistor by Yarmouth | | | 3 | 2 | Frilford | 1 | 6 | 12 | 15 | Prestatyn | 3 | 4 | | |
| Caistor by Norwich | | | 7 | 4 | Gatehampton | | | 5 | 5 | Reading | | 3 | 29 | 13 |
| Canterbury L'market | | | 22 | 4 | Gelligaer | 3 | | | | Reculver | 1 | | 13 | |
| Canterbury Whitefriars | | 1 | 8 | 4 | Gestingthorpe | | | 6 | 2 | Richborough | | 1 | 2 | 1 |
| Carlisle | 3 | 12 | 4 | 1 | Gloucester | 2 | 16 | 3 | | Rivenhall | | | 1 | |
| Carpow | | 2 | | | Glos St Mary | 8 | 14 | | | Sandwich | | | | 2 |
| Carrawburgh | | 4 | | | Glos St Oswald's | 1 | 11 | | | Shelford Quarry | 2 | 2 | 15 | 2 |
| Castleford | 12 | 2 | 8 | | Gosbecks | | 1 | 16 | | Silchester | 23 | 17 | 130 | 12 |
| Chedworth | | 2 | 2 | 1 | Great Chesters | | 1 | | | Slack | 15 | 2 | | |
| Chelmsford | 2 | | 4 | | Great Holt | | | 12 | 1 | South Shields | | 6 | 5 | 5 |
| Chelmsford Mansio | | | 13 | 1 | Grimescar | 4 | 9 | | | Sparsholt | | | 7 | |
| Cheltenham | | 4 | | | Halstock | | | | 3 | Stanton Low | | | 5 | 3 |
| Chester | 17 | 21 | 9 | 1 | Holt | 8 | 19 | 1 | | Tarbock | 24 | 4 | | |
| Chichester | 4 | | | | Housesteads | | 3 | | | Templebrough | | 2 | 3 | 1 |
| Chignall | | | | 1 | Inchtuthill | 1 | | | | Thurnham | | | 18 | 18 |
| Chilgrove 2 | | | | 4 | Lancaster | 4 | 6 | 5 | | Wall | 2 | 5 | 13 | 1 |
| Cirencester | | 17 | 5 | | Layerthorpe | 3 | 4 | | | Wallsend | | 1 | | |
| Colchester Buc Church | 1 | | 7 | 3 | Leicester | 3 | 2 | 4 | | Wantage | | | | 7 |
| Colchester GBS | 4 | 1 | 3 | | Leicester Norfolk St | | | 37 | 27 | Winchester | | | 8 | 1 |
| Colchester Rural | 1 | 1 | 3 | | Leucarum | 2 | 9 | | | Witham | | | 6 | |
| Colchester St Mary's | 4 | | | | London | | | 3 | | Wroxeter | 5 | 27 | 2 | |
| Colchester Urban | 4 | 1 | 5 | 1 | London Dominant H | | 3 | 4 | | Yewden | | | 4 | |
| Cookham | | | 7 | | London Lime St | | | 14 | 1 | York | 12 | 62 | 9 | |
| Corbridge | 5 | 1 | 1 | | London Mansell St | | | 11 | 4 | York Minster | 3 | 5 | 2 | |
| Crookhorn | | 3 | 12 | 4 | Lullingstone | | 1 | 3 | 2 | | | | | |

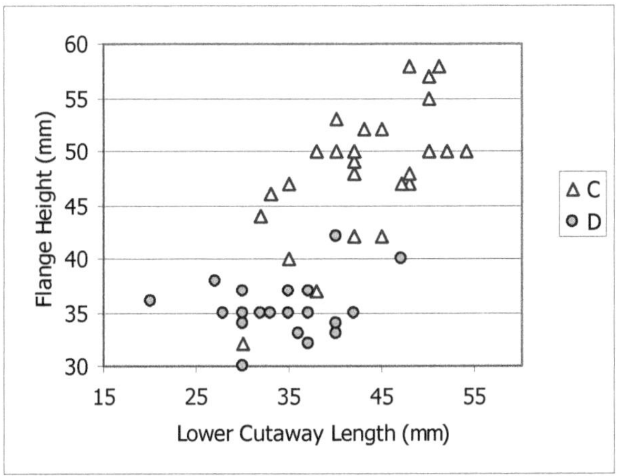


Figure 3.25: Norfolk Street height vs cutaway length

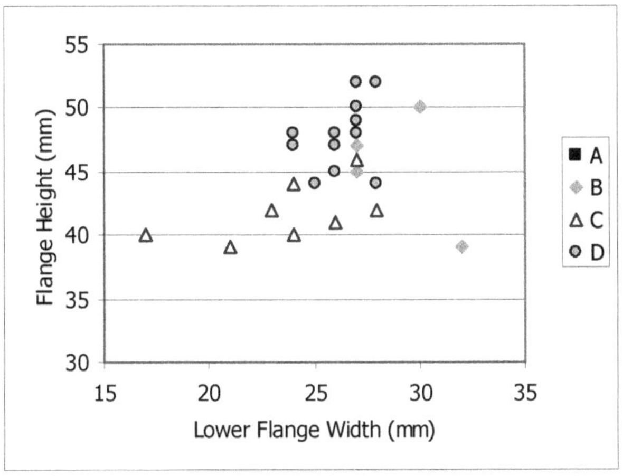


Figure 3.26: Frilford lower flange height vs width

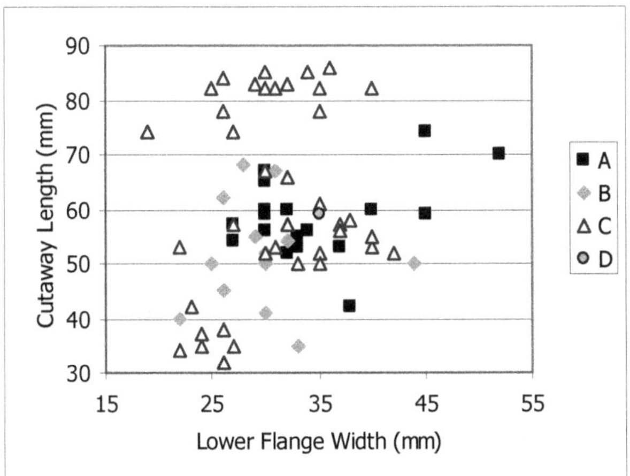


Figure 3.27: Caerwent cutaway length vs flange width

In this section the flange measurements at Chester produced a sharp difference between Group A and B cutaway clusters, those at York a difference between Group B and C cutaway clusters, at Beauport Park between Group C and D, at Silchester between A, C and D, and at Frilford and Norfolk Street between C and D. There were less data on the overall dimensions but these tended to show little difference between Group A and B cutaways but in all cases Group C was distinct. There is therefore a correlation between tegula sizes and cutaway forms, but it is not observed on all sites and on a single site not necessarily between all cutaway forms. The incidence of non-correlation of flange sizes may arise because no correlation exists on those sites/forms or because it is not sufficiently distinct such that the size clusters overlap and/or there is insufficient data to make the distinction clear. It is notable that at Caerleon, where there is no obvious clustering of flange sizes, there is nevertheless a very clear differentiation of Group C in the overall sizes for that site: this suggests that it is the absence of definition rather than the absence of correlation that is confusing the flange data for Caerleon.

The analysis of stamped tiles demonstrated that there appeared to be a range of overall tegula sizes associated with particular flange characteristics and it seems likely that this pattern is being repeated with the cutaway groups, albeit it is not always as sharp on all sites or cutaway forms. The clusters are typically considerably larger than could be explained by differential shrinkage and the implication must be that, where the clusters occur, they constitute several different moulds normally all having similar flange attributes.

## 3.6 Seriation evidence

The majority of Romano-British sites show continuous occupation albeit many do not start until the second or third centuries and many fall out of use before the fourth century. As most buildings were also subject to extension or renovation at least every fifty years, it is possible that the sequence of cutaway forms could be evidenced by seriation analysis. Figure 3.28 shows the number of each cutaway group recorded on each of the 104 sites included in this survey. This shows that the order of cutaways A to D produces continuous occupation across all but six of the sites. It is evident that any other ordering of the sequence, other than the reverse D to A, would produce many more discontinuities. As *Legio* IX tegulae only appear with Group A and B cutaways, and the legion left Britain before AD 120, the sequence must run from A to D.

The probability that the seriation data are a statistical chance can be calculated. Those sites that only have one cutaway form present, or that have all four present, provide no positive evidence about the absence of discontinuity: it is only those sites with two or three cutaway forms, and therefore the opportunity for gaps to appear, that help demonstrate the case. As shown in Figure 3.29 there are six ways that two different cutaway forms could be positioned and three of these would leave a gap; likewise there are four different ways three different cutaways could be positioned and two of these would leave a gap. The probability of a gap arising if the cutaways are randomly located is therefore 50% in each case.

# Figure 3.30: Stamp Seriation Evidence

| | A | B | C | D | | A | B | C | D | | A | B | C | D | | A | B | C | D |
|---|---|---|---|---|---|---|---|---|---|---|---|---|---|---|---|---|---|---|---|
| 2459.1 | | | 1 | | 2459.54 | | | 5 | | 2463.4 | 2 | 1 | | | 2474 | | 1 | | |
| 2459.3 | 4 | | | | 2459.55 | | | 5 | | 2463.6 | 2 | | | | 2476.1 | | 1 | | |
| 2459.4 | 4 | 1 | | | 2459.56 | | | 5 | | 2463.11 | 1 | | | | 2481.7 | | | 2 | 13 |
| 2459.5 | 1 | | | | 2460.9 | | 1 | | | 2463.15 | | 1 | | | 2481.7A | | | | 1 |
| 2459.6 | | 1 | | | 2460.12 | | 7 | | | 2463.18 | 1 | | | | 2481.43 | | 2 | | |
| 2459.7 | 2 | | | | 2460.21 | | 1 | | | 2463.21 | | 2 | | | 2481.75 | | 3 | 4 | |
| 2459.8 | 2 | 1 | | | 2460.23 | | 1 | | | 2463.27 | | 3 | | | 2481.102 | | 5 | 23 | |
| 2459.10 | 1 | | | | 2460.48 | | 1 | | | 2463.28 | | 2 | | | 2481.103 | | 1 | | |
| 2459.11 | | 1 | | | 2460.53 | | 7 | | | 2463.29 | | 3 | | | 2481.108 | | 1 | | |
| 2459.13 | 1 | | | | 2460.54 | | 1 | | | 2463.30 | | 1 | | | 2483 | | 2 | | |
| 2459.18 | 1 | 2 | | | 2460.61 | | 1 | | | 2463.32 | | 2 | | | 2486.1 | | 1 | | |
| 2459.19 | | 2 | | | 2460.62 | | 5 | | | 2463.36 | 4 | 4 | | | 2486.15 | | 1 | | |
| 2459.20 | | 2 | | | 2460.64 | | 1 | | | 2463.38 | | 1 | | | 2488.1 | | 1 | | |
| 2459.21 | 1 | | | | 2460.71 | | 2 | | | 2463.38/39 | 1 | | | | 2489.32 | | | | 1 |
| 2459.24 | 1 | | | | 2460.80 | | 4 | | | 2463.39 | 2 | 1 | | | 2489.40A | | | 1 | |
| 2459.25 | 1 | | | | 2460.81 | | 2 | | | 2463.41 | 1 | | | | 2489.40B | | 2 | | |
| 2459.26 | | 2 | | | 2460.84 | | 1 | | | 2463.43 | | 1 | | | 2489.40C | | | 1 | |
| 2459.29 | | 1 | | | 2460.85 | | 1 | | | 2463.44 | | 2 | | | 2489.42 | | 3 | | |
| 2459.30 | 2 | 4 | | | 2460.90 | | 1 | | | 2463.53 | | 2 | | | 2489.44 | | 3 | | |
| 2459.33 | | | 2 | | 2460.92 | | 1 | | | 2463.54 | | 3 | | | 2489.44B | | | 1 | |
| 2459.36 | | 1 | 10 | | 2462.6 | 1 | | | | 2463.56 | | 4 | | | 2489.44F | | 2 | | |
| 2459.37 | | 1 | | | 2462.7 | 2 | | | | 2464 | | 1 | | | 2489.45A | | 1 | | |
| 2459.42 | | 1 | 1 | | 2462.9 | 3 | | | | 2465.1 | 1 | | | | 2489.45E | | 2 | | |
| 2459.43 | | | 1 | 1 | 2462.9A | 4 | 1 | | | 2467.1 | | 1 | | | 2489.48A | | 3 | | |
| 2459.44 | | 1 | | | 2462.9B | | 2 | | | 2470.1 | 4 | | | | 2489.49 | | 6 | | |
| 2459.45 | | | 2 | | 2463.1 | | | 1 | | 2470.2 | 8 | | | | 2489.71 | | | 1 | |
| 2459.46 | | | 7 | | 2463.2 | | 1 | | | 2472.1 | | | 1 | | 2489.72 | | | 2 | |
| 2459.51 | | | 1 | | 2463.3 | | 2 | | | 2472.2 | | 1 | 2 | | | | | | |

| Two Forms | | | |
|---|---|---|---|
| A | B | | |
| A | | C | |
| A | | | D |
| | B | C | |
| | B | | D |
| | | C | D |
| **Three Forms** | | | |
| A | B | C | |
| A | B | | D |
| A | | C | D |
| | B | C | D |

Fig 3.29: Possible cutaway combinations

Sixty-one of the sites have either two or three cutaway forms present and six of these produced gaps: the statistical probability of this happening by chance is vanishingly small.[21] Moreover, for those sites that show discontinuities there are often explanations, for example all the Group D tegulae from Fishbourne come from an aisled building separate from the palace from which the Group A tegulae arise.[22]

Another approach using seriation is to examine the data based upon different dies. The presumption is that each die represents a single (perhaps extensive) period of production. Thus if the cutaway sequence is correct then most dies should only be represented by a single cutaway form and where two forms are present they should be adjacent in the sequence. This is precisely the observation shown in Figure 3.30. There are 111 different dies but many of these are single examples which provide no proof for this exercise, so ignoring these, there are 58 dies with multiple examples and 15 of these show a cutaway overlap, always between adjacent groups in the sequence.[23] The chances of these 15 adjacent overlaps happening by chance are three in 100,000.[24] The 43 stamps which have multiple examples of just a single cutaway group and the absence of any stamp spanning more than two cutaway groups are also supportive of the notion that each die was restricted to one or two cutaway types.

It would therefore appear that there was a chronological sequencing to the lower cutaway groups and that the dimensions of the tegulae were linked to the cutaway choice. It is also possible that the individual cutaway types that comprise the cutaway groupings could represent sub-divisions of this temporal sequence, but the data are too sparse to investigate this.

## 3.7 Dimensional evolution

The final dimensional aspect to consider is whether the sequence of cutaway groups can provide some explanation for the wide range of tegula sizes that are observed. The average lengths of all tegulae in the survey are shown by cutaway group in Figure 3.31. The number of data points (given after the chart description) used is 521. The cutaway Group "R" refers to the three regional forms noted in Section 2.3(h) and is discussed further in Section 3.9.

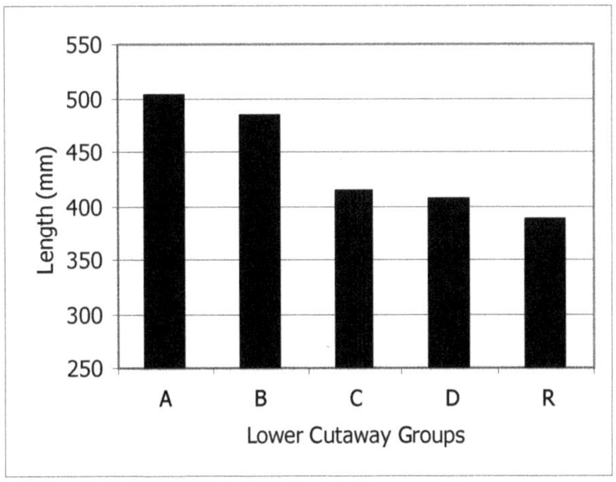


Figure 3.31: Average length by tile numbers (521)

This shows a very satisfactory evolution. However with only 521 measurements the result may be unduly influenced by some of the individual sites with a large number of examples. To remove this potential bias, instead of average values being calculated across all the data, averages have been calculated for each cutaway group at each site and then these site averages taken to produce an overall average of the individual sites. This result is shown in Figure 3.32. Again the number of data points (in this case contributing sites) is stated after the title. A site that has produced an average length for both Group A and B tegulae will be counted as two sites.

---

[21] The probability is the sum of $61!.0.5^{k}.0.5^{(61-k)}/k!.(61-k)!$ for k=0 to 6 which equates to $2.7 \times 10^{-11}$, for example see Sokolnikoff & Sokolnikoff 1941, 502.

[22] Pers Comm: David Rudkin.

[23] Holt 1110 had two different dies (RIB 2463.38 and 39) and a Group A cutaway: there is one other example of RIB 2463.38 which has a Group B cutaway and three other examples of RIB 2463.39, two of which have Group A cutaways and the third has a Group B cutaway. This therefore suggests that both of these dies straddled the Group A/B transition.

[24] The probability is $0.5^{15} = 3 \times 10^{-5}$.

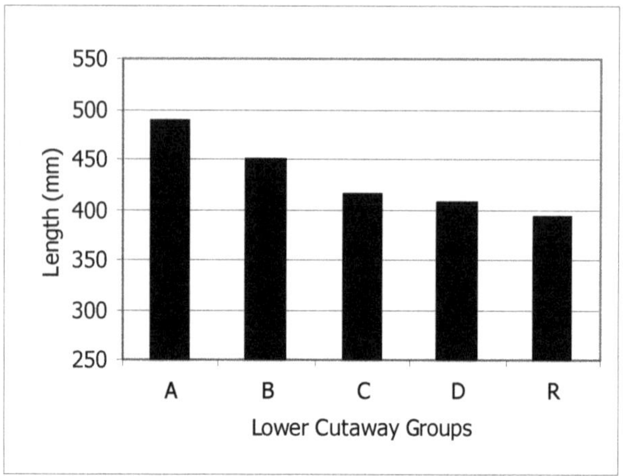


Figure 3.32: Average length by sites (110)

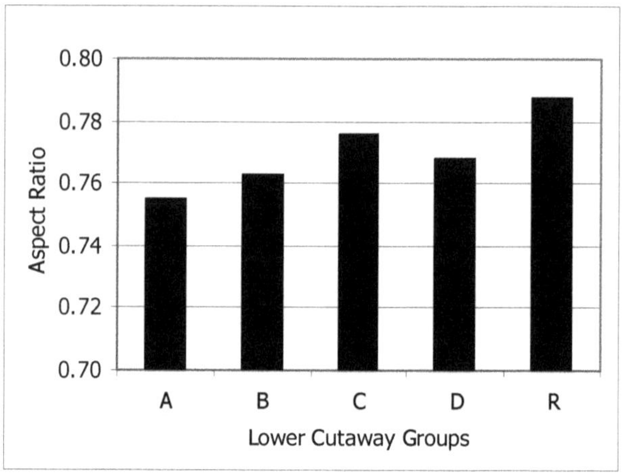


Figure 3.34: Average aspect ratio by sites (91)

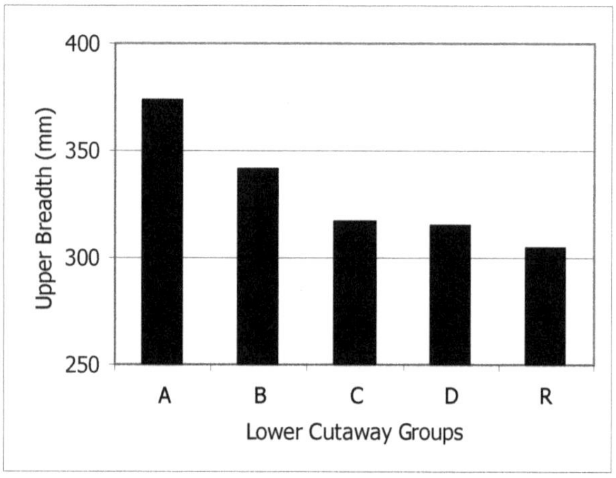


Figure 3.33: Average upper breadth by sites (93)

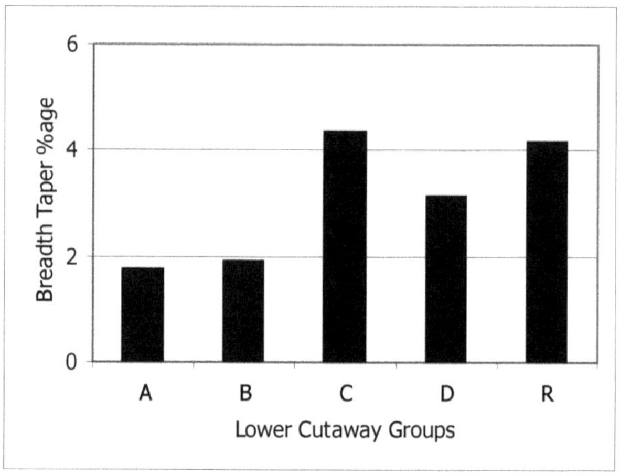


Figure 3.35: Average breadth taper by sites (88)

Average breadth by site also produces a reasonable evolution as shown in Figure 3.33. Given the wide diversity in tegula sizes, even for the same cutaway group, these results are both satisfactory and surprising. They clearly show a typological evolution through the cutaway sequence and therefore demonstrate that the diversity of sizes masks a strong underlying size trend. The result is also consistent with Betts' observation[25] that in London "breadth, and in particular length, shows a progressive decrease throughout the Roman period".

The aspect ratio of tegulae, that is the breadth divided by the length, increases through the cutaway sequence (Figure 3.34) but the breadth taper from the upper to lower end appears to jump upwards part way through the sequence and remain broadly at that level (Figure 3.35). However this analysis is supported by the lowest number of measurements.

Although the foregoing analysis demonstrates the typology, few if any complete tegulae are found in a typical excavation. It would therefore be useful if the typology also worked at a fragmentary level based upon flange pieces incorporating either an upper or a lower cutaway. Figures 3.36, 3.37 and 3.38 show the lower flange height, width and thickness respectively. As there are over 1600 examples of each measurement, the averages have been calculated on total tiles rather than site averages.

[25] Betts 1995, 212.

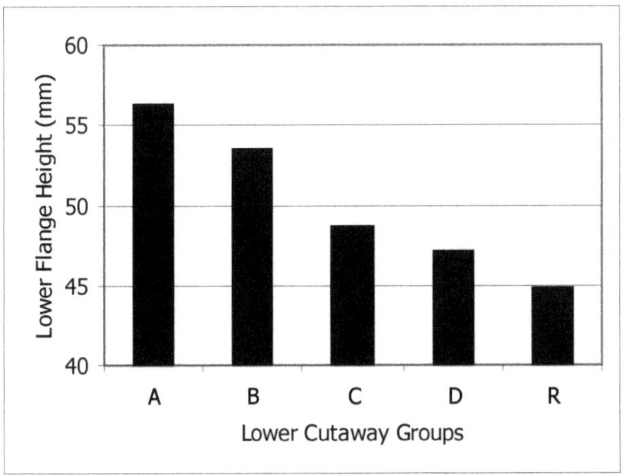


Figure 3.36: Lower flange height by tile numbers (1656)

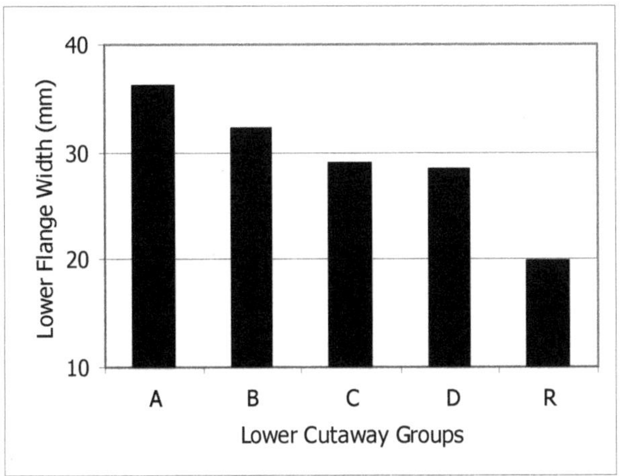


Figure 3.37: Lower flange width by tile numbers (1699)

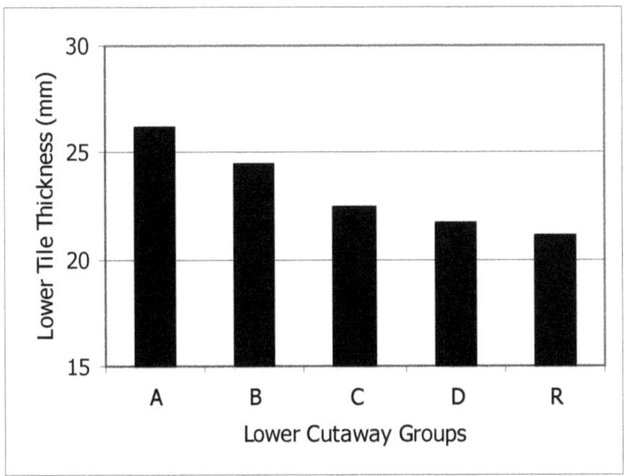


Figure 3.38: Lower tile thickness by tile numbers (1672)

These data look very convincing but, although the differences are statistically very significant, the actual reduction in size between adjacent cutaway groups is less than 10% which is less than 5mm. It would therefore be difficult to match these averages against individual samples from an excavation; instead an average over a dozen or more pieces would be necessary before even tentative conclusions could be drawn. The results are broadly consistent with Drury's observations that in Essex fourth century tegulae tend to have narrow flanges[26] and that in Chelmsford earlier flanges tended to be squat and later ones taller.[27]

The lower cutaway length (again by tile totals) is shown in Figure 3.39. This data is less satisfactory because 16% of the Group C tegulae come from Silchester[28] which also has easily the highest average lower cutaway length. If the Silchester data is ignored then a much more satisfactory result ensues (Figure 3.40).

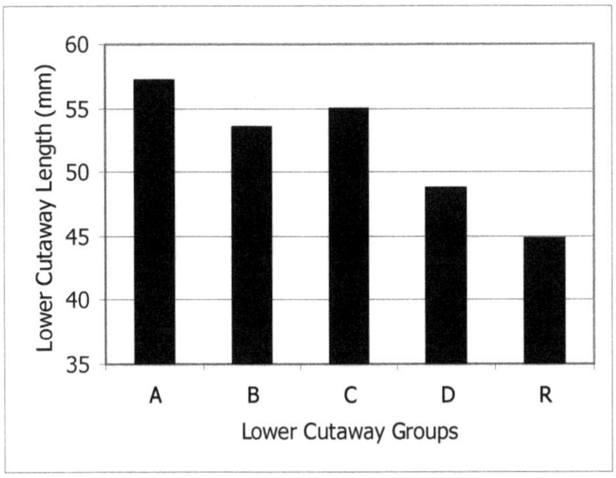


Figure 3.39: Lower cutaway length by tile numbers (1523)

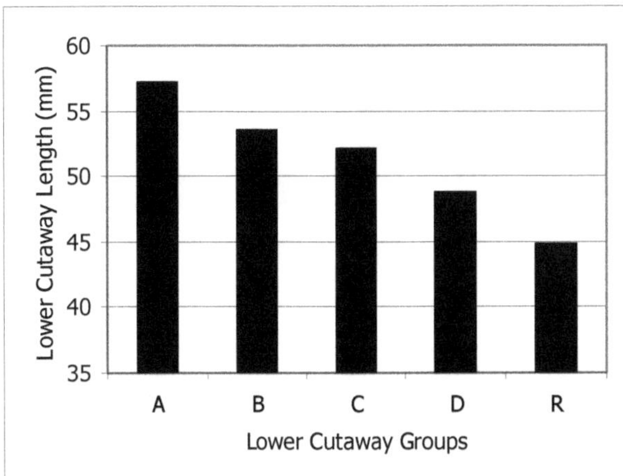


Figure 3.40: Cutaway length excluding Silchester (1415)

[26] Drury 1978, 112.

[27] Drury 1988, 80: the earlier flanges may actually be taller but the wider flanges and thicker bases on the earlier tegulae compared to the thin later ones makes the earlier flanges appear squat and the later ones to project higher above the base.

[28] The majority of Silchester Group C tegulae come from three or four large deposits. These deposits seem to have been carefully placed and comprise predominantly tegulae with relatively few imbrices. The tile would appear to have been moved from elsewhere in the town and its highly homogeneous nature suggests it may have arisen from the stripping of just one or two roofs. If this is correct, then the quantities from these deposits may cause an undue statistical significance relative to the limited diversity of sources from which they have arisen.

The data for the upper flange can only be collected by cutaway form from complete tiles and is therefore less reliable; it is again presented by aggregating the site averages to reduce the bias of the sites with larger numbers of complete tiles. This is shown in Figures 3.41, 3.42, 3.43 and 3.44. The results, with the exception of the upper cutaway, are less convincing than for the lower part of the flange.

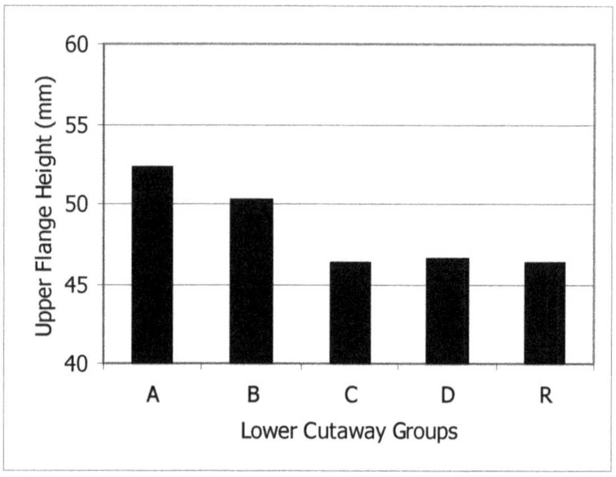


Figure 3.41: Upper flange height by sites (108)

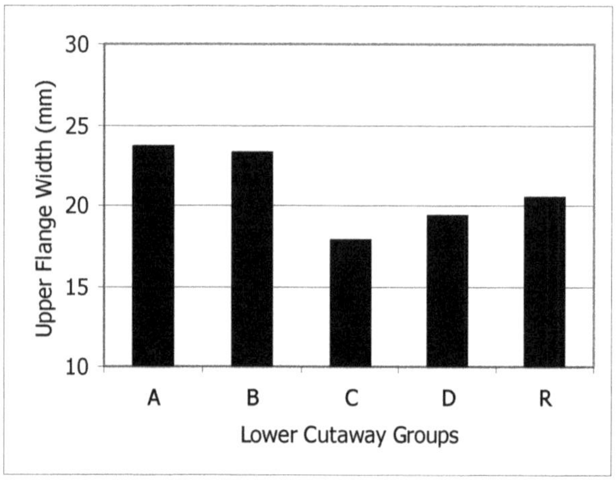


Figure 3.42: Upper flange width by sites (111)

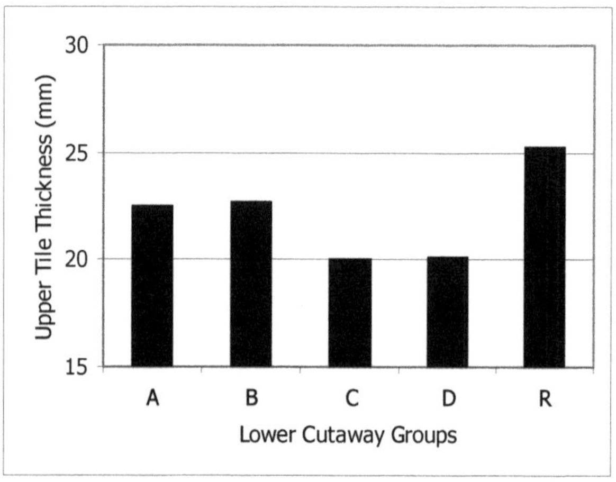


Figure 3.43: Upper tile thickness by sites (108)

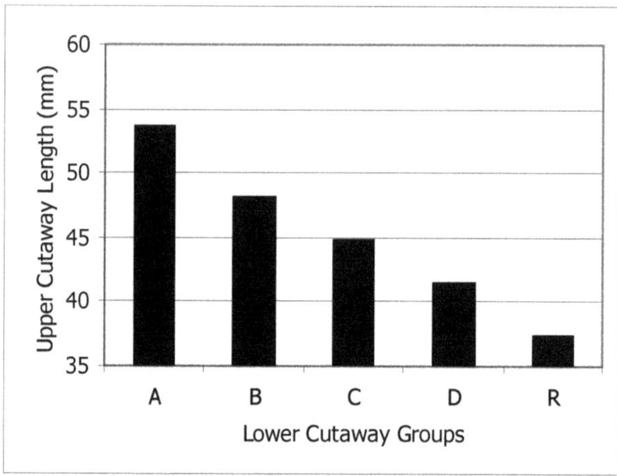


Figure 3.44: Upper cutaway length by sites (104)

As has already been noted in Section 2.3(c), both the height and width of the flanges taper from the lower to the upper end of the tegula. The reduction may be expressed as a percentage of the height or width at the lower end and this is plotted in Figures 3.45 and 3.46. The results, which are dependent on complete tegulae, show no particular trend.

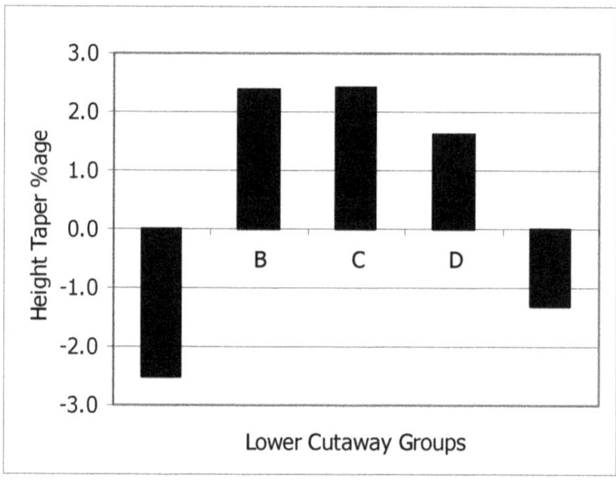


Figure 3.45: Flange height taper by sites (107)

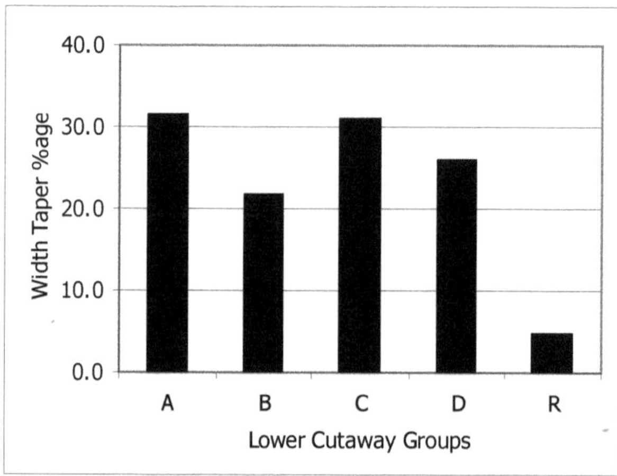


Figure 3.46: Flange width taper by sites (111)

The final piece of data concerns the channels on the base of the tile running adjacent to the flanges. These are normally made by one or two fingers, or very occasionally by three. Figure 3.47 shows that there appears to be a trend with finger channels but, even more so than with the other measurements, this trend hides a wide variety of practice.

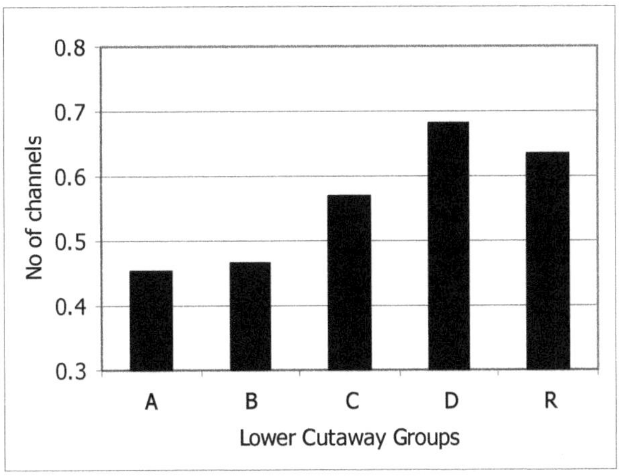


Figure 3.47: Finger channels by tile numbers (1587)

This section has shown that for the overall dimensions, and for the lower flange dimensions where there were significant amounts of data available, there is a very consistent evolutionary pattern emerging that definitely justifies being labelled a typology.

## 3.8 Comparative dimensional analysis

Given the smooth progression of sizes on all dimensions, the relationship between the dimensions is also worth examining. Figure 3.48 shows the ratio of lower flange height to overall length.[29] As can be seen, despite the roughly 20% reduction in overall length between Group A and R tegulae, this ratio is remarkably constant showing that as the length of the tile was reduced the height of the flange was reduced exactly in proportion.

| Cutaway Group | Ratio LH/L |
|---|---|
| A | 0.112 |
| B | 0.114 |
| C | 0.114 |
| D | 0.117 |
| R | 0.125 |

Fig 3.48: Ratio of lower flange height to overall length

In fact the ratio of tegula length to each of the flange dimensions is surprisingly constant as shown in Figure 3.49. The lower cutaway length is almost identical to the lower flange height on all cutaway groups.

[29] Using the site average data for both dimensions.

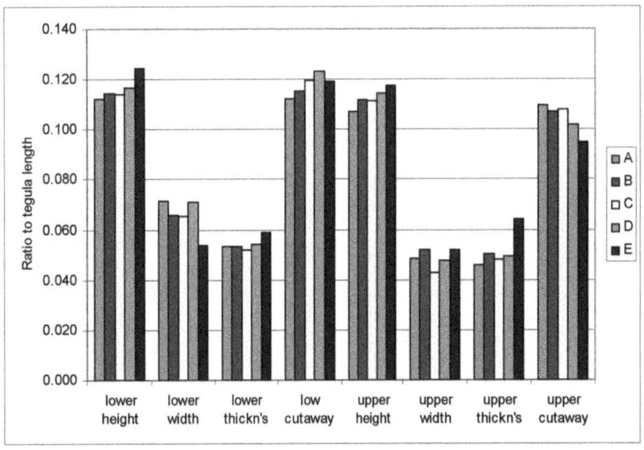


Figure 3.49: Ratio of flange dimensions to tegula length

The attribute that shows the least constant value is the upper cutaway length although it will be remembered that this was the upper flange dimension that displayed the best typology in the previous section. A comparison of upper to lower cutaway length is shown in Figure 3.50 and yields a satisfactory relationship demonstrating that the upper cutaway length reduced at a faster rate than all the other flange attributes. If these data are reliable then it may be possible to predict the lower cutaway form from a fragment incorporating just the upper cutaway.

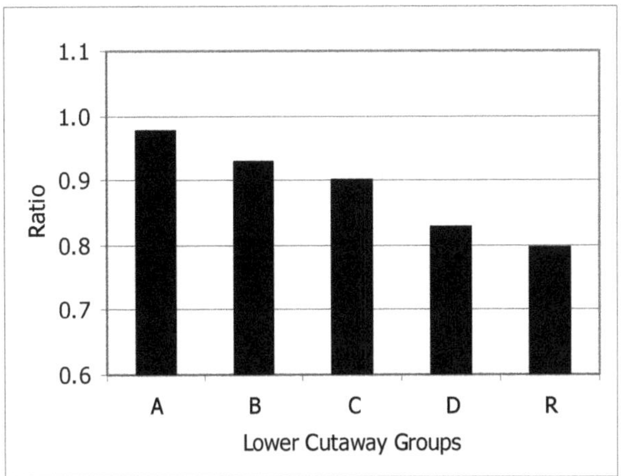


Figure 3.50: Ratio of upper to lower cutaway length (104)

## 3.9 Regional cutaway forms

These data now leave little doubt that there is a meaningful tegula typology that can be unlocked by reference to the form of the lower cutaway. It also shows that the three regional cutaway forms fit naturally into the typology following on from the Group D cutaways. The regional cutaways are, of course, a departure from the single countrywide standard which suggests, at least in this respect, that there was an erosion of common stylistic norms in later Romano-Britain.

In evolutionary terms the Type 7 regional cutaway from the Hampshire area is a natural successor to the Type 1 national form as indicated in Figure 3.13. Plate 3.1 shows that as the Type 1 notch became wider, it removed more and more of the flange until ultimately the whole of the end of the flange was removed yielding the Type 7 form. However there is no obvious evolutionary pathway for the other two regional forms: Type 8 from Arbeia (Plate 2.37) would be best derived from Type 2 but this belongs in Group A and is therefore inconsistent with the typological data, whilst Type 9 from Corbridge (Plate 2.30) represents a radical approach without parallels elsewhere.

Plate 3.1: Evolution from Type 1 to Type 7 cutaway

## 3.10 Conclusion

There is a clear linkage between the average tegula dimensions and the lower cutaway form. Of itself this does not demonstrate any temporal typology because it could be possible that all the cutaway forms were in use contemporaneously and that particular dimensions were simply associated with individual cutaway types: thus for example if the tilemaker chose a Group A cutaway then he would automatically give the tegula large dimensions. However the seriation evidence proves that the cutaways were not simultaneous but sequential production forms; a finding that is also supported by the correlation between stamp dies and dimensions. The dimensional analysis shows that there is a remarkably smooth evolution of tegula dimensions based on the sequence of lower cutaway forms.

The overall size of tegulae reduces through time. The length reduces at a faster rate than the breadth, such that tiles become relatively wider (increased aspect ratio). The flange height, width, tile thickness and lower cutaway length all reduce proportionately such that they maintain the same relative size to each other. The upper cutaway length also reduces but at a faster rate.

Drawing on the evidence from dies and cutaways, at any point in time individual tilemakers were producing several different overall tegula sizes but all utilising the same standard measurements for flange and cutaway dimensions. The typology shows that the average overall dimensions for the range of sizes reduced through time, so one could speculate that each tilemaker produced a standard set of tegula sizes and that each constituent tegula in the set was reduced by the same proportion through time.

However, behind this picture of order and regulated progression derived from average dimensions, there was a wide dispersion of values. In principle, all of the graphs in Section 3.5 should have the data arranged with Group A tegulae furthest from the origin then running through Groups B and C with the Group D tegulae closest to the origin to reflect the smooth reduction in size through time. This is only partly true: the Group A data are never inside the other data but they are often coincident with the Group B (for example York Figures 3.18 and 3.19). The Group D data should always be inside the Group C data as shown at Norfolk Street (Figure 3.25) but not as at Frilford (Figure 3.26) where the Group C is inside the Group D.

A small element of this, particularly the scatter of values, will have been due to differential shrinkage and a further element will have been attributable to accidental variations in what was ultimately a hand finishing process. However the bulk of the variation will surely have come from the fact that there were many hundreds of tegula producers, and although they may all have been striving to comply with the same norms (as witnessed by the fact that they all appear to have progressed though the sequence of cutaway forms), there would inevitably be a significant variation in the moulds they constructed and the clays they used. Moreover, as the data cover nearly 350 years of tile production, many generations of tilemakers were involved. So it should not be surprising that deviations occur and that when new tileries were established, or a new set of moulds was produced, they did not always fit properly within the sequence. Indeed it is not so much the differences that are surprising but that the data, when averaged, produces such a startlingly uniform result.

Other typological groupings of the lower cutaway forms are, of course, possible. For example a typology could be based upon the number of cuts required to make the notch; however, if this or any other alternative typology was adopted then, *inter alia*, the seriation data and size progression would not have yielded the convincing results shown above. The chosen typology does therefore reflect the chronological development of the cutaway forms proposed in Section 3.4 and excludes any alternative hypothesis. Possible explanations for the steady reduction in the size of tegulae are discussed in Chapters 6 and 8.

# 4

# DATING THE CUTAWAY FORMS

## 4.1 Introduction

This chapter addresses the dating of the sequence of cutaway forms developed in the previous chapter. The objective is to assign date ranges to each of the four national cutaway groups and to the regional forms. Only tegulae and fragments of tegulae excavated from datable contexts where the lower cutaway is present are useful for this analysis and these are referred to as *diagnostic* tegulae.

The longer lives of tegulae and their frequent reuse in other structures makes assigning dates to tegulae considerably more difficult than dating pottery. Moreover, whilst it is normally possible to distinguish between different potteries on stylistic grounds, this is much more difficult and uncertain with tegulae. Fabric analysis can help but, as discussed in Section 1.4, this approach appears to be far less satisfactory when applied to tiles than pottery.

In fact the only tegulae that one can be confident come from the same source are military tiles that have been stamped with the same die. However whilst the stamp may confirm the provenance, with the exception of some legionary stamps, it does not give a direct date. Dating has to rely on the association of stamped tegulae with other more datable artefacts but is complicated by the longevity of these tiles and their frequent reuse.

There is only one Romano-British tegula with a stamp incorporating a consular date and even the dating of this stamp is contested, as is discussed further in Section 4.6. Legionary tile stamps bearing honorific cognomina provide the only other primary dating source for tegulae. Six legions are attested in Britain but two of these (*Legio II Adiutrix* and *Legio XIV Gemina*) had left the country before tile stamping was adopted. The datable cognomina of the four other legions are discussed below.

Plate 4.1: RIB 2463.51 *Antoniniana* cognomen[1]

Plate 4.2: RIB 2463.53 possible *Antoniniana* cognomen

## 4.2 Legionary stamped tegulae

### 4.2(a) *Legio* II Augusta

Present from AD 43, based at Exeter, partly Gloucester and ultimately Caerleon.[2] The cognomen *Augusta* was awarded prior to the legion's arrival in Britain and so provides no dating information. The cognomen *Antoniniana* was awarded by Caracalla probably in AD 213 and was in use until AD 222:[3] there are seven different dies with this cognomen and three of them have yielded diagnostic tegulae. The cognomen *Severiana* was awarded during Severus Alexander's reign (AD 222-235) and has two dies but neither has yielded any diagnostic tegulae. Likewise there are no diagnostic tiles from the single die possibly bearing the cognomen *Victoriniana*.

### 4.2(b) *Legio* VI

Present from circa AD120, based at York. A number of different cognomina were awarded to *Legio* VI but the only ones that are datable are *Antoniniana* (AD 213-222), *Severiana* (AD 222-235) and *Gordiana* (238-244).[4] Unfortunately there are no diagnostic tegulae with these

[1] The distance between the sides of the template is 14cm so the plates are just over half life size.
[2] Holder 1982, 104-5.
[3] RIBII.4, 128.
[4] RIBII.4, 148.

stamps.[5] There is a further cognomen *Britannica* which has yielded diagnostic tiles and which has been equated with Severus's adoption of the title in AD 210[6] but there are strong grounds for questioning this attribution which are discussed in Section 4.7.

Plate 4.3: Holt 1104 with *Cohors I Sunicorum* graffito

## 4.2(c) *Legio* IX

Present from AD 43 until circa AD 120, based at Lincoln and then York. *Hispana* is the only cognomen used on these tiles in Britain and had been awarded prior to the legion's arrival here. However, *Legio* IX was replaced by *Legio* VI, which is first attested in Britain circa AD 122,[7] so any *Legio* IX stamped tiles should predate this.

## 4.2(d) *Legio* XX

Present from AD 43, based at Colchester, possibly Gloucester, then Wroxeter and finally Chester.[8] The principal cognomen used by *Legio* XX was *Valeria Victrix* which was awarded for its part in the defeat of Boudicca.[9] As tiles were not being stamped by the legions in Britain until around AD 100,[10] this cognomen does not help dating. The title *Antoniniana* was adopted for the period AD 213-222[11] and there are two dies with lettering "LEGXXANTO"[12] (Plate 4.1), but the seven examples on tegulae are all non-diagnostic. There is a third die, originating from Holt, where only the final letters "VVA" (Plate 4.2) remain: the final "V" and "A" are ligatured and the two "V"s inverted which RIB tentatively interprets as *V[aleria] V[ictrix] A[ntoniniana].* [13]

There are several reasons to doubt this interpretation, the most important being that, whilst the "VVA" die was found at Holt, none of the sixteen examples of the "LEGXXANTO" dies were found by Grimes in his report on the excavation of Holt,[14] nor have any subsequently come to light.[15] Recent research has confirmed the absence of evidence for third century domestic occupation at Holt.[16] Grimes himself found no evidence for third century occupation of Holt other than this presumed *Antoniniana* tile and another tile[17] bearing a graffito by a soldier of the *Cohors I Sunicorum* (Plate 4.3) which was believed to be of third century date.[18] The suggested dating of this graffito is based upon an inscription of *Cohors I Sunicorum* which relates to AD 198-209[19] but, as this unit appears on diplomas as early as AD 122[20] and is not attested elsewhere, there is no reason to favour the later date.[21] Indeed Holder[22] states the graffito to be second century. Thus the argument that Holt production continued into the third century based upon the attribution of the "VVA" stamp as referring to "Antoniniana" is essentially reduced to circularity and is inconsistent with the other dating evidence from the site. There is only one other *Legio* XX stamp with the ligatured "V" and "A" which reads "LEGXXVAV" (the ligature in this case being with the first "V") where the "VAV" is interpreted as *VA[leria] V[ictrix].*[23] In the circumstances it seems more likely that the "VVA" example is a mistaken rendition of "VAV" (perhaps caused by inverting the two Vs) particularly as the "LEGXXANTO" stamps do not adopt the *Valeria Victrix* cognomen at all.

[5] RIB 2460.41 mistakenly records nine tegulae stamped with the *Severiana* cognomen in the Yorkshire Museum but in fact these are *bipedales* tiles.
[6] RIBII.4, 148.
[7] RIB 1427.
[8] Holder 1982, 106-7.
[9] RIBII.4, 175.
[10] RIBII.4, 125.
[11] RIBII.4, 175.
[12] RIB 2463.51 and 52 but in fact there are three different dies of this type (See Plate 5.7)

[13] RIB 2463.53, RIB pairs this with another tile where only the starting letters "LEGX" are visible to produce a suggested interpretation of "LEGXXVVA".
[14] Grimes 1930, 143 notes that "ANTO" stamps have been attributed to Chester *and* Holt in the past, but he found no such examples at Holt and none are recorded in RIBII.4, 189.
[15] In fact the "LEGXXANTO" dies and the "VVV" dies discussed below, together with RIB 2463.7, are the only dies listed in RIB where more than two examples have been found in Chester but none found at Holt.
[16] Ward 1998, 64-5; Shotter 1998, 71 notes that no coins dated between AD 192-235 have been found at Holt and Swan & Philpott 2000, 62 believe the absence of ceramic and coin evidence implies that there was no military activity at Holt after AD 150.
[17] RIB 2491.96.
[18] Grimes 1930, 52.
[19] RIB 430.
[20] de la Bedoyere 1999,79.
[21] Contra Stephens 1989, 224-6 who argues for a date not much earlier than AD 235 based, *inter alia*, on the association with "Deciana" tiles which he incorrectly assumes are dated AD 250–252 (see later discussion) and some loose associations of secondary material. It is also worth noting that the graffito tile is stamped with RIB 2463.15 whilst at Segontium, where *Cohors I Sunicorum* were based, the only stratified tile with a stamp (RIB 2463.36) comes from a Trajanic or earlier context (Davies 1993, 231).
[22] Holder 1982, 121.
[23] RIB 2463.20.

There are two further tentative imperial cognomina that have been proposed for *Legio* XX: *Victoriniana* (AD 268-70) on stamps lettered "LXXVVV" and "LEGXXVVV", and *Deciana* (AD 249-51) on stamps lettered "LEGXXVVDE".[24] The first of these attributions has been comprehensively undermined by the discovery of stamped tiles with "VVV" lettering at the same probable kiln site at Tarbock as the sole stamp with a consular date[25] (Plate 4.4). This stamp has lettering "VERO COS III" which could fit two different senators who achieved their third consulships in AD 126 and AD 167 respectively.[26] The excavators believe that tile production was restricted to a very short period in the second century[27] and so it seems inconceivable that production could have stretched for at least the further hundred years that would have been necessary to encompass *Victoriniana* production. The very strong inference is therefore that the third "V" stands for Viducius who is named on the consular dated tile and was presumably making the tiles under contract to the legion in the second century.

Plate 4.4: Tarbock 265 – RIB 2463.59 Consular stamp

The attribution of "LEGXXVVDE" to *Deciana* is also improbable because, again, there is no evidence for Holt being in production in AD 250 where these stamps have been found and presumably originated; moreover they have also been discovered in late second or early third century contexts in Deanery Field, Chester.[28] The interpretation of this stamp is discussed further in Section 5.7.

## 4.3 Other stamped tiles

### 4.3(a) Auxiliary Units

Although no tiles stamped by the auxiliary units can be directly dated, they offer the potential for indirect dating because many of these units are attested on other dated inscriptions. However, there is evidence that many units were divided and stationed at different places and that many garrisons were formed of mixed units,[29] so conclusions based solely on the presence or absence of a particular unit should not be treated as incontrovertible.

[24] RIBII.4, 175.
[25] Swan & Philpott 2000, 61.
[26] See Section 4.6.
[27] Cowell & Philpott 2000, 114.
[28] Newstead & Droop 1936, 40-41.
[29] Hodgson & Bidwell 2004, 142.

RIB adopts the practice of implicitly dating such stamps by reference to the inscriptions identifying these auxiliary units, but this gives just a single date identifying the presence of a particular unit which may have been at that base for a hundred years or more. Whilst there is an expectation that auxiliary units in general were regularly moved in the first and second centuries to adapt to the changing frontier requirements before adopting a more settled existence after the last of the campaigns in AD 211, there should be no presumption that individual units moved unless there is evidence to support it. So in the subsequent analysis it is assumed that an auxiliary unit remains at the same base between its first and final epigraphic attestations unless it is also confirmed at another location or another unit is known to have arrived at the base. The consequence of this approach is that it generates longer potential site occupations and therefore limits the usefulness of this data, albeit hopefully making it more accurate.

### 4.3(b) Non-military officially stamped tiles

The only datable stamps are the Neronian imperial dies from Silchester but these have not yielded any diagnostic tegulae. The incuse imperial stamps from Carlisle[30] do occur on diagnostic tegulae but the stamp itself is undated. Procuratorial stamped tiles from London and municipal stamped tiles from Gloucester offer further dating potential. Both of these groups of stamps are believed to be of first and second century date but as neither offer any direct dating evidence based on the stamps themselves, they are treated on their merits based upon the site or kiln evidence available.

### 4.3(c) Civilian stamped tiles

No direct dating is possible on civilian stamps so they can only be indirectly dated from associated artefactual material.

## 4.4 Methodology

Each of the 105 sites[31] that produced diagnostic tegulae has been researched to attempt to establish dates for the different cutaway forms, but only just over half of the sites have actually provided usable evidence. This evidence is catalogued in Appendix 4. Unfortunately relatively few excavation reports provide contextual information for unstamped tegulae and even fewer identify the lower cutaway forms. Moreover, on many sites that might have been expected to provide useful dating evidence very little tile, and in some cases no tile, has been retained. To this must be added the difficulty

[30] RIB 2483.
[31] Tegulae from Braughing, Herts, although not examined by the author, are also included on the basis of the excellent excavation report of that site (Partridge 1977).

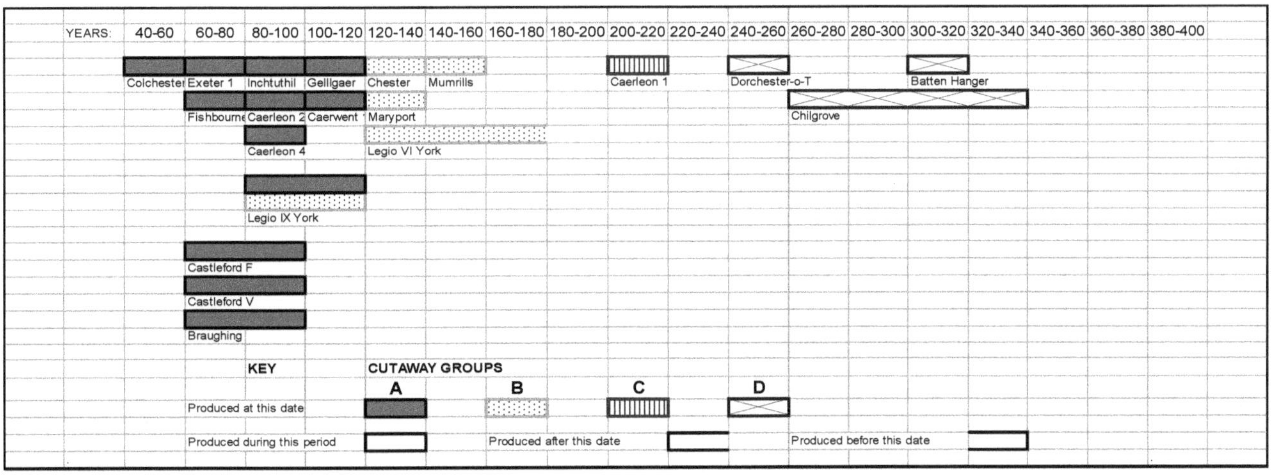


Figure 4.1: Compilation of individual sites with *strong* dating

associated with the regular reuse of roofing material in secondary contexts.

Ideally the analysis should consider the *terminus post quem* or *terminus ante quem* for each diagnostic tegula. However this would make the analysis somewhat unmanageable for a first iteration of the dating so, to make the problem more tractable, a number of simplifying assumptions are made in the dating analysis in Appendix 4. It is assumed that where tiles can be tied to a particular phase of a building's construction, then the tiles were new when placed on the roof, although manifestly this will not always be correct. If only one phase of construction has been identified then the tegulae will be assumed to come from this initial construction phase even if the site had an extended duration. A further assumption made is that the sequence in which the cutaway forms occur in Chapter 3 is correct.[32] Only those tegulae with a relatively restricted date range are useful. Those where the date range is in excess of sixty years, or is open-ended, have largely been ignored except where the dating has a significance that goes beyond this study, as is the case with stamped tegulae from the auxiliary units.

Clearly the more tegulae a site produces the more reliable the conclusions will be. In some cases damage to the tile means that the assignment to a particular cutaway group is not assured and likewise not all the stamp impressions can be determined with certainty. Site dating may also be subject to a degree of error, as may be the identification of the contexts from which the tiles come. To recognise the variability of the evidence and the use of the simplifying assumptions, the conclusions have been subjectively graded according to their reliability into three levels: *strong, probable* and *possible*. *Strong* evidence would normally consist of direct evidence based upon a stamp or good contextual information. *Probable* evidence would be where a reasonable deduction has been made but reliance has been placed on either the assumption that the cutaway sequence is correct or that the tegulae were new when placed upon the roof (or both). *Possible* means the evidence is weak and tendentious.

The conclusions drawn from each site are stated at the end of its entry in Appendix 4 using the following nomenclature: [B=125] would signify that tiles with Group B cutaways were *possibly* in production circa AD 125; B=120-140 would indicate that production *probably* took place some time in the period AD 120-140; whilst <u>B=120+</u> would show that there was *strong* evidence for Group B production sometime after AD 120. The analysis includes all the possible dating evidence acquired in this survey, even when that data are doubtful and unhelpful.[33]

## 4.5 Consolidation of the dating evidence

Figure 4.1 pulls together all the dating evidence from Appendix 4 that was regarded as *strong, inter alia*, placing no reliance on the cutaway sequence established in Chapter 3. Twenty year intervals are used for the dating. A solid box indicates that the site in question was producing tiles of the relevant cutaway within that period. If the evidence only points to production after a specific date but with no indication of when this ceased then the box is left open at the right hand end. Likewise if the evidence shows that production must have taken place

[32] For example, if a tile kiln were known to have been operating continuously between, say, AD 100-150 and only Group A and B tegulae could be sourced back to the kiln, then the assumption that the cutaway sequence is correct means that it may reasonably be inferred that Group A cutaways were being produced in AD 100 and that Group B cutaways were being produced in AD 150.

[33] The numbers of tegulae identified on individual sites will not always tally with those shown in Figure 3.28 because some tegulae have come from undated contexts or unstratified material. On sites with abundant material not all the tegulae will necessarily have been recorded in this survey, so totals will not always equate with those in excavation reports

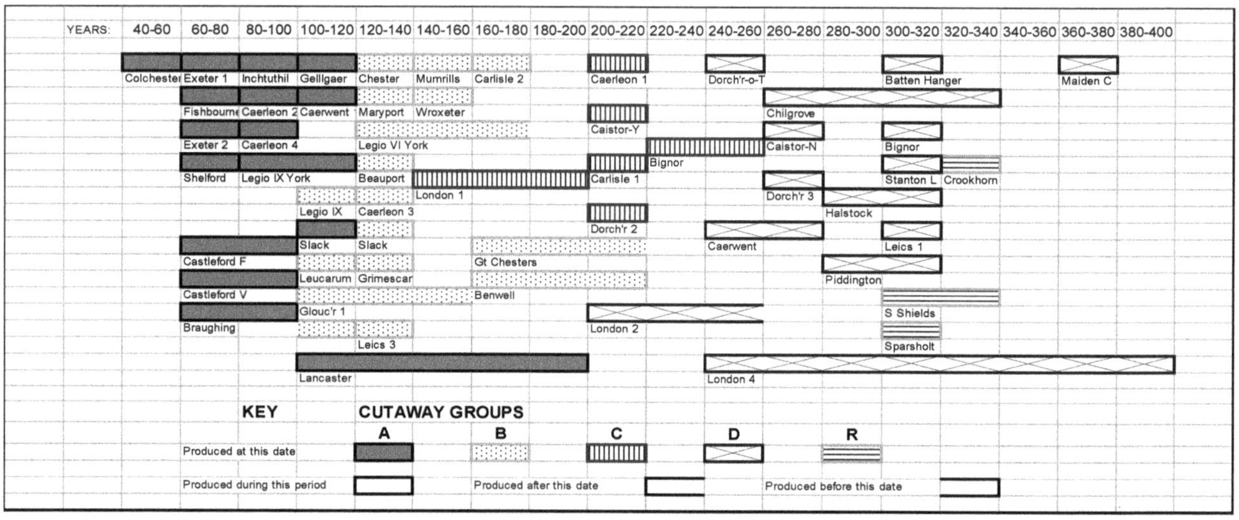


Figure 4.2: Compilation of individual sites with *strong* or *probable* dating

before a given date then the box is unclosed at its left hand end. The site name is given directly below the box: thus the table shows that Colchester was producing Group A tegulae in the period AD 40-60 and Exeter 1 was producing them in the period AD 60-80.

Figure 4.1 shows that the dates of the different forms cluster together very satisfactorily and demonstrate that the cutaway sequence predicated in Chapter 3 must be correct. The only possible conflict is the Group B tegulae stamped by *Legio* IX but if one assumes that the legion switched from Group A to B in AD 110 then there need be no inconsistency with the other data.[34]

The preponderance of Group A data simply reflects the absence of pre-existing cutaway forms on these sites thus making the evidence much more straight forward. Later sites suffer from the complexity of disentangling earlier data unless they were newly established on undisturbed land. The difficulty is most acute with Group C which sits in the middle of the Romano-British period and therefore has both earlier and later forms to contend with, whereas the later third and fourth century sites often only have Group D tegulae present.

Inclusion of the *probable* data should help fill some of the gaps in Figure 4.1. The dating of these sites either relied on evidence that was slightly less certain than those classified as *strong* or were dependent on the correctness of the cutaway sequence. As the accuracy of the cutaway sequence has been demonstrated, it is now possible to add the *probable* data and this is shown in Figure 4.2.[35]

[34] If the granularity of the table is reduced from twenty years to ten than this apparent problem disappears (a fact that remains true even when the additional data is added in the subsequent Figures 4.2 and 4.3).

[35] The *Legio* IX data from York has been amended to take advantage of the acknowledgement of the cutaway sequence which means that the Group A tegulae can be assigned to the start of the AD 90-120 date range whilst the Group B can be assigned to the end of that range.

This data all reinforces the pattern of the previous table and presents no conflicts. The regional cutaways are now included and follow on after Group D as expected. If it were not for Maiden Castle temple, the Group D range could end in AD 320 which would then create a minimal overlap with the regional cutaways. However there is no equivocation about the dating in Wheeler's excavation report nor about the temple having had clay roof tiles, so the only doubt surrounds whether new tegulae were used. Only three complete tegulae have been retained from the excavation: two of these have an upper cutaway insert but the third shows no sign of an insert. Although coming from at least two different moulds, the tiles are sufficiently dimensionally similar not to be obviously different manufacture. Nevertheless the possibility that the tiles were originally installed on another building or buildings prior to coming to the temple should not be discounted.

Finally Figure 4.3 adds in all the remaining *possible* data. Use of the indicative data helps populate the more difficult Group C range and continues to support the overall date ranges with four exceptions: Sandwich, Beauport Park, London 3 and Leicester 2. The Group D tegulae at Sandwich are dated as AD 100 consistent with the methodology that tegulae are assumed to be contemporaneous with the original construction of the building unless there is evidence of modification which was absent from Sandwich's severely truncated foundations. However, as the excavators thought that tiles could have come from the collapse of the porch in the late third or early fourth century,[36] even if this was an original feature, it is extremely unlikely that these would have been the original tiles as roofs would have needed to be renewed at least once every hundred years. The fact that tegulae were found in association with late third or fourth century pottery[37] also adds weight to a later dating.

[36] Parfitt 1980, 236.

[37] Parfitt 1980, 244.

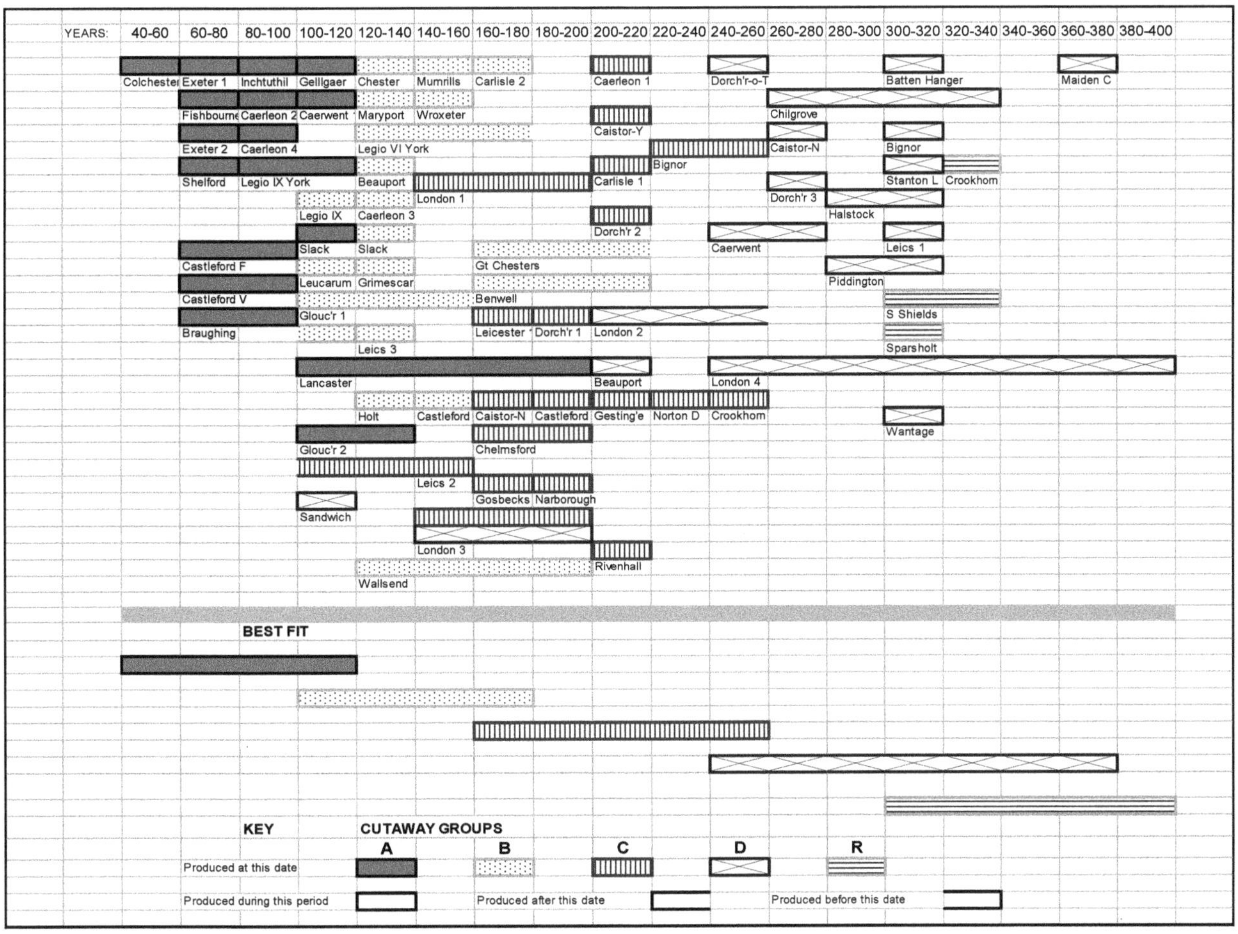


Figure 4.3: Compilation of all site dating

The Beauport Park Group D tegulae were stamped with RIB 2481.7 and have been dated by comparison with the same die from Phase III of the fort at Dover. However, almost all the examples of this die at Beauport Park were found on tegulae[38] whereas almost all the examples at Dover were found on brick,[39] so it is possible that the assumption that the use of the die at the two sites was contemporaneous is incorrect. Moreover, Williams suggests caution should be exercised in using the Dover tile stratigraphic evidence.[40]

The attribution of some Group D tiles from London 3 as prior to AD 200 is dependent upon fabric dating, which as previously discussed, is problematic. In this case, Betts[41] identifies the same fabric as going through to AD 300 (which would fit well) but also differentiates the quality of the mould sand used: the coarser sand being placed as pre AD 200 and the finer as later. There must be a possibility that the consistency of the mould sand was not always maintained.

The final conflict arises with Leicester 2 where tegulae used in a corn-dryer have been assumed to be secondary material and therefore to pre-date the corn-dryer by at least twenty years. If this assumption were wrong and the tegulae were actually new (which would be supported by the quality of the tiles and the fact that they mainly appear to have been complete when installed), then this could place these Group C tegulae as AD 180 which would fit well within the range.

Best fit date ranges for the cutaway groups are suggested in the second half of the table. The dates[42] are as below:

| | |
|---|---|
| Group A | 40-120 |
| Group B | 100-180 |
| Group C | 160-260 |
| Group D | 240-380 |
| Regional | 300 onwards |

The Group A dating is very secure. The start of Group B seems equally secure but the ending is less certain with

[38] 170 out of a total of 200 per RIB 2481.7.

[39] 256 out of a total of 309 per Williams 1981, 124.

[40] Williams 1981, 127.

[41] Betts 2000, 341-2.

[42] Crummy 1984, 299 records Group B and C tiles from Boudiccan destruction levels at Lion Walk, Colchester. This appears to be an isolated observation not matched at the Boudiccan destruction site of St Mary's Hospital discussed in Appendix 4, nor observed on any of the other early sites across the country.

civilian sites appearing to switch to Group C earlier than military ones, some of which in the north might arguably go a few years beyond the postulated AD 180 finish date. Group C presents the greatest difficulty and the date range of AD 160 – 260 must be regarded as tentative. Whilst it seems unlikely that Group C persisted beyond AD 260, it is possible that adoption on some civilian sites took place before AD 160.[43]

Group D clearly starts in the third century but without the evidence of Maiden Castle an earlier end date than AD 380 might have been appropriate. However the regional forms, which appear to commence circa AD 300, are not found throughout the province and thus an earlier end date for Group D would effectively imply an earlier end date for tegula production outside of those areas producing the regional forms. Whilst some areas had switched to stone roof tiles by the fourth century (and some much earlier) there is no evidence for this occurring in the southeast where a replacement regional form is absent.

Despite these qualifications, there can nevertheless be no doubt about the basic structure of the dating. This allows the dating of three further sites which apparently contradict these date ranges to be discussed in the next three sections of this chapter: the three sites are Tarbock from where the consular dated stamps appear to originate, Carpow which has yielded stamps with the cognomen *Britannica*, and Templebrough which has tegulae stamped by the *Cohors IIII Gallorum*.

## 4.6 The date of the consulship of Verus

Seven tegulae have been found with a stamp[44] referring to the third consulship of Verus (Plate 4.4). Two of these were found in Chester but the remaining five come from Tarbock in Merseyside which the excavators thought was a probable rural kiln site operating in the mid-second century, *inter alia,* supplying tile to *Legio* XX in Chester.[45] Unfortunately all of these tiles were too fragmentary to be diagnostic.

The excavators believed that production from the site was very short-lived and placed it as centring around AD 167,[46] no doubt influenced by the then established date for the tile stamp.[47] They noted that tile was introduced onto the site prior to the erection of the principal structure because it was used to pack some of the post holes.[48] The post holes also contained probable Holt pottery which they argued must date to the active period of Holt production between AD 90 and AD 130.[49]

However, the dating of the Holt pottery is difficult to reconcile with a very short-lived duration for the site centring around AD 167. Twenty-eight (unstamped) tegulae with lower cutaways were recovered from the site of which twenty-four were Group A and just four were Group B which suggests they were deposited close to the date of transition from Group A to B and therefore nearer to say AD 120.

More recently, it has been argued[50] that the consul referred to in the inscription is M. Annius Verus and that it should therefore be dated to AD 126. This dating fits the occurrence of the Group A and B tegulae much better and also fits with the Holt pottery.

This conclusion can be further tested by reference to the distribution of tegulae cutaway lengths found at Chester compared with those from Holt and Tarbock. The Chester data are broadly bi-modal for both lower and upper cutaways as shown in Figures 4.4 and 4.5.

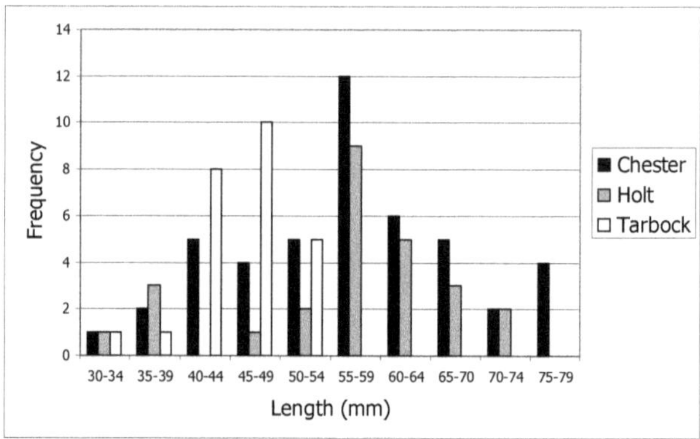


Figure 4.4: Chester/Holt/Tarbock lower cutaway lengths

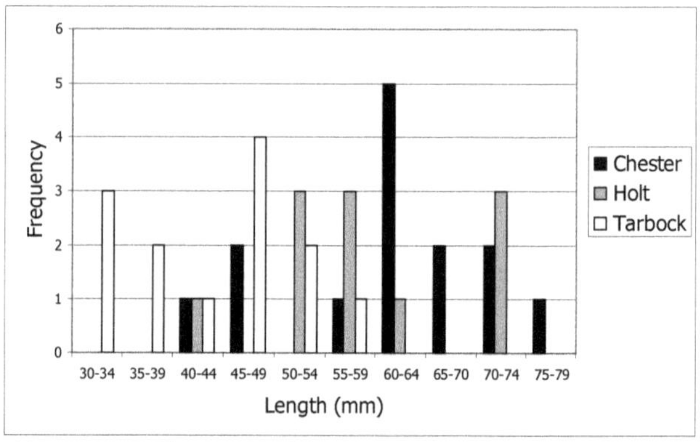


Figure 4.5: Chester/Holt/Tarbock upper cutaway lengths

[43] The large assemblages of tegulae from the forum/basilica at Caerwent and Insula IX at Silchester have representatives of all of the national cutaway groups but proportionately there are few Group B. This could simply reflect a reduction in construction activity on these sites during the period of Group B production or it could possibly imply that on these sites the use of Group B tegulae was very short-lived such that the conversion to Group C took place much earlier than postulated above. This issue is considered further in Sections 4.11 and 6.13.

[44] RIB 2463.59.

[45] Cowell & Philpott 2000, 67-116.

[46] Cowell & Philpott 2000, 114.

[47] Hassall & Tomlin 1978, 476 No 16 note 29.

[48] Cowell & Philpott 2000, 112.

[49] Cowell & Philpott 2000, 72.

[50] L'Annee Epigraphique 2000, 322 No 831.

The distribution of cutaway lengths of the tegulae found at Holt coincides with one of the Chester modes and the Tarbock tegulae with the other. However, because second century building work at Chester had essentially stopped by AD 130,[51] the Tarbock tegulae must have been supplied prior to that date in order to fit with the distribution of Chester tegulae shown on the graphs.

The attribution of the tile stamp to M. Annius Verus and the date AD 126 must therefore be preferred.

## 4.7 Dating the *Britannica* Cognomen

### 4.7(a) The Carpow fortress

Tegulae bearing the stamp "LEG VI VIC BPF" (Plate 4.5) where the "B" is interpreted as the cognomen *Britannica* have been excavated at Carpow, a fortress on the Tay estuary. The site has been traditionally regarded as a Severan establishment and the "B" has therefore been equated to the adoption of the title of *Britannicus* by Severus and his sons in AD 210 and the subsequent award to *Legio* VI for its part in their northern campaign.[52] However the two tegulae from Carpow held at the National Museum of Scotland both have clear Group B cutaways which, on all the evidence so far presented in this chapter, could not have been in production as late as AD 210.

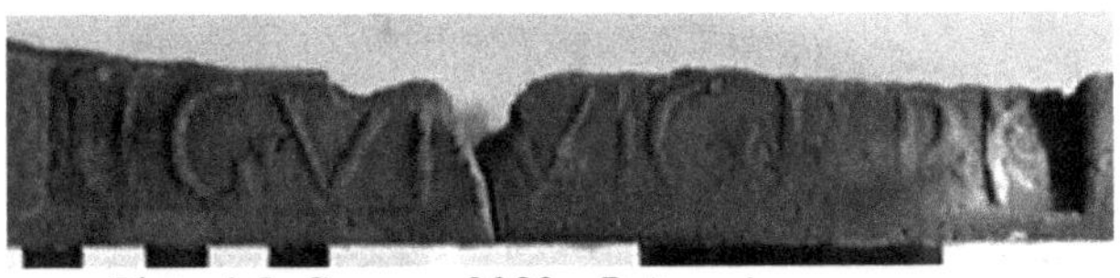

Plate 4.5: Carpow 2083 –*Britannica* cognomen

Dore & Wilkes summarise their comprehensive report on Carpow by stating "Structural evidence suggests that the occupation was confined to a single period, of short duration, and ceramic and numismatic evidence suggests that this occurred between c AD 180 and c AD 220".[53] Within this date range the pottery report favoured a Severan occupation whilst expressing caution about the small number of stratified finds on which it was based.[54] The coin report[55] reached a similar conclusion without being able to rule out a presence during the reign of Commodus.[56] It also suggested that an occupation much beyond AD 211 was unlikely.[57]

There are a number of possible reconstructions: Birley has suggested that the thirty acre fortress could have been constructed after the initial success against the Caledonians in AD 209[58] and its size and solid construction could have meant that it was not completed until perhaps AD 212. Alternatively construction of the fortress could have been started shortly after AD 200 when the campaigning appears to have begun and the fortress been abandoned in AD 211 when Caracalla and Geta returned to Rome. A more radical proposal is that the fortress was built by Commodus and continued in use for the Severan campaigns. Each of these propositions have arguments both for and against so, in order to better assess these and thereby to fix the date for the *Britannica* cognomen, the various sources of evidence are listed in the next subsection and then tested against these three possible reconstructions in the subsequent subsections

### 4.7(b) Evidence for the dating of Carpow

First: the inscription from the east gate which Wright reconstructed to refer to Caracalla as sole emperor and therefore dated as AD 212 or later.[59] Mann & Jarrett[60] have argued that the reference to "IMP" in this inscription does not preclude a reference to other joint rulers later within the text. Although Wright[61] roundly rejects this possibility for want of sufficient space for the text, his own interpretation based on just six letters in a twenty-nine letter reconstruction, must be open to doubt.

Second: *Britannicus* was a title first adopted by Commodus in AD 184 and subsequently by Severus and his sons in AD 210.[62] If Severus granted the title to *Legio* VI in recognition of its success in his Scottish campaign then it would presumably not have been before he adopted the title himself. Heil has dated Severus's adoption of the title to 31 March 210,[63] so it is possible that the award could have been passed onto *Legio* VI in time for it to appear on that summer's tile production.

Third: Dio[64] and Herodian,[65] whose accounts are not necessarily objective, state that the campaign amounted to little more than a series of skirmishes with Caracalla and Geta returning home promptly on the death of their father.[66] Birley goes further stating that "there is ample evidence on record that Caracalla withdrew the northern garrisons before he left for Rome in 211"[67] and Frere notes that the sons were back in Rome in AD 211.[68] Dio also says that, following the completion of the campaigns, the Maeatae (who are assumed to have previously been compliant) rebelled. The Caledonians

[51] Strickland 1997, 107.
[52] RIBII.4, 148.
[53] Dore & Wilkes 1999,481.
[54] Dore 1999, 537-552.
[55] Holmes 1999,528-535.
[56] Holmes 1999, 531.
[57] Holmes 1999, 532.

[58] Birley 1962-3, 199.
[59] Wright 1965, 223.
[60] Mann & Jarrett 1967, 64.
[61] Wright 1974, 289-292.
[62] Birley 1962-3, 197.
[63] Heil 2003, 268-271.
[64] Dio LXXVII, 13.
[65] Herodian III, 14, 10.
[66] Herodian III, 15, 7.
[67] Birley 1962-3, 197.
[68] Frere 1999, 170.

joined this revolt,[69] but Severus died before he could make war on them in person. As Severus died in February AD 211, this implies that this new revolt took place in AD 210.

Fourth: St Joseph[70] identified two further large military enclosures by aerial photography which preceded the excavated fortress on the Carpow plateau: one is most probably Flavian but the later polygonal one does not match any British type. There may, however, be an exemplar at Alteburg, a *Classis Germanica* base on the Rhine, which has a similar polygonal form expanding towards the river.[71]

Fifth: Swan[72] has noted the absence of Ebor ware pottery at Carpow, which was being produced by *Legio* VI in the late second and early third centuries. She believes that the African troops responsible for the production of Ebor ware arrived as reinforcements for *Legio* VI following Commodus's war.[73] This could suggest that the legion was at Carpow in one of its earlier structures but not during Severus's campaign.

Sixth: two coins of AD 208; one of Severus showing a permanent bridge and the other of Caracalla showing a boat bridge. Reed suggests that the permanent bridge was sited at Carpow crossing the Tay and the boat bridge crossing the Forth,[74] thereby providing a line of march via Cramond and Carpow to the north.

Seventh: there is a wide spread of coins from the site but all the Severan coins have little or no wear and the latest are dated AD 207.[75]

Eighth: Holmes has concluded that the archaeological record at Cramond, some fifty kilometres south of Carpow, does not support any period of abandonment between AD 160 and 208[76] which has obvious ramifications for the interpretation of Carpow. This has always been a possibility: Frere has highlighted the difficulty of interpreting Dio's reference to the storming of the Wall[77] in the uprising circa AD 180 as being Hadrian's Wall rather than the Antonine Wall because of the lack of evidence of any consequential repairs to Hadrian's Wall in the following twenty-five years.[78] Salway has speculated that following the uprising one possibility was that the Antonine Wall was reoccupied and maintained through to the time of Severus.[79]

Finally: the east gate inscription only bears the symbols of *Legio* II[80] but not those for *Legio* VI, although, as the right hand part of the inscription is missing, it is possible that that side carried the *Legio* VI symbols.[81] Conversely, none of the 216 stamped tile fragments that were recovered from the *principia* and *praetorium*[82] mention *Legio* II. These two most important buildings on the site would therefore have been clearly badged to *Legio* VI. At Carlisle, where vexillations of *Legio* II and *Legio* XX were under joint command at around the same time,[83] tiles stamped by both legions occur. These tiles, uniquely for both legions, have incuse stamps and appear to be of identical manufacture suggestive of a joint tile-making operation.

In addition to the 216 tiles at Carpow, two of the thirty-two tiles listed in RIB under third century cognomen for *Legio* VI have the *Britannica* cognomen, but none of the 136 similar *Legio* II tiles have the *Britannica* cognomen, nor do any of the sixteen *Legio* XX tiles.[84] This would suggest that the title was only awarded to *Legio* VI.

### 4.7(c) Construction of the fortress started AD 209 and finished after AD 212

This proposition would be consistent with Wright's interpretation of the inscription and allows time for buildings to be constructed after AD 210 using tiles with the *Britannica* cognomen. It would also be consistent with a single phase of archaeology. However, the position of the fortress in such a defensive position is difficult to comprehend except as part of a wider scheme to reinstate the whole Antonine system, for which there is no evidence on other sites. Moreover, if the fortress was not started until AD 209 then the bridge across the Tay built for the campaigns of AD 207/8 would have had no protection.

The proposition is inconsistent with the pottery evidence and literary sources and it probably stretches the date of the unworn coins too far. It also fails to explain the absence of *Legio* II tiles from Carpow and the omission of the *Britannica* cognomen from their abundant third century tiles.

On balance this proposition is difficult to sustain.

[69] Dio LXXVII,15.
[70] St Joseph 1973, 220-222.
[71] Bogaers & Ruger 1974, 169.
[72] Swan 1992, 5.
[73] Swan 1992, 7.
[74] Reed 1975, 93.
[75] Holmes 1999, 531.
[76] Holmes 2003, 151.
[77] Dio LXXIII, 8.
[78] Frere 1999, 152-3.
[79] Salway 1981, 206.

[80] Wright 1965, 223.
[81] Wright 1974, 291.
[82] Birley 1962-3, 200.
[83] Hassall & Tomlin 1989, 331 note 9, inter alia, suggest a date of AD 219 based upon an inscription under the same joint command at nearby Netherby. However, at this date both units were stamping their tiles with the *Antoniniana* cognomen, so it is possible that they had already been under joint command at Carlisle for some time, perhaps as early as AD 200 as is suggested by the use of incuse dies (see Section 5.15).
[84] RIBII.4.

## 4.7(d) Construction of the fortress started circa AD 200 and finished soon after AD 210

There are two sub-options with this proposal: that building continued into AD 212 which tallies with Wright's interpretation of the inscription but disagrees with the literary sources and stretches the coin evidence; or that the fort was abandoned by AD 211 which fits with the literary sources but conflicts with Wright's interpretation of the inscription.

Either way there is a difficulty with the construction of the principal camp buildings: as previously noted the *praetorium* and *principia* were tiled with tegulae stamped with *Britannica* cognomen which cannot in this proposal have occurred before AD 210. It seems most unlikely the fortress existed without these buildings for some ten years, so the stamped tiles must represent a re-roofing presumably from either thatch or shingle (replacing good unstamped tegulae with stamped tegulae seems most unlikely). This re-roofing would not simply be a matter of stripping off the old cladding and replacing it with the new because the tiled roof would require a different angle of pitch and greater supporting beams. Structural alterations would most likely have been required to accommodate a new roof but the excavators found no evidence for this. Moreover, if Dio is accurate in reporting the renewal of resistance shortly before Severus's death, one would have expected the refurbishment of the camp buildings to be low in the legions' priorities in AD 210.

The location of the fortress is also awkward because if it was built as a base for *offensive* operations then a more logical siting would be on the north bank of the Tay rather than the south. Severus had marshalled a large armada including detachments from the Rhine and Danube fleets[85] so establishing such a bridgehead on the north bank must have been within their compass and indeed was necessary before a bridge could be built.

It also fails to explain the pottery and the absence of *Britannica* stamped *Legio* II tiles both at Carpow and more widely.

On balance this proposition is also difficult to sustain.

## 4.7(e) The fortress was built circa AD 180 and modified after AD 200

This proposition stretches the short duration single phase proposed by Dore & Wilkes but is not outwith their overall possible date range. It is, moreover, consistent with almost all the other evidence. In particular, because it was reutilising an existing facility, it explains why Severus's fortress was in a defensive position on the south bank of the Tay estuary rather than the north where it would have otherwise have been better located for offensive operations.

[85] Dore & Wilkes 1999, 570.

Assuming that *Legio* VI occupied the fortress circa AD 180 when it constructed the principal buildings, it explains why there was no Ebor ware pottery (which post dated *Legio* VI's occupation) and also why there were no stamped *Legio* II tiles found at Carpow. Under this scenario Commodus will have awarded the Britannica title to *Legio* VI but probably not to the other legions so it explains the absence of the *Britannica* tiles amongst the prolific third century *Legio* II tiles and also from *Legio* XX tiles. The fact that only two tiles[86] with the *Britannica* cognomen have been found elsewhere other than at Carpow is not inconsistent with the relatively low rate of stamped tile production (or low rate of stamped tile recovery) of *Legio* VI during this period.[87] It may also have been impacted by the *damnatio memoriae*[88] that Commodus suffered after his death in AD 192.

The earlier occupation of Carpow also fits better with the evidence from Cramond which appears to have been occupied throughout the second century and could clearly not have continued in isolation.

This reconstruction provides no resolution to the conflict between the literary sources and the dating of the inscription. However, because this proposition does not require construction to continue into at least the summer of AD 210 in order to explain the *Britannica* roof tiles, it makes it much easier to suggest that Wright's dating is wrong.

## 4.7(f) An alternative interpretation of the Carpow inscription

Almost all of the evidence, historical, numismatic, pottery and archaeological, conflicts with Wright's reconstruction of the inscription to Caracalla which should now be examined in more detail. On the basis of the size and spacing of the lettering, and the room available, Wright's reconstruction was as below:

IMP E[T D N M AVR ANTONINUS]
[PIV]S F[ELIX]…..

"IMP" implied a sole emperor which, in view of the Severan date for Carpow, had to be Caracalla who became sole emperor in AD 212. The formula "IMP ET D N" was not matched elsewhere but represented a reasonable evolution of the formula "IMPERATOR DOMINUS NOSTER", a version of which was first attested in AD 155.

[86] RIB 2460.75, both found at York.
[87] RIB records only one tile with the *Antoniniana* cognomen for *Legio* VI (RIB 2460.70) compared with over hundred examples for *Legio* II.
[88] Mann 1988,134.

However Commodus was also a sole emperor after AD 180. Between AD 180-190 he was named Marcus Aurelius Commodus and in AD 183 adopted the title "Pius".[89] The formula "DOMIN N" has already been attested for Commodus[90] so the next stage "IMP ET D N" would not be unreasonable if it was to reach its fully standardized form of "IMP D N" by AD 213 under Caracalla.[91]

The word "Commodus" has only eight letters compared to the nine of "Antoninus" but offsetting this "Commodus" has two "Ms" which take up more space than an average letter and has no "I" which takes up less.[92] Thus an alternative reconstruction, which is virtually identical to Wright's and matches all of his stipulations, could be:

IMP E[T D N M AVR COMMODUS]
[PIV]S F[ELIX].....

This inscription would date to AD 183 or later.

## 4.7(g) A revised history for Carpow

Using this revised reconstruction of the inscription it is now possible to propose a sequence of events that fits much better with all the known evidence. Construction of the fortress was originally undertaken by *Legio* II circa AD 180 in reaction to the tribes crossing the Wall. They built the turf rampart and stone gateways (reminiscent of the walls at Verulamium and Cirencester of the same period). However, before the fortress was completed, the legion was moved. Perhaps it may have been *Legio* II to which Dio was referring in his account when a general was slain at the head of his troops; and perhaps it was *Legio* VI which was drafted in and secured the victory for which it was awarded the title *Britannica* in line with Commodus's own adoption of the title in AD 184. Either way *Legio* VI took possession of the fortress and completed the construction including the *principia* and *praetorium*[93] which they roofed with tiles bearing their new cognomen.

Carpow was then maintained as the last link in a line of outposts running through Newsteads and Cramond, policing the separation of the friendly Maeatae from the Caledonians.[94] Carpow could be supplied by sea and presumably Cramond was as well. The line of outposts and the absence of a hostile threat in this area would explain why the forts at the eastern end of Hadrian's Wall do not appear to have been repaired in the twenty five years leading up to the campaigns of Severus.

If Holmes is correct that Cramond was continuously occupied between AD 160-208 then it seems likely that Carpow was as well. Perhaps the early post-Antonine occupation was restricted to the polygonal enclosure which was maintained as a relatively limited *horrea classis*. When the tribes crossed the Wall, which on this proposition must be the Antonine Wall,[95] a more substantial base was required which led to the construction of the fortress. Carpow would also have fitted logically with the extra chain of forts (if these were also occupied) running up Strathallan and Strathearn to the Tay to secure the good agricultural land in the salient between the Forth and the Tay estuaries.[96]

The extent to which the fortress continued to be occupied after Ulpius Marcellus's campaign is obscure,[97] but when Severus launched his campaigns just fifteen years later there must have been a sufficient military structure still extant at Carpow for him to choose this as his northern base over establishing a new base on the northern bank of the Tay estuary.

## 4.7(h) Further dating evidence

Section 4.5 established the date range for Group B cutaways as AD 100-180 and for Group C cutaways as AD 160-260 whilst Chapter 3 established that there appears to be a uniform rate of average tegula size reduction through time. It is therefore worth examining where the single complete Carpow tile[98] fits into the size distributions of Group B and C tegulae. These are shown in Figures 4.6 and 4.7.

From these figures it can be seen that the Carpow tegula falls neatly between the modes of the Group B and C tiles both in respect of length and breadth, which would suggest that it was manufactured close to the transition date from Group B to Group C tiles: that is circa AD 170-180. This analysis should be regarded as highly speculative as it depends upon the dimensions taken from a single tegula but it nevertheless tends to support manufacture of the *Britannica* tiles in the later second century rather than the third.

89 Cagnat 1976, 203-4.
90 CIL viii 10570.
91 CIL xiii 7338, which may also cast doubt on Wright's "IMP ET D N" for the previous year.
92 In Times New Roman font they are identical in length: ANTONINUS
COMMODUS
93 At Inchtuthil the *praetorium* was the last building to be started and had not been constructed before the fortress was abandoned.
94 And perhaps monitoring the older demarcation between the philo-Roman Votadini and their more hostile neighbours the Selgovae.
95 Mann 1988, 134 believes that the Antonine Wall was the northern frontier at this time
96 Ardoch, Strageath and Bertha: Jones & Mattingly 1990, 121-126.
97 However some measure of continuous occupation seems likely because any period of abandonment would have shown up in the archaeology. A continuing occupation by *Legio* VI could also fit with the evidence of Samian potters' stamps from York which show a marked decline after AD 180 (Dickinson & Hartley 1971, 129 Figure 18) although this in part mirrors the overall pattern of Samian production.
98 Dore & Wilkes 1999, 536

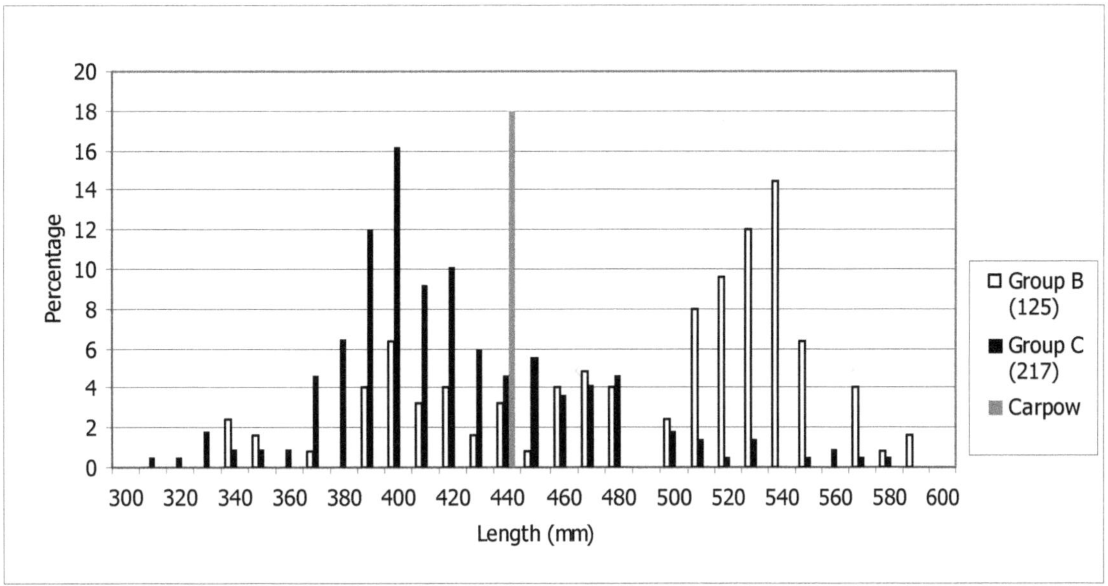


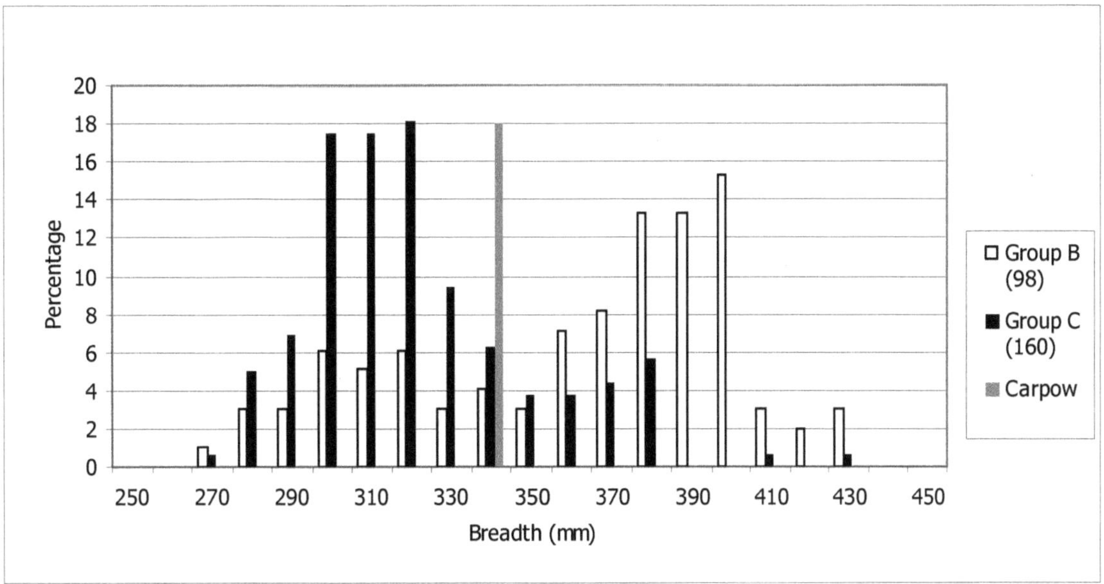


Figures 4.6 and 4.7: Carpow length and breadth compared to national distribution

## 4.8 Dating *Cohors IIII Gallorum's* occupation of Templebrough[99]

Templebrough represents a series of three successively smaller forts starting in AD 54 and in use up to circa AD 175.[100] Ten tegulae stamped by *Cohors IIII Gallorum* (Plates 4.6a and 4.6b) have been recovered from the site,[101] however only three of these were diagnostic and in every case the cutaway was damaged such that it was not possible to be certain whether they were from Group B or C. Three other unstamped diagnostic tegulae have been retained from the 1916 excavation, one of which was Group B and the other two Group C.

Any consideration of Templebrough cannot be undertaken in isolation from the fort at Castleford where two further tegulae stamped by *Cohors IIII Gallorum* were found in 1922.[102] (As discussed in Appendix 4, the military occupation of Castleford lasted until AD 100, but the vicus continued into the fourth century). One of the Castleford tegulae has the same RIB 2472.2 die as the majority of the Templebrough tiles. The other Castleford die, although the same overall size, has much thinner lettering (Plate 4.6c) and could represent a different (new)

[99] May's 1922 excavation report uses the spelling Templebrough although later references adopt the spelling Templeborough.
[100] May 1922, 11.
[101] RIB 2472.1 and 2.

[102] Castleford Museum Accession Book.

die or, more probably, an earlier version of RIB 2472.2 which was subsequently recut thereby broadening the letters.[103] Only one of these tegulae was diagnostic but it was damaged and, again, it was impossible to be certain whether it came from Group B or C.

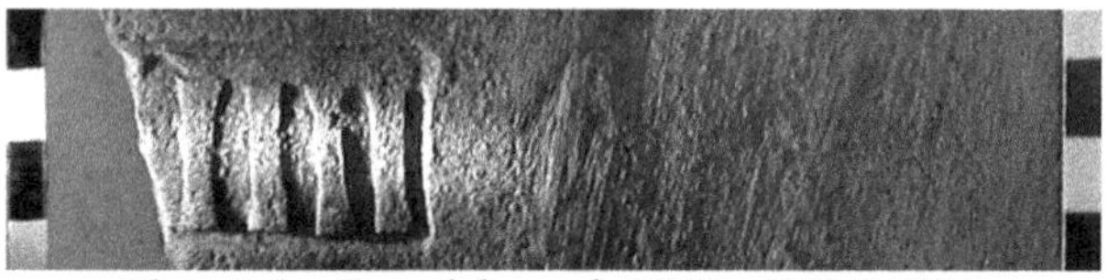

Plate 4.6a: Templebrough 2054 – RIB 2472.1

Plate 4.6b: Templebrough 2051 – RIB 2472.2

Plate 4.6c: Castleford 2309 – RIB 2472.2

Based on the existence of the stamped tegulae, it has been suggested that *Cohors IIII Gallorum* was the garrison during the first century military occupation of Castleford.[104] If this were the case then the stamped tiles would be expected to be similar to the unstamped Group A tegulae from Castleford which have been excavated from confirmed first century contexts. Figure 4.8 below compares the dimensions of these Castleford Group A tiles against all the other tegulae from both Castleford and Templebrough.

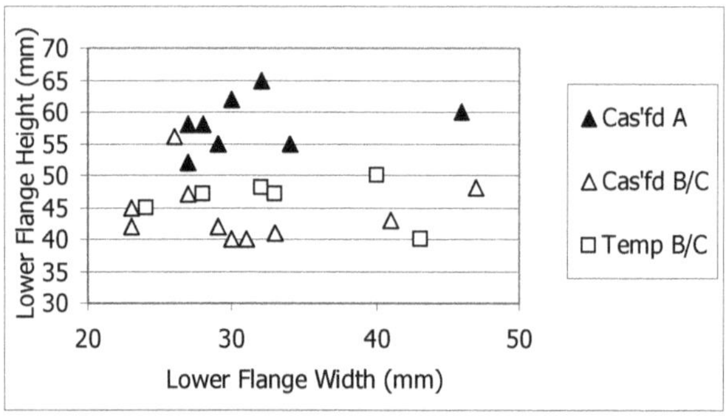


Figure 4.8: Castleford and Templebrough dimensions(1)

It can be seen, with the exception of one unstamped Castleford Group B/C tegula, that the Group A tiles are dimensionally different to all the other tiles, stamped and unstamped, from both sites. Thus it seems most unlikely, even without the evidence of the cutaways, that *Cohors IIII Gallorum* was the first century garrison at Castleford,[105] moreover they are not attested on a diploma in Britain until AD 127.[106] However, as there was no second century military presence at Castleford, an alternative reason is required for the presence of the stamped tegulae.

The probable explanation is that the same tilery was supplying both the military at Templebrough and the civilian establishment at Castleford. This would make sense because, although roller die stamps are known to have travelled,[107] there is no evidence for the same military tile stamp being used at two separate locations. Indeed all the evidence, for example from the *Classis Britannica,*[108] is to the contrary.

Plate 4.7: Templebrough antefix

Plate 4.8: Castleford Antefix

[103] The Castleford tiles are not listed in RIB.
[104] Betts 1998, 231.

[105] A better argument could be constructed for Cohors IIII Breucorum being the garrison as a tile of theirs was also found at Castleford and they were at nearby Slack in the early second century. However, whilst the stamp is identical to those appearing at Slack, it appears to have had a red paint or slip applied which was a feature noted on a number of the later tiles at Castleford but was not noticeably present on the tiles excavated at Slack (or Grimescar). If it was slip this presents something of a conundrum because it would have been applied prior to firing and so would make it difficult to argue that this was simply residual material quarried from Slack at some later date in lieu of manufacturing new tiles.
[106] De la Bedoyere 1999, 75.
[107] Betts, Black & Gower 1997, 52.
[108] Peacock 1977, 245.

A common source is also suggested by the similarity of unusual antefixes found at Castleford and Templebrough which take the form of a cresting mounted on top of the open end of an imbrex (Plates 4.7 and 4.8) instead of being the normal stopping for the open end of a column of imbrices.[109] The fact that this antefix design appears not to have been reported elsewhere,[110] the superficial similarity of the fabrics,[111] and the incidence of *Cohors IIII Gallorum* tiles on both sites make a compelling case for the two sites sharing the same tile manufacturing unit. Presumably the source lay somewhere between the two forts which were 35km apart, although rather more if travelling via Doncaster as there is no known Roman road between Castleford and Templebrough.

The antefixes had different motifs: the Templebrough ones had a military insignia[112] and the Castleford one a sun ray design which suggests that production for the two sites was deliberately differentiated. It could therefore be that the stamped tiles were sent to Castleford in error when they were included by mistake in a consignment of other tiles.

On the assumption that the source of the second century tile to the sites is the same, the combined assemblage can be examined to attempt to fix a date for the stamped tegulae. This is shown in Figure 4.9 in terms of lower flange width versus cutaway length. Castleford tegulae are represented by triangles and Templebrough by squares. Group B tegulae are shown as unfilled symbols and Group C as black. The uncertain stamped tegulae are shown in grey.

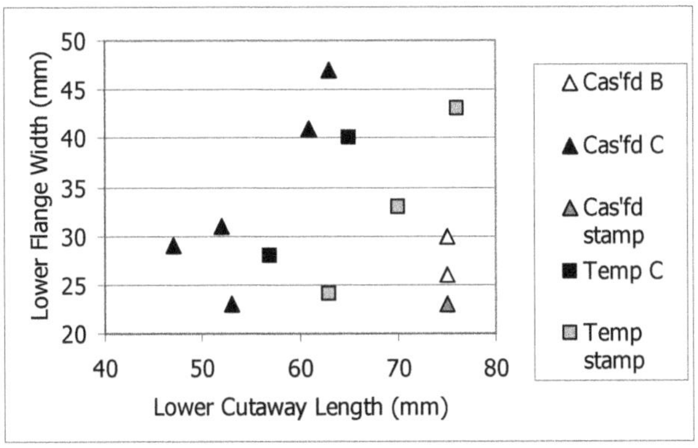


Figure 4.9: Castleford and Templebrough dimensions(2)

This suggests that the Castleford stamped tegula should be attributed to Group B but leaves the Templebrough stamped tegulae ambiguously placed between the Group B and C clusters. An alternative approach (Figure 4.10) is to look at flange height versus cutaway length (having now allocated the Castleford stamp to Group B). This indicates that one of the Templebrough stamped tegulae is Group B, one Group C and leaves one which could fit into either cluster. In total this means there are two Group B, one Group C and one unallocated tegula stamped *Cohors IIII Gallorum.* The likelihood is therefore that the tegulae originate from around the transition from Group B to Group C which would give a date centred around AD 170.[113]

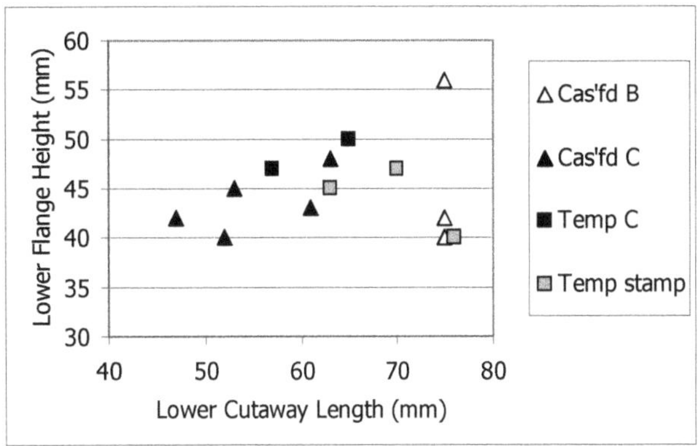


Figure 4.10: Castleford and Templebrough dimensions(3)

*Cohors IIII Gallorum* is believed to have been the Trajanic garrison at Templebrough based upon the stylistic evidence from two tombstones.[114] The unit is subsequently attested at Castlesteads, perhaps during the reign of Hadrian,[115] Castlehill on the Antonine Wall,[116] Risingham[117] and at Vindolanda in AD 212[118] where it appears to have remained. The Risingham inscriptions refer to NUMINIB[US] AUGUSTOR[UM] which could fit with the joint reigns of AD 161-9, AD 176-80 or AD 198-211.[119] As Risingham lies beyond Hadrian's Wall, the most likely dating is AD 161-163 during the Antonine II occupation which would also be consistent with the absence of the AUGG abbreviation[120] in the inscription which would otherwise have suggested a third century date.[121] *Cohors IIII Gallorum* cannot have reached Vindolanda much before AD 212 because the fort served as a probable refugee camp[122] during Severus's campaign.

This therefore leaves a gap between the unit's departure from Risingham, possibly in AD 163, and its arrival at

---

109 RIB 2458.9 and 2458.10 illustrate two *Cohors IIII Gallorum* antefixes of conventional form.

110 Neither Toynbee 1964, 428-431 nor Brodribb 1987, 29-31 mention it in their reviews of the subject.

111 Betts 1985, 261 states that the tile from the Castleford vicus appears to be of the same fabric as the stamped *Cohors IIII Gallorum* tegulae.

112 Templebrough have an "88" motif which May 1922, 7 suggests relates to the auxiliary unit.

113 This allows a possible timeline as follows: RIB 2472.1 (Templebrough-B) AD 160, RIB 2472.2 first die (Castleford-B) AD 170, RIB 2472.2 recut die (Templebrough-C) AD 180.

114 RIB 619 and 620.

115 RIB 1979, 1980 however Stephens and Jarrett 1985, 77-80 argue that the Castlesteads altars are third rather than second century.

116 RIB 2195.

117 RIB 1227 and 1249.

118 RIB 1705.

119 Wright 1965, 223.

120 Stephens & Jarrett 1985, 78.

121 In any event *Cohors IIII Gallorum* must have left Risingham by AD 208 when *Cohors I Vangionum* are attested there: RIB 1234.

122 Birley 2002, 159-161.

Vindolanda some fifty years later. Given that the stamped tiles appear to have been manufactured in the years around AD 170, it may therefore be suggested that *Cohors IIII Gallorum* was reposted back to Templebrough during this period. Alternatively the unit could have continuously occupied Templebrough with only detachments of the unit serving in the north.

The dating of their first posting to Templebrough is dependent upon the dating of the two tombstones which had been utilised in a drain.[123] As May had defined only three phases for the site ending in AD 175,[124] these tombstones had to come from the first or second phase and been reused in the third. However, third and fourth century coin finds from the first 1877/8 excavation, and subsequent finds from the site,[125] suggest that there may have been a further phase of construction and occupation, possibly of a civilian nature, when reutilisation of the tombstones may have been more probable. If this is correct, then it is possible that the post AD 163 assignment of *Cohors IIII Gallorum* was in fact their first and only posting to Templebrough.[126]

## 4.9 Dating the upper cutaway inserts

Of the sites considered in this chapter unambiguous upper cutaway inserts have been noted at Batten Hanger, Castleford, Chilgrove, Crookhorn, Dorchester, Maiden Castle, Sparsholt and Stanton Low. With one exception at Dorchester, they have all occurred on tegulae with Group D or regional cutaway forms, or where the tegulae have not had a lower cutaway themselves, they have been associated with other tegulae with these cutaway forms. The presence of an upper cutaway insert can therefore be regarded as evidence of a late third or fourth century tegula. As discussed in Section 2.4(g), these tiles were made in an inverted box mould which must therefore have been introduced into the tiler's repertoire in the later third century.

Intriguingly, tegulae from Leucarum with triangular flanges but without upper cutaway inserts must have been made in inverted box moulds by a military unit in the early second century, but the practice was not adopted elsewhere[127]. However, a small proportion of the Caerwent tegulae with square flanges and smooth undersides but without upper cutaway inserts were also made in inverted box moulds. Some of these are Group C and some may possibly be Group B (but were too damaged to be certain) suggesting a local transfer of technology or perhaps a continuation of the same tilery.

## 4.10 Evidence from the Continent

Although a study of tegulae from continental sources is outside the scope of this work, it is still possible to draw some inferences. Tegula cutaway variations occur throughout the Roman world and in *Britannia* the typology is equally applicable to both military and civilian manufacture. When *Legio* VI arrived at York in circa AD 120 it was immediately producing tegulae with Group B cutaways, consistent with the rest of *Britannia*, and had presumably been producing this form in Lower Germany prior to its transfer. In fact, despite the regular movement of auxiliary units and legionary vexillations around the Empire, there is no evidence for any discontinuities or aberrations in the cutaway typology in British military tegulae in this survey. This strongly suggests that the cutaway typology was not unique to Britain.

Anecdotally it may also be noted that in Italy tegulae from Pompeii have Group A cutaways.[128] In Germany two Group C tegulae stamped by *Legio VIII Augusta* come from Miltenberg[129] which was part of the new *limes* established at the end of the reign of Antoninus Pius[130] and was overrun by the Alemanni no later than AD 259. Epigraphic evidence shows that the neighbouring *limes* fort at Osterburken had work carried out by *Legio* VIII in AD 185-192,[131] so it is possible that a similar dating should apply to the *Legio* VIII tiles at Miltenberg but, in any event, these Group C tegulae must date between AD 160-260. In Brittany Group B and C tegulae have come from Le Yaudet where there was late Antonine activity[132] although the main presence on the site was late third century.

## 4.11 Demonstration of the chronology

An interesting demonstration of the cutaway chronology is provided by an analysis of the tegulae from successive seasons of excavation of Silchester Insula IX. Figure 4.11 shows the mix of cutaway forms found each season whilst the number of cutaways found is stated beneath the year designation. It can be seen that as the excavation went deeper, the Group D tegulae diminished and faded away, whilst the Group A tegulae began to emerge from

[123] May 1922, 127 and 129.

[124] Frere 1999, 147 shows Templebrough as one of the forts occupied in the period AD 163-196 to maintain control of the Brigantes which, if correct, would suggest that the military occupation went some twenty years beyond May's AD 175 date and would give more room for the dating of the tegulae suggested here.

[125] pers comm. Karl Noble of Rotherham Museum.

[126] However Stephens & Jarrett (1985, 77) argue strongly that the formulas used in the inscriptions could not be later than Trajanic. Moreover the presence of two different forms of antefix at Templebrough, both of which are probably associated with *Cohors IIII Gallorum*, is suggestive of two separate periods of occupation.

[127] There are some undatable triangular flanged tegulae at Carlisle which are dimensionally similar to those at Leucarum and could therefore have been made by the same unit after it had been moved north.

[128] Rook 1977, 107 note 12.

[129] Pers Comm: Hermann Neubert, Museum der Stadt Miltenberg.

[130] Schonberger 1969, 168. Epigraphic evidence shows that the old Odenwald and Neckar limes was still operational in AD 148.

[131] Schonberger 1969, 173.

[132] Pers Comm: Barry Cunliffe.

the lower levels. The 2004 season presents a hiatus caused by some very large deposits of tegulae introduced into the *insula* as levelling material. The extreme left-hand column shows the composition of the complete tegulae from the Victorian excavations held at the Reading Museum. This would suggest that the Victorian excavations either did not go very deep or that complete tegulae were more likely to survive from the later buildings.

However the virtual absence of Group B tegulae from the more recent excavation years raises the possibility that their period of use on this site was a good deal more abbreviated than suggested by the postulated date range of AD 100-180 and that the Group C were introduced earlier than AD 160.

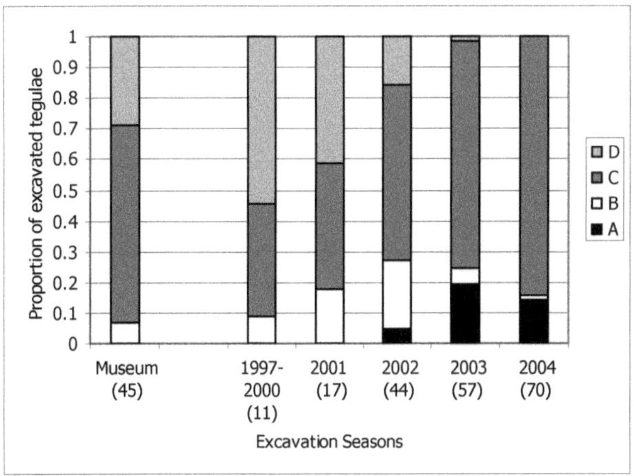


Figure 4.11: Silchester *Insula* IX tegulae by season

# 5

# STAMPS, SIGNATURES AND TALLY MARKS

## 5.1 Introduction

This chapter explores how tile stamps were used, when they were used, who they were used by and the interpretations that can be made from the manner of their use. In particular it seeks to establish whether each legionary cohort had its own stamp, whether they stamped all of the tiles that they made, and the extent to which contractors were used to supplement their output.

It examines the practices of other military units, of the procuratorial and municipal tilemakers, and of the private tilemakers. It considers whether incuse and ansate framed stamps were limited to particular producers or periods of production.

It uses the example of Classis Britannica stamps to explore whether particular stamps can be associated with particular phases of construction and how such buildings were subsequently maintained.

Finally it looks at signatures and tally marks to determine who made them and their significance. However, before any of these questions can be addressed, it is necessary to develop an estimate of the average die life because this will be crucial to the interpretation of all the other evidence.

## 5.2 Estimate of die life

By 1992, when the RIB listing of tile stamps was published, an average of 77 dies had been identified for each of the three principal British legions. As will be discussed later in this chapter, legionary tiles appear to have started being produced around AD 90 and finished around AD 260: so simplistically this works out at roughly one new die per legion every two years. If none of the dies were contemporary such that the dies actually had an average life of two years then it would be extremely rare to find instances of the same die appearing on tegulae with different cutaway forms because only those dies in production around the time of the switch to new cutaway forms in AD 110, 170 and AD 250[1] would be affected. On the other hand if many dies were in use at the same time and the dies had lives of fifty years then there should be very few dies that did not appear with two different cutaway forms because only those that commenced production in the years AD 110-120 or AD 170-200 and therefore were replaced in the years AD 160-170 and AD 220-250 would miss the switch to the new cutaway forms in AD 110, 170 and AD 250.

This principle allows the life of a die to be estimated using the seriation data in Figure 3.30. There are 77 different legionary stamps listed in this table[2] and eleven appear with two cutaway forms. However 35 of the entries have just single examples of dies which could not therefore show two different forms of cutaway. Eliminating those single examples leaves 42 different dies with multiple examples and eleven of these show a cutaway overlap. This is 26% of the total.

Using a theoretical model where, for convenience, new dies are issued every year,[3] it is possible to test different die life assumptions to see which one most closely fits the 26% of cutaway overlaps shown by the seriation evidence. For example, if dies had an average life of twenty years then the die starting production in AD 90 should cease production in AD 109 which is just before the switch from Group A to B in AD 110 and therefore that die should not produce any overlaps. The die starting production in AD 91 should go out of use in AD 110 and therefore the final year's production should be Group B. Production starting in each of the following 19 years should also show a cutaway overlap until the die starting its life in AD 110 which went out of use in AD

[1] Although Chapter 4 indicated that the switch from one form to the next may have taken place over a period of twenty years when both forms may have been concurrent, to simplify the argument it is assumed that the switch occurred at an instant in time at the start of AD 110, AD 170 and AD 250 respectively.

[2] RIB numbers 2459.1 to 2463.56.

[3] If new dies were issued every other year, or at longer intervals, the same result would obtain but the argument would be less clear.

129 which would use exclusively Group B cutaways. No further cutaway overlaps would occur until the die starting its production life in AD 151 which would have its final use in AD 170 with Group C cutaways. Continuing this logic for dies being introduced every year until AD 260 yields a total of sixty dies that should have cutaway overlaps out of a total of 170 years' production. That is 35% of the dies would have overlaps compared with the actual observation of 26% overlaps which suggests that the first guess of a twenty year life is too great. Adopting this same logic for a fifteen year life produces 45 overlaps in the 170 years, or a proportion of 26%, which exactly matches the observed occurrence.

However this model could be an underestimate of the die life because in reality there may be insufficient examples excavated to reveal the overlapping cutaways in all cases where they actually exist. For example, if only three copies of a particular die with a life running between, say, AD 100-115 had been excavated then it is possible that all of these could have been from Group A simply because Group B examples were still buried. Figure 5.1 below, shows that there are eight ways in which three examples of such a die with potentially two cutaway forms could emerge.

| First excavated | Second excavated | Third excavated | Cutaway overlap |
|---|---|---|---|
| A | A | A | X |
| A | A | B | √ |
| A | B | A | √ |
| A | B | B | √ |
| B | A | A | √ |
| B | A | B | √ |
| B | B | A | √ |
| B | B | B | X |

Figure 5.1: Ways cutaway overlaps can occur

Manifestly an overlap will fail to appear only if three Group A or three Group B tegulae are excavated. In this example the period of production was fifteen years and the switch from Group A to Group B took place in AD 110 which provided ten years of Group A production and five years of Group B production. Assuming production took place at a uniform rate and there was no bias in the way the tiles were excavated, then the probability of the first tile excavated being Group A is 10/15 and the probability of all three being Group A is $(10/15)^3$. Likewise the probability of the first tegula being Group B is 5/15 and the probability of them all being Group B is $(5/15)^3$.

The probability of no overlap showing even though the production period spans two forms is therefore $(10/15)^3 + (5/15)^3$ which equals 0.33. This can be generalised to $(F/T)^n + ((T-F)/T)^n$ where n is the number of examples of the die, T is the total number of years it was in production and F is the number of years before there was a change of cutaway form.

The average number of examples found in each of the eleven cutaway overlaps in Figure 3.30 is actually five rather than the three so far assumed. So, using the calculated fifteen year life, the formula can be used to assess the probability of an overlap failing to appear between, say, AD 109-124. This yields a probability of 0.7. As might have been expected it is far more likely that there would be no "visible" cutaway overlap if the date range only just crosses the cutaway date change in AD 110. On the other hand if the date change was in the middle of the production date range (eg AD 103-118) then the probability of no overlap showing in five examples falls to 0.07. Repeating the calculation for all possible positions of the date range yields an average probability with five examples that a cutaway overlap will fail to be revealed where one exists of 0.29.

This average probability can then be used to estimate the number of "concealed" cutaway overlaps where eleven have already been revealed which yields an answer of 4½.[4] Rounding this down to four gives a total of fifteen revealed and concealed cutaway overlaps out of a total of 42 dies and therefore a revised proportion of 36%. This equates to the proportion of overlaps that would be expected with a twenty year die life as previously calculated.

There are number of simplifications made in this analysis which is itself based upon limited data, so no pretence of accuracy should be assumed. Nevertheless this approach does indicate that the most likely average life of the legionary dies was between fifteen and twenty years. Within this fifteen to twenty year average, some dies will have had longer lives and others shorter ones. Given that an average of 77 different dies have so far been excavated for each legion, or roughly one for every two years of stamped tile production, then a twenty year average life would suggest that each legion must have had around ten different dies in concurrent use throughout the period.

This analysis has focussed on legionary dies where there is sufficient evidence to draw tentative conclusions which is not the case with the other dies. Nevertheless, the absence of any overlaps in the eight dies of private tilemakers that have multiple examples is suggestive of much shorter die lives than for the legionary output.

## 5.3 The date of the *Legio* II dies

The fifteen to twenty year life of legionary tile stamps means that where two different cutaways have been observed on the same die then the die is likely to have been in use either side of the cutaway transition. For example if Group A and B have been observed then the

[4] If "x" is the total number of overlaps then (1-0.29).x overlaps will be revealed and this number is known to equal 11. The unrevealed overlaps will equate to 0.29.x which equals 11x0.29/(1-0.29)=4.5.

die was probably in use between AD 100-120, but if only Group A is observed then AD 80-100 is more likely, although use as late as AD 120 still remains possible. Based on this structure, the following approximate dating has been adopted.

| Cutaways present | Date range |
|---|---|
| A only | 80-100 |
| A and B | 100-120 |
| B only | 120-160 |
| B and C | 160-180 |
| C only | 180-240 |
| C and D | 240-260 |
| D only | 260 + |

Figure 5.2: Approximate cutaway date ranges

Boon[5] categorised the *Legio* II dies based upon their size and style. He then discussed possible dates for each of his stylistic groups based upon the excavation evidence. These have been crystallised into suggested dates in the table below, although it should be appreciated that these date ranges may give Boon's conjectures the appearance of greater certainty than he ever intended.

| Boon Type | Description | RIB Number 2459.xx | Suggested Dating |
|---|---|---|---|
| Ai | Large serif letters | 3-6 | 90-100 |
| Aii | Bordered ansate label | 7-14 | (100)-140 |
| Aiii | Plain ansate label | 15-16 | c150? |
| Aiv | Small plain label | 17-19 | c120? |
| Av | Large Pi numeral | 20-22 | 90-140 |
| Avi | Tall letter | 23 | 175-250? |
| Avii | Small block letter | 24 | c130-150 |
| Aviii | Broken letter | 25 | 90-130? |
| Aix | Retrograde serif | 26 | c140 |
| Ax | Crude block letter | 27-28 | c150 |
| Axi | Retrograde Pi numeral | 29-30 | c140? |
| Axii | Crude letter | 31-32 | c200? |
| Axiii | Irregular letter | 33 | none |
| Axiv | Pinched label | 34 | c150? |
| Axv | Mostly retrograde | 36 | c140 |
| Axvi | Bordered plain label | 37-38 | c210 |
| Axvii | Orthograde II for E | 39-40 | none |
| Axviii | Retrograde II for E | 41-42 | c200 |
| Axix | Nailed die | 43-45 | c200 |
| Axx | Large thin letter | 46-47 | c200 |
| Axxi | Retrograde crude block | 48-51 | c200 |
| Axxii | Retrograde clumsy | 52-53 | c200 |
| Bi-vi | Antoninian | 54-60 | 213-222 |
| Ci-ii | Severan | 61-62 | 222-235 |
| Di | Doubtful suffix No 1 | 63 | 200+ |
| Dii | Victoriniana? | 64 | 269-271? |

Figure 5.3: Boon's *Legio* II die classification and dating

Boon's data may be compared with that derived from the cutaway dating and also with well stratified finds from Zienkiewicz's excavation of the Caerleon Museum site[6] which was completed after Boon's analysis. The results for Boon's first two die groups (Plate 5.1) are shown in Figure 5.4.

Plate 5.1: Types Ai and Aii (RIB 2459.4 and 2459.13)

| RIB 2459 | Boon type | Boon dating | Cutaway dating | Examples in RIB |
|---|---|---|---|---|
| 3 | Ai | 90-100 | 80-100 | 67 |
| 4 | Ai | 90-100 | 100-120 | 33 |
| 5 | Ai | 90-100 | 80-100 | 21 |
| 6 | Ai | 90-100 | 120-160 | 14 |
| 7 | Aii | (100)-140 | 80-100 | 23 |
| 8 | Aii | (100)-140 | 100-120 | 21 |
| 9 | Aii | (100)-140 | | 18 |
| 10 | Aii | (100)-140 | 80-100 | 32 |
| 11 | Aii | (100)-140 | 120-160 | 16 |
| 12 | Aii | (100)-140 | | 33 |
| 13 | Aii | (100)-140 | 80-100 | 16 |
| 14 | Aii | (100)-140 | | 3 |

Figure 5.4: Comparative dating of Type Ai and Aii dies

This table shows a complete correlation between the four examples where there is both a Museum Site date and cutaway date. As Boon did not actually specify a date for the start of his Type Aii, there is also no inconsistency with his data. Four of the dies have Group B cutaways which suggests that they were in production after AD 110 and possibly as late as the AD 140 suggested in Boon's date range. Two of these dies (RIB 2459.3 and 2459.10) also came from the construction deposits of Phase IVa of the Museum Site[7] which is dated as AD c130-140.[8] As these same dies have also been found in pre AD 100 contexts it suggests that either Phase IVa employed secondary material or these dies had a life of at least thirty years.

The Vindolanda tablets, which are broadly contemporaneous with these tiles, indicate that troops were allocated to the kilns.[9] If the practice in this auxiliary unit is accepted as a model for the legion, then it is possible that each cohort had to serve its turn in the kilns and presumably produce the tiles for its own buildings.[10] If each cohort had its own stamp then ten

[5] Boon 1984, 23-36.

[6] Zienkiewicz 1993, 127-129.

[7] Zienkiewicz 1993, 129.

[8] Zienkiewicz 1993, 28.

[9] Bowman 1994, 106.

[10] It would be interesting if different dies could be associated with different barrack blocks.

different stamps would be required which could explain why so many dies appear concurrently.

In fact there are twelve different dies in Boon's Types Ai and Aii listed in Figure 5.4 above. However there are only three examples of RIB 2459.14 listed in RIB compared to a minimum of fourteen for every other die, so it seems likely that this die was lost or broken shortly after being made and that one of the other Aii dies was made as a replacement.[11] This leaves eleven different dies each represented by a significant number of examples. It is possible that the first double strength cohort had two dies or perhaps another die represented by a lower number of examples was also lost. Thus it seems probable that each cohort had its own die, the first four such dies were made to one design (Boon Type Ai) and the next seven to the new design (Boon Type Aii). All the dies were manufactured prior to AD 100 and many remained in use through to Hadrianic times. This is remarkably consistent with the life of dies suggested in Section 5.2 and satisfactorily explains why so many of these early dies occur with both Group A and B cutaways.

It may also be relevant that this series of tile stamps (both Type Ai and Aii) are the only Legio II dies that appear with any frequency in the outlying forts with six dies at Brecon, five at Cowbridge, three at Usk and four other locations with single dies represented. Only five of the remaining fifty-nine dies (ignoring Carlisle) are found outside Caerleon and in every case just at a single location.

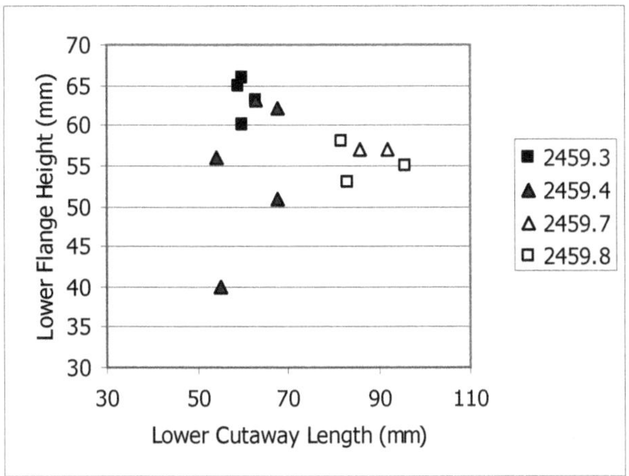


Figure 5.5: Differences between Ai and Aii dies

Boon was reluctant to place his large serif letter Type Ai as contemporaneous with his bordered ansate label Type Aii on stylistic grounds[12] and whilst they do now appear to be contemporaneous, there are nevertheless dimensional differences between them as shown in Figure 5.5, suggesting perhaps that both the stamps and moulds were made by different carpenters.

The next set of dies for which there is cutaway dating evidence is shown in Figure 5.6. The conflict with the Museum Site dating of RIB 2459.20, 29 and 30 is not significant because these dies came from Phase V demolition debris which included some of the earlier stamps featured in Figure 5.4 and, as Zienkiewicz says, they should "therefore be taken as representative of the varied composition of a roof at the fortress about AD 200".[13]

| RIB 2459 | Boon type | Boon dating | Museum site dating | Cutaway dating | Entries in RIB |
|---|---|---|---|---|---|
| 17 | Aiv | c120? | | | 41 |
| 18 | Aiv | c120? | 130-140 | 100-120 | 30 |
| 19 | Aiv | c120? | | 120-160 | 10 |
| 20 | Av | 90-140 | 200 | 120-160 | 40 |
| 21 | Av | 90-140 | | 80-100 | 4 |
| 22 | Av | 90-140 | | | 2 |
| 24 | Avii | c130-150 | | 80-100 | 16 |
| 25 | Aviii | 90-130? | | 80-100 | 21 |
| 29 | Axi | c140 | 200 | 120-160 | 15 |
| 30 | Axi | c140 | 200 | 100-120 | 57 |

Figure 5.6: The second set of *Legio* II dies

Plate 5.2: Dies from Figure 5.6 (RIB 2459.18, 20 and 30)

Although Boon's type Aiv and Avii could possibly have been copies of the same exemplar, none of the other Boon types have much in common with each other (Plate 5.2). It would be convenient to attach weight to the fact that there are ten dies in the table above suggesting again ten cohorts but this would probably be wrong as two of them are represented by only a handful of examples. An alternative hypothesis could be that as five of the first group of Type Ai and Aii dies in Figure 5.4 are not attested by cutaway evidence into the Hadrianic period, and five of this second group of dies in Figure 5.6 are pre AD 110 according to the cutaways, then these could have been replacements for the previous five Type Ai and Aii

[11] Alternatively they could just possibly be a mis-attribution of RIB 2459.10.

[12] Boon 1984, 15.

[13] Zienkiewicz 1993, 128.

dies. In other words five of the first batch of dies needed replacing before AD 110 as they came to the end of their anticipated fifteen to twenty year average life. As these wore out at different times they were not replaced as a group and so the replacement dies do not reflect any consistent style. The remaining six dies continued into Hadrianic times before being replaced.

The remaining stamped tegulae with cutaway dates are considered in Figure 5.7 below.

| RIB 2459 | Boon type | Boon dating | Museum site dating | Cutaway dating | Entries in RIB |
|---|---|---|---|---|---|
| 26 | Aix | c140 | 200 | 120-160 | 9 |
| 33 | Axiii | none | | 180-240 | 3 |
| 36 | Axv | c140 | | 160-180 | 97 |
| 37 | Axvi | c210 | 200 | 120-160 | 20 |
| 42 | Axviii | c200 | | 160-180 | 37 |
| 43 | Axix | c200 | | 240-260 | 2 |
| 44 | Axix | c200 | | 120-160 | 5 |
| 45 | Axix | c200 | | 180-240 | 11 |
| 46 | Axx | c200 | | 180-240 | 52 |
| 51 | Axxi | c200 | | 180-240 | 5 |
| 54 | Bi | 213-222 | | 180-240 | 7 |
| 55 | Bii | 213-222 | | 180-240 | 22 |
| 56 | Biii | 213-222 | | 180-240 | 46 |

Figure 5.7: Later *Legio* II dies

This table continues the conflict with the debris from Phase V of the Museum site which can now all be taken to be secondary material. There are also a number of inconsistencies with Boon's suggested dating which, in part, may be due to the longevity of the use of these stamps. Boon's Type B are the dies with the *Antoniniana* cognomen and all had Group C cutaways[14] which are consistent with the AD 213-222 dating. However the three dies from Boon's Type Axix are interesting; these dies are unmistakable in having the impression of a large nail head in the middle and Boon treated all of these together (Plate 5.3). The cutaway evidence suggests that they may represent a sequence of three consecutive dies perhaps spanning the hundred years AD 150-250 where the same stamp tool and over large nail were used to retain each of the dies.

Plate 5.3: RIB 2459.44 and 2559.43 – large retaining nail

[14] There was one tegula with an RIB 2459.56 stamp that had a damaged cutaway that was tentatively identified as Group D but this was probably erroneous.

The last dies are Boon's Type C and D. The discovery of two examples of RIB 2459.61 in Phase IV of the Museum site[15] demonstrates that the reading of this die as containing an "S" for the cognomen *Severiana* is wrong and undermines the case for the attribution of RIB 2459.62 to *Severiana*. The other possible late cognomen is a single stamp which has been read as "[..]VG.VI" and has been tentatively interpreted as *Victoriniana,* but the discovery of the "LEGXXVVV" tiles in a second century context at Tarbock[16] must now make this attribution even more doubtful.

The latest *Legio* II die is therefore likely to be RIB 2459.43 (Plate 5.3) which has a Group D cutaway that could have been contemporaneous with the last known inscription from Caerleon of AD 253-5.[17]

Three further dies of *Legio* II have been found that are not attested at Caerleon: two at Carlisle and one at Seaton in Devon. The Carlisle dies are the only incuse examples from *Legio* II and probably date circa AD 200 as will be discussed in Section 5.15. Boon classified the die from Devon[18] as his type Axviii for which he postulated a date of circa AD 200. This dating was confirmed by the cutaways used on the other die of this Boon type.[19] Williams[20] has examined the fabric of the Devon example which corresponds with other unstamped tile from the same site but differs from the Caerleon fabrics. The tile is also very unusual in having the stamp impressed *over* the print of a nailed shoe. The evidence points to this being a legionary detachment of sufficient size to justify its own kiln, albeit not operated as efficiently as the Caerleon tilery. If the circa AD 200 dating is correct[21] then perhaps the detachment was engaged in restoring order following Albinus's abortive attempt at the purple.

## 5.4 The *Legio* VI dies

*Legio* VI dies are characterised by groups that have similar nomenclature and broadly similar design. These are summarised in Figure 5.8. With the exception of the "LEGIONIVI" dies which have produced no diagnostic tegulae, all the others have been found with Group B cutaways and should therefore date AD 120-180. Given the relative paucity of examples per individual die, some dies may be missing from the table but the overall impression is of new die designs being issued in batches of at least ten. However, unlike *Legio* II where the first dies were almost certainly produced by the same two

[15] Zienkiewicz 1993, 128.
[16] See Section 4.2(d).
[17] Boon 1984, 36 (RIB 334).
[18] RIB 2459.41.
[19] RIB 2459.42.
[20] Williams 1987, 72-3.
[21] As suggested by Holbrook 1987, 67, the dating is also supported by the use of stone defences.

hands, in most instances the *Legio* VI dies were produced by several hands incorporating the same lettering.

It is surprising that there are so many different dies found on Group B tegulae and means that these stamps could not have had a twenty year life if only one die per cohort were concurrently in use. One possibility that would allow the proposition of one die per cohort to be maintained is that the changes in the nomenclature were the result of official decisions to adopt new forms of appellation. This would appear to have happened about once every ten years and on each occasion all the existing dies were withdrawn before they had worn out. The fact that none of the nomenclatures support more than ten dies tends to support this theory.

| Lettering | RIB 2460. | No. of dies | No. of examples |
|---|---|---|---|
| LEGVI | 5-14 | 10 | 27 |
| LEGIOVI | 32-36 | 6 | 19 |
| LEGIONIVI | 37-40 | 4 | 38 |
| LEGVIV | 48-57 | 10 | 103 |
| LEGVIVIC | 63-67 | 5 | 14 |
| LEGVIVPF | 80-89 | 10 | 21 |

Figure 5.8: *Legio* VI die groups

The "LEGVIV" dies are particularly interesting because four of them, RIB 2460.48-51, were probably produced by the same hand and these are the only relatively common *Legio* VI dies (other than Carpow) that are not found at York but have been found at ten sites all associated with Hadrian's Wall. Betts has examined the fabric of some of these tiles and was unable to differentiate the fabric from that used in tiles found at York[22] but he nevertheless concluded that they probably originated from somewhere closer to the Wall, perhaps Corbridge.[23] Whether or not they originated from York, the dies strongly suggest that a vexillation of four cohorts of *Legio* VI was assigned to the Wall area in this period.

Plate 5.4: *Legio* VI *Severiana* and *Gordiana* cognomina

[22] Betts 1985, 257.

[23] This exemplifies the difficulties of fabric analysis.

No stamped tegulae with Group C cutaways have been observed at York although unstamped tegulae from the Minster, and therefore presumably from the fortress, have been. There is no reason to doubt the attribution of four dies to *Severiana* and a further four to *Gordiana* which dates to AD 238-244[24] (Plate 5.4) so tile stamping would appear to end around AD 250 in line with *Legio* II practice.

## 5.5 The *Legio* IX dies

It seems unlikely that *Legio* IX issued individual stamps to each cohort because only a total of sixteen dies are known and only eight of these dies have been found in York. Moreover two of these York dies account for three quarters of all known examples of *Legio* IX stamps.

*Legio* IX stamps have been found in the bathhouse at Castleford dating to AD 85-100[25] which suggests that the legion started stamping its output circa AD 90 in line with the *Legio* II practice. Stamped tegulae are found with both Group A and B cutaways showing that *Legio* IX continued to produce tiles up to its departure in circa AD 120.

## 5.6 The *Legio* XX dies

There are eight dies from *Legio* XX that have Group A cutaways and half of these also occur with Group B cutaways as shown in Figure 5.9 suggesting that this legion started stamping tiles at a similar time and manner to *Legio* II.

| RIB 2463 | Lower Cutaways | No of examples |
|---|---|---|
| 6 | A | 10 |
| 11 | A | 4 |
| 18 | A | 6 |
| 41 | A | 1 |
| 4 | AB | 29 |
| 36[26] | AB | 60 |
| 38 | AB | 2 |
| 39 | AB | 9 |

Figure 5.9: Early *Legio* XX dies

The dies with just Group A cutaways have little in common (Plate 5.5). Stylistically three of the stamps in the Group AB set have the same lettering: one in a square frame (RIB 2463.4), one in an ansate frame within a square frame (RIB 2463.36) and one just with an ansate frame (RIB 2463.38). The fourth die (RIB 2463.39) in the Group AB set has different lettering but was contemporary with the others because it has been found with RIB 2463.38 on the same tegula at Holt (Plate 5.6). This is, of course, also excellent evidence for different

[24] RIBII.4, 148.

[25] Abramson et al 1999, 295.

[26] Davies 1993, 231 records this die at Segontium in a context that was Trajanic or earlier.

military units within the same legion having different dies.

Plate 5.5: RIB 2463.4, 11, 18, 36 and 39 from Figure 5.9

Plate 5.6: Holt 1110 stamped with RIB 2463.38 and 39

*Legio* XX has yielded less than half as many examples of stamped tiles as *Legio* II so it is not surprising that the evidence for individual unit stamps is less compelling, but even if one cannot be sure that individual dies were allocated to each cohort in *Legio* XX, one can be sure that several stamps were in contemporaneous use.

RIB lists two dies with clear *Antoniniana* cognomina[27] but there were in fact three as shown in Plate 5.7. All of the examples of the *Antoniniana* cognomen were non-diagnostic, but there is one diagnostic tegula with an incuse stamp[28] and Group C cutaway from Carlisle and other unstamped tegulae from Chester that have Group C cutaways, so it seems certain that *Legio* XX or its contractors were still producing tegulae into the third century.

Plate 5.7: *Legio* XX *Antoniniana* dies RIB 2463.51/52/N

No *Legio* XX dies have been discovered with *Severiana* or *Gordiana* cognomina and the dies which were thought to have the *Victoriniana* cognomen can now be dismissed (see Section 4.2(d)). So the only hope of demonstrating that *Legio* XX continued stamping tiles until circa AD 250 in parallel with observations for *Legio* II and *Legio* VI are the tiles with a possible *Deciana* cognomen which would date to AD 249-251.[29] However these also fail, *inter alia*, because these stamps have been discovered in late second or early third century contexts in Deanery Field, Chester.[30] Moreover all the diagnostic examples have Group B cutaways datable to AD 100-180. There is therefore no evidence for *Legio* XX production after AD 222.

## 5.7 The "LEGXXVVDE" dies

The "LEGXXVVDE" stamps warrant further consideration. The lettering of the stamps (Plate 5.8) is unusual in having curiously shaped "G"s, fishtail "E"s[31] and in the case of RIB 2463.55 an additional inverted "V" detached from the main lettering (but this is not visible in the photograph). Figure 5.10 tabulates the other occurrences of these features. With the exception

[27] RIB 2463.51 and 52 but note the discussion in Section 4.2(d) concerning RIB 2463.53.

[28] RIB 2463.1; see Appendix 4 for suggested dating of circa AD 200.

[29] RIB 2463.54 and 55.

[30] Newstead & Droop 1936, 40-41.

[31] RIBII.4, 191 makes a similar point but identifies fewer dies.

of RIB 2463.65,[32] all the dies are in ansate frames where sufficient of the die has been recovered for this to be determined.

Plate 5.8: *Legio* XX dies listed in Figure 5.10 and text (RIB 2463.44, 54, 56, 58, 60, 61 and 65)

| RIB 2463. | 14 | 44 | 54 | 55 | 56 | 58 | 60 | 65 |
|---|---|---|---|---|---|---|---|---|
| Two stroke "G" | √ | √ | √ | | | √ | | |
| Curly "G" | | | | √ | √ | | √ | √ |
| Fishtail "E" | √ | √ | √ | √ | | √ | | |
| Extra "V" | √ | | | √ | | | | |

Figure 5.10: Associated *Legio* XX dies

As these features are not found on any of the other 57 *Legio* XX dies there are grounds for suspecting some relationship between the dies in the table above. RIB 2463.54 and 2463.55 are the dies with the "DE" nomenclature. RIB 2463.56 and 2463.65 are dies with "VVV" nomenclature and the only other example (RIB 2463.57) contracts "LEG" to "L" thereby omitting the "G" that might have included it in this table. RIB 2463.58 and 2463.60 are two of the four *Legio* XX examples of dies with wording that goes beyond the simple legionary designation; one of the other two examples is RIB 2463.59 which is the consular dated stamp that was found at Tarbock in association with "VVV" stamps and the other is RIB 2463.61, which is only fragmentary, includes no letters "G" or "E" but the lettering is such as to suggest that the fishtail "E" might well have been used. There is no obvious relationship with the remaining two dies RIB 2463.14 and 2463.44.[33] Only three of these dies have produced diagnostic tegulae which were all Group B.

Given the known relationship at Tarbock between the consular dated stamp and the "VVV" die, and the similarity of the lettering with the other stamps with extended scripts, it is reasonable to group these together and suggest that Viducius may have been involved in the production of all of these dies[34]. However the same lettering extends to the "DE" stamps so arguably Viducius could have been involved with the production of these dies as well. This is supported by the extra "V" within the ansa of RIB 2463.55. It has been suggested that this is simply a correction for the omission of the second "V" in the main text "LEGXXVDE"[35] but given the strength of the other relationships, an attribution to Viducius should not be ruled out.

Another similarity relates to signatures found on the "VVV" and "DE" tiles. Two of the unstamped Tarbock tegulae had a letter "V" incised onto their upper surface[36] which is rare and RIB records only one other site that has produced "V" graffiti:[37] the obvious inference is

[32] This die is not actually recorded in RIB so it has been arbitrarily assigned the next available number.

[33] A more remote linkage can be made by examining the seven dies (out of a possible 64) that were found in the walls of the Holt tile kilns (RIB 2463.15,21,44,49,54,55 and 60 compiled from Grimes 1930,139-143, references in RIB 2463 and identifications in the Grosvenor Museum store). Presumably only scrapped tile was used for this purpose so normally little weight would be given to such a relationship. However the fact that the seven dies found in the kiln walls include four of the dies with unusual lettering could be more than coincidental.

[34] That is RIB 2463.56-61 and the unlisted "RIB 2463.65".

[35] RIBII.4, 191.

[36] Jones 2000, 93.

[37] Castleshaw (RIB 2491.77). There is also one tile from Ashtead with a "V" signature in this survey.

therefore that the "V" stands for Viducius, the tile contractor. There are also two tiles with a graffito incorporating the letter "D", one an imbrex found at Holt[38] and the other a tegula found at Prestatyn[39] (Plate 5.9). The use of a letter "D" on tiles is extremely rare, indeed none is recorded in RIB[40] although it does record a tile with a name starting with a "D" that was also found at Holt.[41]

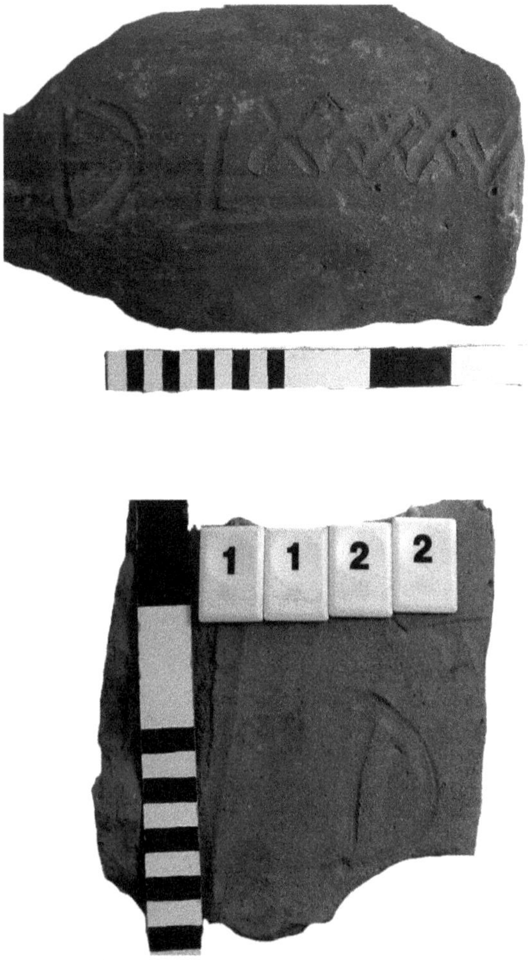


Plate 5.9: *Legio* XX tiles with "D" graffiti

The tegula has a simple letter "D" made with the finger on a tile that is also stamped with RIB 2463.36 whilst the imbrex has a ligatured "DE" followed by a frame containing "XXXXV" or possibly "XXVV" all produced with a finger. RIB 2491.49 interprets the imbrex graffito as the number DLXXXXV putting weight on the "D" being barred to indicate a numeral. However, RIB's "L" actually appears to be part of a frame that surrounds all four sides of the numerals. Although there is no doubt that the numerals as written are XXXXV, their context within a frame at Holt make it possible that the intention was to write XXVV. It is also worth considering the parallel from Piddington villa that shows the relationship between signatures and stamps: Plate 5.10 shows a stamp probably reading PRO[42] and a signature (found on a number of tegulae) which is clearly intended to replicate the stamp.

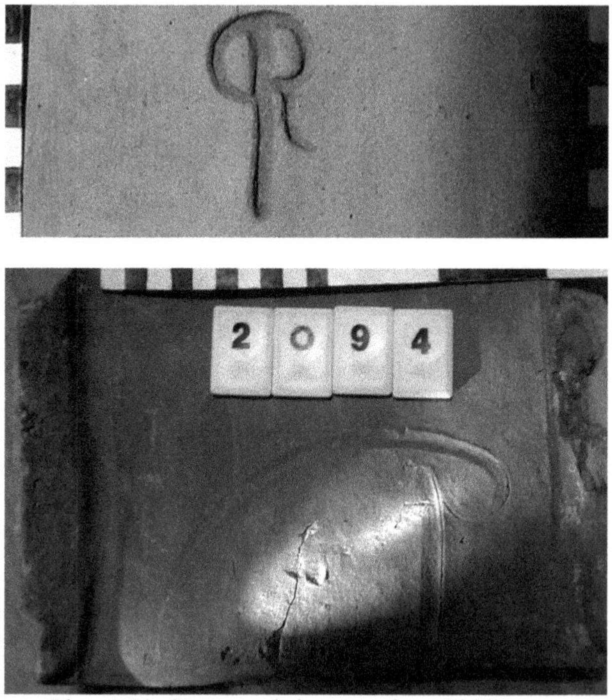


Plate 5.10: Stamp and signature from Piddington

All of these similarities tend to suggest that the "DE" tiles were made by a contractor who may have been connected to Viducius or simply operated in a similar manner to him and appended his output with his own "DE" initials. It certainly seems most unlikely that if the "DE" were the cognomen *Deciana* that it would simply be recognised by a piece of graffito.

One objection to the contractor theory is that the "DE" tiles originate from the legionary tile works at Holt rather than from an assumed private contractor's site such as Tarbock. However, the inference from the tegula with stamp RIB 2463.15 inscribed by Julius[43] Aventius, soldier of the first cohort of Sunici[44] which was found in the wall of tile kiln No 4 at Holt (Plate 4.3) must be that his unit were producing tiles at Holt in the second century.[45] The other stamps found in the wall of that kiln were RIB 2463.44, 49 and 55[46] and it is a remarkable coincidence that two out of these three are tiles with the unique lettering. So if the legion were no longer exclusively operating the site, then contractors as well as auxiliaries could have been involved either

[38] RIB 2491.49, in National Museum of Wales store at Cardiff.
[39] Tile 1122 from the National Museum of Wales store at Cardiff.
[40] RIB 2491.67-77.
[41] RIB 2491.93.

[42] Ward 1999, 62. This stamp is not in RIB.
[43] Is it just coincidence that Julius is also the name on the die RIB 2463.61?
[44] RIB 2491.96.
[45] This is Holt 1104 which has a Group B cutaway.
[46] Grimes 1930, 140/141.

| Cutaway group | A | | B | | C | | Total | |
|---|---|---|---|---|---|---|---|---|
| | None | Stamp | None | Stamp | None | Stamp | None | Stamp |
| Caerleon | 0 | 2 | 1 | 2 | 2 | 6 | 3 | 10 |
| Chester | 8 | 2 | 4 | 3 | 2 | 0 | 14 | 5 |
| York | 5 | 4 | 19 | 30 | 6 | 0 | 30 | 34 |
| Total | 13 | 8 | 24 | 35 | 10 | 6 | 47 | 49 |

Figure 5.11: Stamped and unstamped complete tegulae

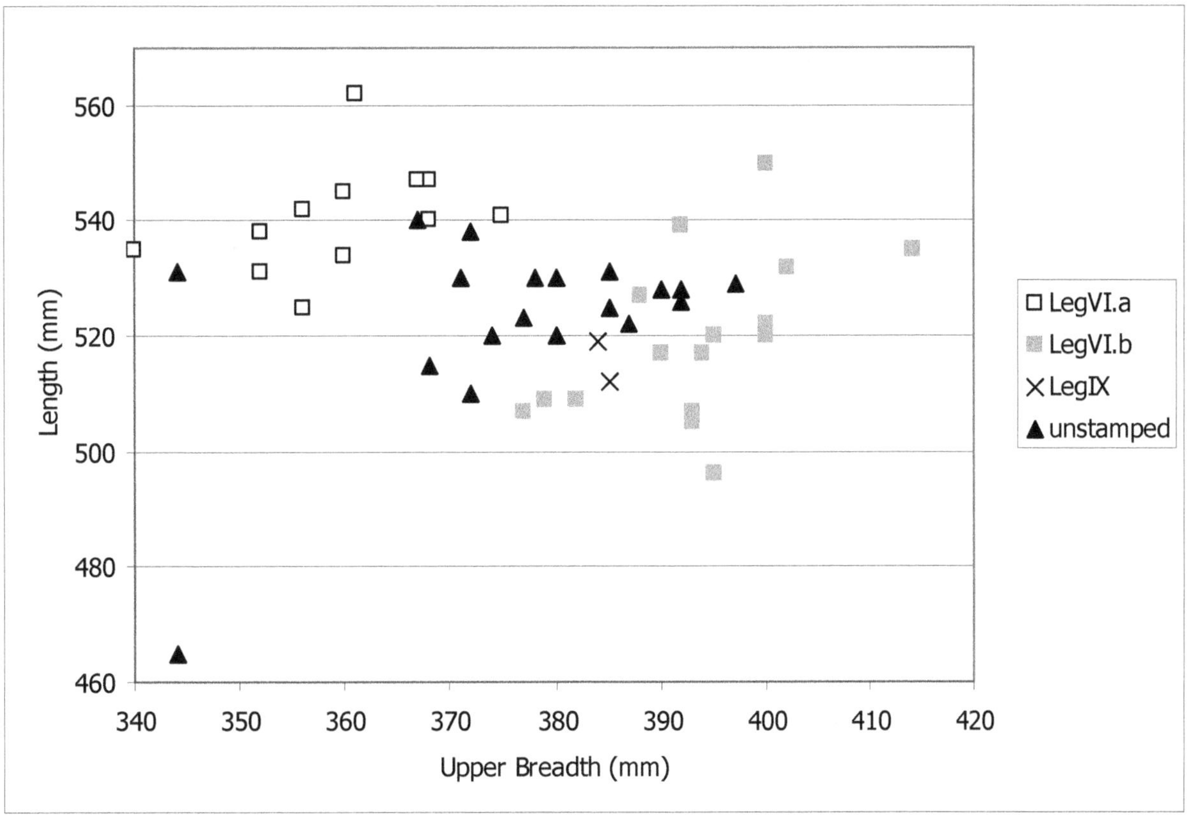


Figure 5.12: Dimensions of York Group B tegulae

contemporaneously, or more likely, at different times.[47]

The definite existence of a contractor to *Legio* XX in Viducius at Tarbock and the probable contractor "DE" using the facilities at Holt, coupled with continental parallels, suggests that the other British legions should also be examined to see if they adopted a similar practice. *Legio* II has some dies[48] that could possibly have a contractor's initial but the reading is uncertain, whilst there is nothing in *Legio* VI and IX dies *per se* to indicate that they were using contractors.

[47] Greene 1977, 125 comments that it was unthinkable that civilian potters could have been working at Holt which was thoroughly military in character. This accurately reflects the pre AD 130 site but perhaps not the later period.

[48] RIB 2459.61, 62, 63 and 64.

## 5.8 The frequency of use of tile stamps

Conventional wisdom has held that only a proportion of tile production was stamped.[49] Figure 5.11 tabulates all the complete[50] tegulae found in Caerleon, York and Chester/Holt which would suggest that, overall, one tegula in two was stamped. This, however, probably significantly overstates the proportion that was stamped because these tegulae will have been differentially retained by antiquarian excavators to the detriment of unstamped tegulae. Conversely the Group A data may be

[49] For example RIB II.4, 127. At the Dover forts, Williams 1981, 123 speculates that the *Classis Britannica* were stamping between one in ten and one in five hundred of the tiles they produced. This would be very surprising, especially in the light of the very high stamping rate at Beauport Park. Unfortunately the tiles from the Dover excavations are still not available for public study, so it is not possible to test this result.

[50] Complete in this context really does mean having the whole upper surface of the tegula intact such that the presence or absence of a stamp is certain. This means that several tegulae with just a complete length or breadth have been ignored even when a stamp was visible, because to include them would have been to bias the result against unstamped tegulae where the whole tile has to be examined to confirm the absence of a stamp.

understated because some of the unstamped tegulae may relate to the period before tile stamping began in Britain.

Easily the largest group of complete stamped and unstamped tegulae are the Group B ones from York and Figure 5.12 explores the dimensions of these in more detail. All of these tegulae have much the same surface area, so the differences shown in the chart are not so much in terms of size as the aspect ratio of the tegulae. It would appear that the moulds used to make these tegulae were all intended to be the same size or suite of sizes but they ended up marginally different. Nevertheless it is clear that there are three separate clusters of tegulae each with slightly different aspect ratios. Two of the clusters are stamped and the third is unstamped. The two stamped tegulae clusters comprise different dies with the left hand group made up of RIB 2460.21, 23, 53 and 80 whilst the right hand group is made up of RIB 2460.12, 62, 64, 81, 85 and 92.

The unstamped tegulae cannot be part of the same population from which the stamped tegulae were drawn because, if one tegula in two was chosen at random to be stamped, then the distribution of their dimensions should be the same as for the stamped tegulae. This strongly suggests that *all* of the tegulae from the legionary works were stamped and that the unstamped tiles come from another source.[51] As most of the tegulae came from the York tile tombs, which were presumably composed of secondary material, it is possible that some of the unstamped tegulae were not of legionary origin. However, this is unlikely, especially as they are all of legionary rather than domestic size.

Swan and McBride[52] have opined that *Legio* VI may have had two or more tileries operating in the vicinity of York and the clusters in Figure 5.12 may therefore reflect this.[53] However, it seems likely that all three clusters used the same clay source because Bett's fabric analysis identified no difference in the *Legio* VI fabrics stamped and unstamped.[54]

It is just possible that some cohorts did not stamp any of their tegulae but still used the legionary works; but even if this were the case, it is still clear that those units that stamped their tegulae must have stamped all of them. This observation is also consistent with the 216 stamped tiles from Carpow[55] which most probably result from stamping every tile.

Although each stamped cluster is comprised of different dies, each cluster has dies with a range of nomenclatures as identified in Figure 5.8 and in particular each cluster has a die with the nomenclature "LEGVIVPF". It was argued in Section 5.4 that every time *Legio* VI decided to change its style of appellation each cohort was required to produce a new die in conformance with this instruction. If this were correct, then the different aspect ratios of the two clusters could be interpreted as the same moulds continuing in use in each cluster whilst the dies were changed to conform to the new style. The data therefore seems to reinforce the interpretation put forward in Section 5.4.

If all the tegulae were stamped then the stamping is likely to have occurred as part of the production process because it would have been physically difficult to reach across to stamp every tegula once they had been laid out *en masse* for drying (although this would offer an alternative explanation for why so many legionary tiles have boot marks). It could suggest that the stamp had the dual purpose of identifying legionary property and acting as a "piece-work" receipt. Each man assigned to the tilery may have been given a fixed number of tiles to produce: as each one was completed it was taken to the officer who stamped it and entered a tick against the man's name. By stamping the tile it made it more difficult for him to recycle the same tile and thus cheat on his output. It would also have enabled the officer to keep track of the day's production. Such a system would also help explain why the cohorts did not seek to differentiate their dies, especially Boon's Aii group of *Legio* II dies, because the stamp was for record keeping in the tilery rather than differentiation once the tegulae were on a roof.

Two of the Group B tegulae in Figure 5.12 above were from *Legio* IX which appear to conform with the *Legio* VI findings. The Group A complete tegulae for *Legio* IX are shown in Figure 5.13, but the sparseness of the data and the possibility that some of the tegulae pre-date stamping makes it unwise to draw any conclusion.

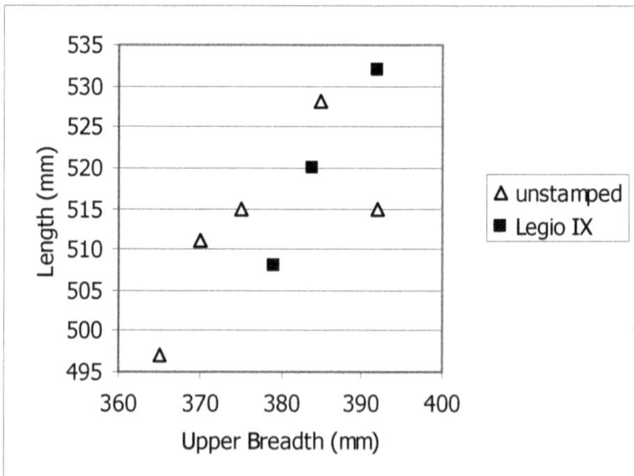


Figure 5.13: Dimensions of York Group A tegulae

The same information for complete tegulae from Chester and Holt is displayed in Figures 5.14 and 5.15: the Group

[51] Figure 6.9 shows a similar result for the unstamped imbrices from Beauport Park.
[52] Swan & McBride 2002, 193.
[53] See also Figure 3.4.
[54] Betts 1985, 242.
[55] Birley 1962-3, 200.

A data are inconclusive but the Group B data tend to support the idea that the unstamped tegulae were drawn from a different population.

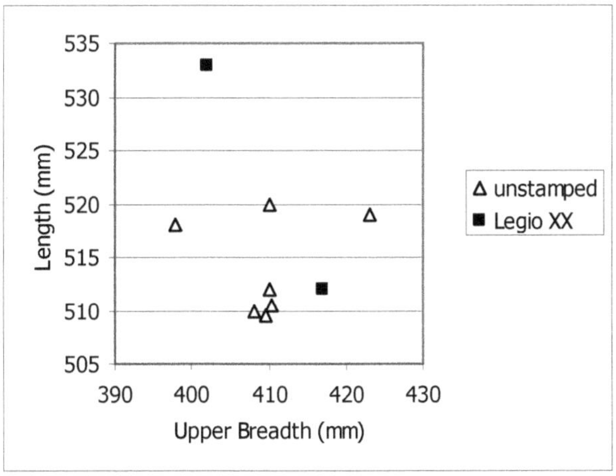


Fig 5.14: Chester Group A

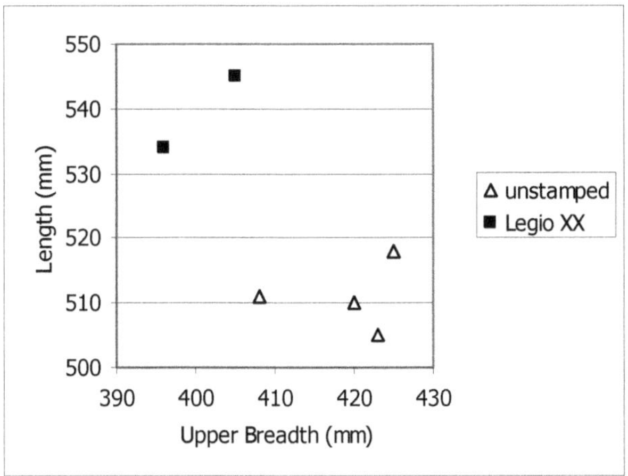


Fig 5.15: Chester Group B

Only one unstamped tegula from Groups A and B was recorded at Caerleon (which may itself be indicative) but the Group C data are shown below. Again this supports the conclusion that the unstamped tegulae came from a different population.

It would therefore seem likely that the practice in all the British legions was to stamp all of *their own* production of tegulae and to supplement that production with the (normally) unstamped output of contractors. It has already been suggested in Section 5.7 that *Legio* XX was using contractors in the first part of the second century and it is quite possible that all their Chester requirements were contracted from then on, perhaps including the *Antoniniana* tiles which do not appear at Holt. On this evidence *Legio* VI and *Legio* IX must also have been using contractors, perhaps from their arrival in Britain, whereas *Legio* II does not seem to have used contractors until around AD 200 when unstamped tegulae begin to occur at Caerleon.

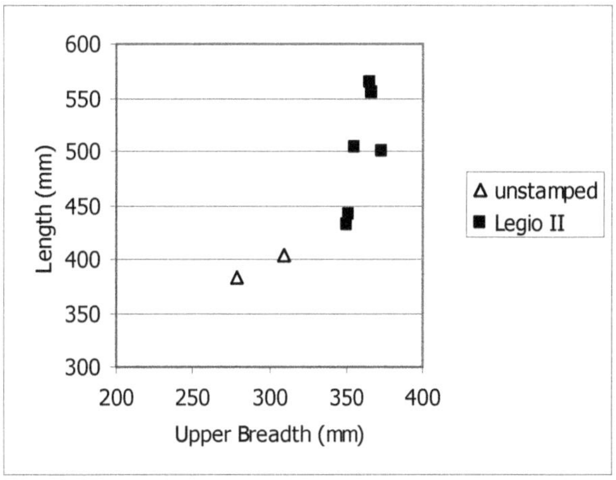


Fig 5.16: Caerleon Group C

## 5.9 Application of stamps to products

An analysis of the legionary stamps listed in RIB II.4 shows that tegulae were the most common tile to be stamped across the legions and brick[56] the least, moreover, based on the dating of the dies, the stamping of bricks appears to reduce or die out towards the end of the second century. However there was a significant variation in legionary practice: *Legio* XX produced only three examples of stamped imbrices whereas almost half the dies of *Legio* VI were used with imbrices and half the examples of their stamps were on imbrices.

If a typical military roof required twenty columns of imbrices on each slope then nineteen columns of imbrices would be required to bridge the gaps between the columns. Thus, in principle, if tegulae and imbrices were both stamped at the same rate then, assuming an equal survival rate, there should be 5% fewer stamps found on imbrices. This is consistent with the *Legio* VI data but suggests that *Legio* XX either did not stamp its imbrices[57] or subcontracted their manufacture from an even earlier date than the tegulae.

RIB II.4 only records three examples of legionary stamps used on box-flue tiles: one from *Legio* II and two from *Legio* VI.[58] There is also one die from *Legio* XX that was found on half box flues at Prestatyn.[59] Depending upon the height of the walls and the extent of the jacketing, there could be as many box-flue tiles used in a

[56] RIB II.4, 127 effectively classifies all tiles that are not tegulae or imbrices as brick, so this definition includes wall, floor and *pilae* tiles but only very rarely box flue tiles. However identification cannot always be certain and antiquarian records will not always have been reliable.

[57] Of the three stamped imbrices one was found at Carlisle which was presumably produced in association with *Legio* II (see discussion in Appendix 4), one has RIB 2463.58 die which may be a contractors stamp (see discussion in Section 5.7) and one is unprovenanced.

[58] RIB 2459.4 2460.26 and 2460.55.

[59] Newstead 1937, 226: RIB 2463.39 (iv-viii) mistakenly records these as tegulae.

bathhouse as tegulae on its roof, but there would be relatively few bathhouses compared to other tiled military buildings. Thus only a small proportion of the stamped tiles would be expected to be box-flues, but the exceptionally low number of such tiles suggests that the legions must have generally employed specialist contractors for this difficult task.

## 5.10 Distribution of legionary stamps

There are distinct differences in the distribution of legionary tiles: 22% of *Legio* IX stamped tiles are found outside York (mainly on brick) and 20% of *Legio* XX stamped tiles have been found outside of Chester and Holt, whereas only 4% of *Legio* II stamped tiles are found outside of Caerleon. Indeed, if the Carlisle production is ignored then, after the first batch of stamps (Boon's Type Ai and Aii), there are only six *Legio* II examples that have been found beyond Caerleon. Almost half the *Legio* VI tiles are attributed to Carpow and a further significant number come from the vexillation sent to Hadrian's Wall discussed in Section 5.4, but excluding these, the number of examples found outside York falls to 3%.

A partial explanation for this disparity could be that all the legions provided tile for the outposts until towards the end of the Hadrianic period after which the outstations were expected to make their own arrangements. *Legio* IX was posted away and *Legio* XX stamped output appears to have almost ceased after circa AD 130 whereas *Legio* II and VI production continued strongly into the third century. As a result of this continuing production the proportion that the early tiles sent to the outposts bear to the total for *Legio* II and VI is much lower than for *Legio* IX and XX. Moreover, as much of the later production was roof tiles (perhaps for maintenance and replacement after the main phase of construction was complete), this could also explain why *Legio* II and VI appear to have produced a lower proportion of brick than the other two legions.

## 5.11 The auxiliary unit dies

Very few examples of stamps from auxiliary units have been found so in the main there is no evidence to prove or disprove the allocation of individual dies to subunits in the fashion adopted by the legions. However at Benwell over sixty examples of stamps of *Ala I Asturum*[60] have been found, represented by a single die and at Slack/Grimescar over one hundred stamped tiles have been found, represented by just two dies of *Cohors IIII Breucorum.*[61] In contrast, at South Shields *Cohors V Gallorum* has produced nine different dies, all in similar style, from just fifty examples of stamped tile.[62] All of the diagnostic tegulae with these *Cohors V Gallorum* dies are regional Type 8 dating to circa AD 300.[63] It is tempting to associate these different dies with the unit's six centuries and four *turmae* although it would be surprising if the *turmae*, which probably comprised just thirty men,[64] warranted their own dies.

Based upon their cutaway forms, *Cohors IIII Breucorum* tegulae and possibly the single example of *Ala Sebusiana*[65] from Lancaster date to circa AD 100; the majority of auxiliary stamps are second century; *Classis Britannica* and *Cohors IIII Gallorum* have Group C examples and only *Classis Britannica* and *Cohors V Gallorum* were producing stamped tegulae in the later third or fourth century. This analysis is more consistent with continental practice and contrasts with Hassall's belief[66] that the auxiliary units in Britain mainly stamped their tiles in the third century.

The virtual absence of unstamped Group A tegulae from the sites of *Cohors IIII Breucorum* could suggest that this unit stamped all of its tegulae during the period that it employed stamping. This was also the practice of the *Classis Britannica* as evidenced at Beauport Park. A similar argument could apply to *Ala I Asturum* and it may therefore be possible that *all* military units stamped all of their own output.

## 5.12 The Procuratorial dies

RIB lists ten dies all found in the London area taking the form "P P BR LON" or just "P P BR" which may be expanded as *procuratores provinciae Britanniae Londini.*[67] Betts has subsequently added a further three dies[68] to this list. RIB notes that the earliest excavation dates are early second century but a similar mortarium stamp had been dated to the first century and therefore suggests that tile stamping with these dies started towards the end of the Flavian period.[69] Betts, drawing on slightly more evidence, suggested that production probably started in the last twenty years of the first century but ended around AD 120 because no new dies have emerged from buildings which post-date the major Hadrianic fire that swept through London.[70]

Unfortunately none of the accessible stamped tegulae from London has proved to be diagnostic, but the tegula illustrated in Betts' paper[71] has Group B cutaways and should therefore date after AD 110.

[60] RIB 2464.
[61] RIB 2470.1 and 2470.2.
[62] RIB 2473.1-9.
[63] Appendix 4.
[64] Hodgson & Bidwell 2004,134.
[65] RIB 2465.1.
[66] Hassall 1979, 265.
[67] RIBII.5, 30.
[68] Betts 1995.
[69] RIB II.5 31.
[70] Betts 1995, 222.
[71] Betts 1995, 213.

## 5.13 The Municipal dies

The municipal tilery at St Oswald's Priory, Gloucester produced stamped tiles with lettering "RPG" which may be expanded as *res publicae Glevensium*[72] and also stamps naming the civic officials. Twenty-three different dies carry the names of the *Duoviri* and four the names of the *Quinquennales*[73] (Plate 5.11). One pair of *Qinquennales* is represented by two dies, but assuming all the other dies (some of which are incomplete) represent unique pairs of magistrates, then these stamps must cover a minimum of 26 years. RIB lists 111 examples of these stamps with a minority being represented by up to thirteen examples, but nine dies have only single examples and five only a pair of examples.[74] The frequency of singleton examples would suggest that many more different dies were produced than have so far been excavated. This should perhaps be moderated by the fact that the singleton examples tend to be the most fragmentary and may therefore not constitute quite as many different types as suggested.

Plate 5.11: *Quinquennales* stamp RIB 2488.1

The tilery probably started production in the late first century but tile stamping did not commence until the early second.[75] There are three diagnostic tegulae[76] all from Group B which, if representative of all the stamped production, would suggest that the stamping was limited to the sixty years AD 110-170. Given that the magistrate stamps so far recovered must cover at least 26 years, as well as the high number of singleton examples, a production span of sixty years therefore seems reasonable. However, continuation of tile stamping into the third century as implied by Heighway and Parker[77] seems unlikely as, assuming the magistrates continued to produce annual stamps, at least three quarters of the dies would therefore still remain undetected.

## 5.14 The civilian tile stamps

The majority of examples of tiles stamped by private tilemakers have originated from Gloucestershire where all of the diagnostic tegulae examined have been from Groups B and C. Crookhorn in Hampshire has a die which, as previous authors have noted, is very similar in size and design to some of the Gloucestershire dies and examples of this are from Group B. Piddington has produced three dies and the two that were diagnostic had Group C cutaways, whilst Dorchester has produced a die with a Type 15 cutaway which represents the transition from Group C to Group D and should therefore date circa AD 250. Stanton Low in Buckinghamshire has not yielded any stamped diagnostic tegulae but it seems fairly certain that unstamped Group D fragments can be matched with non-diagnostic but stamped fragments: these stamps were applied in relief to the outside of the flange via the mould (see Plate 2.4). No examples of civilian tile stamps with Group A cutaways have been observed.

Some of the Gloucestershire stamped tiles appear to originate from Minety in Wiltshire from where unstamped tiles also arise.[78] This might suggest, contrary to the military practice, that civilian tilemakers only stamped a proportion of their output. However, Minety may be regarded as a *manufactory* site in Peacock's terminology[79] where a number of independent artisans came together to share some common facilities. Thus it is possible that those civilian artisans who stamped their output did in fact stamp all of it.

## 5.15 Incuse dies

There are only three legionary dies with incuse stamps: two from *Legio* II and one from *Legio* XX.[80] All of the examples have been found in Carlisle. Two complete tegulae (one from each legion) have been examined and it seems almost certain that they were the product of the same mould and kiln. These tegulae have Group C cutaways. They should perhaps be considered alongside the imperial die[81] which is also incuse and where the majority of examples have again been found in Carlisle, although RIB also lists five bricks from Bainbridge in North Yorkshire and an isolated example from Housesteads. These tiles were much more deeply fired and had Group B cutaways. As discussed in Appendix 4 the imperial stamps have been found in a stratified context dating from AD 170-183 which must provide a *terminus ante quem* for the start of the life of this die. It is possible that this imperial die should be linked directly with Commodus rather than simply being the product of an imperial estate. The dating of the legionary dies is less certain but, as discussed in Appendix 4, circa AD 200 may be reasonable, especially as a wide separation in dates between these unique incuse dies seems unlikely.

There are three auxiliary units that have yielded incuse dies: *Ala Sebusiana* at Lancaster, *Cohors I Aquitanorum*

[72] RIB II.5 41.
[73] RIB 2487 and 2488.
[74] RIB 2487 and 2488.
[75] Heighway & Parker 1982, 31.
[76] Two of these have "RPG" stamps and the third is a *Quinquennales* stamp RIB 2488.1.
[77] Heighway & Parker 1982, 31.
[78] Darvill 1979, 318.
[79] Peacock 1982, 9.
[80] RIB 2459.1, 2 and RIB 2463.1.
[81] RIB 2483.

from Norfolk and *Ala I Asturum* from Benwell. There are no diagnostic examples of the first two units although RIB suggests an early third century date for *Cohors I Aquitanorum*. As discussed in Appendix 4 *Ala I Asturum's* occupation of Benwell could date from AD 170 onwards. The tegula had a Group B cutaway.

A small minority of the *Classis Britannica* dies are incuse but the majority of these are on known French fabric. In view of the very limited number of examples found in Britain compared to the dies in relief, it is probable that all the incuse *Classis Britannica* dies are of French origin. No diagnostic examples of the incuse dies have been observed in this survey.

There are 69 known dies from the municipal tile-works in Gloucester and all but one[82] of these are incuse. The stamping of tiles is believed to have started early in the first century and continued into the third,[83] but as discussed in Section 5.13, this may overstate the duration of the practice. There are three diagnostic examples all from Group B.

RIB lists 68 dies, or groups of dies, of private tilemakers of which 31 are incuse.[84] To this may be added three incuse dies from Piddington that post-dated RIB, which means that almost half the civilian dies are incuse. Six of these dies have yielded diagnostic cutaways all from Group B or C which includes two sets of dies from Gloucestershire that had both Group B and C cutaways but on different members of the die set.[85]

Drawing this evidence together, almost all the military dies are in relief and, outside of Carlisle, only three auxiliary units used incuse dies. The Carlisle dies perhaps date AD 150-200 and this range could fit with the auxiliary units as well. Civilian incuse dies could also fit this date range but the municipal dies from Gloucester almost certainly start closer to AD 110.

## 5.16 Ansate framed dies

Dies with ansae or ansate frames have been regarded as second century in line with similar dating of ansate pottery stamps.[86] However tegulae from both the first[87] and last groups[88] of *Legio* XX production had ansate frames indicating that their use extended from the end of the first century into the third.

*Legio* XX was unique in its use of ansae with almost half[89] of their dies being ansate compared to none for *Legio* IX, eleven out of 72 for *Legio* II (and eight of these were Boon's group of Aii dies) and seven out of 94 for *Legio* VI. Five of the 34 auxiliary dies illustrated in RIB are ansate. Outside of this only one of the 119 *Classis Britannica* dies is ansate, two[90] of the 69 Gloucester municipal stamps and three of the 68 private tilemakers stamps have ansae.

The use of ansae or ansate frames was therefore primarily restricted to the army, and within the army, mainly to *Legio* XX which appears to have applied the practice throughout the (mainly second century) period that it stamped its production.

## 5.17 The use of *Classis Britannica* stamps in buildings

RIB lists 119 different dies for the *Classis Britannica*[91] although at least seven of these were in French fabric which would have been imported from the fleet's depot in Boulogne. There are 2272 examples of these dies of which the vast majority have come from Beauport Park (1459) and Dover (694). However RIB omits 179 tiles from Dover which were too fragmentary to firmly allocate between the three derivative forms of the RIB 2481.7 stamp;[92] adding these back gives 873 examples for Dover[93] which is the number that is used in the rest of this section.

Philp determined that the second Dover fort had had three periods of construction: the first period started circa AD 125-140 and the final demolition took place some time after AD 210.[94] By way of comparison, Brodribb and Cleere suggested that the life of the Beauport Park bathhouse was from the last quarter of the first century to the second quarter of the third century.[95] This proposition was based primarily on coins found on the adjacent iron-working site but which were not found in the bathhouse. The earliest stamps from other units are those of *Legio* II and IX dating to circa AD 90. The circumstantial evidence for *Classis Britannica* stamping from Dover, where no stamped tiles were found in the first fort which dated to circa AD 120, but were present in the first phase of the second fort,[96] points to an introduction of stamped tile by the fleet no earlier than AD 125. So unless the bathhouse was entirely re-roofed and all the old tile carted away from the site (for which there is no evidence), then a first century date for Beauport Park seems unlikely. However decorated samian ware finds from the bathhouse itself gave a more

---

82 RIB 2487.23 is the only die in relief.
83 Heighway & Parker 1982, 31.
84 RIB 2489.
85 RIB 2489.40A and B, and 2489.44B and F.
86 Dickinson & Hartley 1971, 138-9.
87 RIB 2463.41.
88 RIB 2463.51 which has the *Antoniniana* cognomen.
89 Thirty out of sixty-four.
90 One of which is the unique RIB 2487.23 die in relief.
91 RIB 2481.
92 Williams 1981, 135.
93 888 are mentioned as the total from the fort (Williams 1981, 123) and 17 from the house with Bacchic murals (Molenkamp 1989, 57).
94 Philp 1981, 96.
95 Brodribb and Cleere 1988, 243.
96 Williams 1981, 127.

limited time span from AD 130 to 200[97] and this seems to provide at least a more probable construction date. If this is so, then the life of the bathhouse could be very similar to that of the second *Classis Britannica* fort at Dover – both start around AD 125, both have three phases of construction and both fall out of use in the third century.

Of the 119 different *Classis Britannica* dies (ignoring sub-classifications), 48 are represented by single examples: 33 from Beauport Park, 11 from Dover and four from other sites. Around half the Beauport Park roof tiles have been recovered[98] so there is a reasonable probability that somewhat more than half of the dies, even ones represented by single tiles will have been found. By contrast, at Dover only a small proportion of the tile remained which meant that, whilst many examples of the more popular dies would have been excavated, dies originally represented by few examples were far less likely to have been found. This would be consistent with the evidence above. Ignoring the singleton examples which are therefore likely to be misleading because a higher proportion will have been found at Beauport Park than Dover, the dies found at Beauport Park and Dover are summarised in Figure 5.17 below. It can be seen that just over half the remaining 71 dies with multiple examples were present on each site and that both sites had an identical distribution between dies that were unique to their own site and common British types.

| | Beauport Park | Dover |
|---|---|---|
| Unique | 14 | 14 |
| Present at both | 23 | 23 |
| Present at this site and a third site(s) | 4 | 4 |
| **Total** | **41** | **41** |
| Gaulish fabric not included above | | 7 |

Figure 5.17: Dies found at Beauport Park and Dover

Next, Figure 5.18 examines the top six dies from both sites which shows that these represented 80% of all the tiles found at Beauport Park whereas at Dover the top six represented 70% of the total. So, even though the bulk of the Dover stamps were not from roof tiles, both sites had the same basic distribution of dies as shown in Figure 5.17 and a broadly similar top six profile. Peacock's fabric analysis[99] has shown that all the British *Classis Britannica* tiles came from the same source of Fairlight clay just outside Hastings.[100] So given all these similarities, it seems highly probable that the same tile supply and construction processes occurred on both sites.

| | Beauport Park | Dover |
|---|---|---|
| RIB2481. | **Top 6** | |
| 75 | **296** | 1 |
| 102 | **267** | 2 |
| 7 | **200** | 327 |
| 63 | **173** | 0 |
| 8 | **133** | 144 |
| 72 | **98** | 11 |
| | **1167** | |
| Share of all 1459 examples | **80%** | |
| RIB2481. | | **Top 6** |
| 7 | 200 | **327** |
| 8 | 133 | **144** |
| 40 | 0 | **53** |
| 61 | 0 | **36** |
| 39 | 2 | **29** |
| 92 | 41 | **19** |
| | | **608** |
| Share of all 873 examples | | **70%** |

Figure 5.18: Top six dies at Beauport Park and Dover

Both sites had three phases of construction and therefore presumably three main phases of tile production and supply. The Beauport Park top six dies represent three stamps that were found on tegulae (RIB 2481.7, 2481.75 and 2481.102) and three stamps that were found on imbrices (RIB 2481.8, 2481.63 and 2481.72). It is therefore possible that these tegulae and imbrices can be divided into three pairs to yield the three phases of construction. In fact, because RIB 2481.75 and 2481.102 both appear with Group B and C cutaways, it is likely that these were contemporaneous and used in a single construction phase circa AD 170. This would suggest that very few of the tegulae from the initial construction survived although there were two large Group B tegulae with die RIB 2481.43. Similar conclusions cannot be drawn from the Dover material because the majority of the stamps have come from bricks rather than roofing tiles.[101]

Two of the top six tiles appear on both the Beauport Park and Dover lists: these are RIB 2481.7 and 2481.8 found (at Beauport Park) on tegulae and imbrices respectively. This suggests that one phase of construction was common to both sites[102] but that the other phases of construction did not overlap. One of the Beauport Park top six was not found at all at Dover (or indeed anywhere else) and two of the Dover top six were not found at Beauport Park which reinforces the proposition that different construction periods were represented by different stamps.

If the most popular dies represented the principal phases of construction then an explanation is required for all the

[97] De la Bedoyere 1988, 244.
[98] Brodribb and Cleere 1988, 267.
[99] Peacock 1977.
[100] This is some five miles from Beauport Park but thirty-five miles from Dover. Peacock 1977, 242 also mentions Romney Marsh as a "just possible" alternative source of the clay. This would be equidistant from Beauport Park and Dover.

[101] Williams 1981, 124: only RIB 2481.8 and 2481.61 at Dover were mainly comprised of imbrices.
[102] However there is a potential mis-match on the dating of RIB 2481.7 between Beauport Park and Dover which is discussed further in Section 8.7.

other dies that were found less frequently on those sites. The most obvious explanation is that these "infrequent" dies represented tiles that were used in the maintenance and repair work that would have been necessary over the life of the buildings. At Beauport Park there were 40[103] singleton dies and 13 doubletons which comprised 32 dies found on imbrices, 17 on tegulae, two on brick and two that were unattributed.[104] If each imbrex die represented a different period of production, then this would imply at least 32[105] separate episodes of very minor repair assuming that the tegulae dies did not represent further separate episodes. Alternatively a repair team could have serviced all the *Classis Britannica* sites in rather less frequent maintenance episodes utilising a stock of variegated second hand tile which they had accumulated. Such a process would help to explain the proliferation of different dies that appear on the *Classis Britannica* sites.

The virtual absence of brick within the singleton and doubleton examples makes it clear that, as would be expected, almost all the maintenance was required to the roof. However there are almost double the number of "infrequent" dies found on imbrices compared to on tegulae which would imply that the imbrices required more maintenance. One possible explanation could be that, because the imbrices lie on top of the tegulae on the roof, they would be more prone to slipping off the roof and breaking.

## 5.18 Signatures

Around eighty percent of the complete tegulae in the survey had signatures and roughly eighty percent of these signatures were semi-circular (including arcs and similar curves). There was no significant chronological variation in the forms. These signatures are such an obvious feature on tiles that they have attracted much speculation. Some subscribe to the belief that they were for decoration or for good luck whilst others have preferred more functional explanations ranging through being a quality indicator, size indicator, dryness test or even to show which way up the tile should go on the roof. Many of these propositions quickly fail when faced with the fact that only a proportion of tegulae were signed, that they were signed when the tegulae were wet and other types of tile, which may have been invisible, were also signed albeit in far lower proportions than for tegulae.[106] However the key point is that signatures can be correlated with tilemakers and tile factories making it much more likely that their purpose was to identify the tilemaker, the factory or possibly the guild. The evidence for this is discussed below.

Signatures were almost certainly applied by the original tilemaker because they were applied when the tile was still wet and they were almost invariably at the lower end. This was the end most accessible to the tilemaker, whereas if they had been applied later then, even if they were all intended to be at the lower end, a proportion of mistakes would have been expected. At Beauport Park one signature (Q1) is unambiguously associated with an individual tilemaker because all of the tiles with this signature had undercut flanges. Instead of running his fingers down the profile of the flange in the smoothing process, the tilemaker had dug into the flange with one finger producing the distinctive undercut flange.[107] At Lime Street in London, Betts observed that differences in the manufacture of the tegulae correlated with the two different signatures on the majority of those tiles[108] thereby supporting the proposition that they were applied by the tilemaker.

| Die | Signature[109] | No. of examples[110] | Cutaway Groups | Production period |
|---|---|---|---|---|
| 2481.75 | B1 | 5 | BC | 160-180 |
| 2481.102 | A1 | 9 | BC | 160-180 |
| 2481.102 | M1 | 7 | BC | 160-180 |
| 2481.102 | Q1 | 7 | BC | 160-180 |
| 2481.7 | B1/SS1 | 4 | CD | 240-260 |
| 2481.7 | SL3/A3 | 2 | D | 260+ |

Figure 5.19: Beauport Park stamps and signatures

On the assumption that all the signatures at Beauport Park can be associated with individual tilemakers, then the following picture emerges from Figure 5.19 (using the codes for signatures illustrated in Appendix 3). Three tilemakers were involved with the manufacture of tegulae with RIB 2481.102 dies circa AD 160-180. Probably at the same time a fourth tilemaker was producing tegulae with RIB 2481.75 dies. Then, in the mid-third century, at least two different tilemakers[111] were producing tegulae with RIB 2481.7 dies. Clearly the second century B1

[103] This differs from the 33 singleton examples referred to earlier because that total excluded singletons at Beauport Park where the same die had also been found on other sites.

[104] RIB 2481 listings.

[105] Brodribb & Cleere 1988, 267 estimated that around half of the Beauport Park tegulae had been recovered, so this would imply that there was a 0.5 probability of recovering a die represented by a single example on the roof. If the die originally had two examples then there would be a 0.75 probability of recovering at least one example, and a 0.875 probability if there were three examples originally. This would suggest rather than potentially 32 episodes of minor repair that the figure would have been nearer 60. (Following the same logic there were probably an additional 44 dies that were not recovered, so the 74 that were recovered would have represented roughly two-thirds of the original complement at Beauport Park).

[106] Betts 1985, 200. However why brick is less frequently signed still needs to be considered. Was it because tegulae were the most difficult to make and therefore most worthy of being identified with the maker?

[107] Perhaps as a result of arthritis.

[108] Betts 1985, 191-195.

[109] See Appendix 3 for descriptions of these signatures.

[110] The total number of Beauport Park tegulae in this survey was 86.

[111] There is little difference between the small semi-circle SS1 signature and the small more elliptical B1 signature; likewise there can be relatively little difference between the three arcs of A3 signatures and three small semi-circles of SS3 signatures. Brodribb 1979, 155 distinguishes rather more variants than in this analysis but the significance of such small variations in these signatures is arguable.

signature cannot have been made by the third century B1 user, but perhaps they may have been related. There are no other signatures on the tiles with these dies which suggests a maximum of four tilemakers were involved with manufacturing tegulae for the bathhouse in each phase.

There are too few multiple examples of individual dies to apply this approach more generally. At Caerleon, however, the six copies of RIB 2459.30 with signatures are represented by three different types, and the three examples on RIB 2459.46 and RIB 2459.54 are both represented by two different signatures. At York the six signed copies of RIB 2460.12 have four different signatures but other *Legio* VI dies show less diversity, whilst almost every example of *Legio* IX RIB 2462.9 and its derivatives have a different signature. On the sparse data for *Legio* XX different signatures are the exception rather than the rule. This tends to suggest that the legions did not have established tilemakers whose signatures appeared regularly on tegulae, but a much wider group of makers who used a variety of signature devices. This could be consistent with each cohort contributing to the tile-making, and individuals from the unit being assigned to making tiles when appropriate. In contrast the greater consistency in the tile making team within *Legio* XX could perhaps be linked with its greater use of contractors even for making stamped tile.

Plate 5.12: RIB 2489.49 stamp with signature

All the above examples tend to suggest that the purpose of the signature was to denote the work of the individual tilemaker, but this is not necessarily always true. Section 5.7 has already commented on the "V" and "D" signatures found on some Tarbock and Holt tegulae which were most probably used to indicate the contractor's work rather than the tilemaker's. A similar interpretation must be made with the "PRO" signature and matching stamp at Piddington (Plate 5.10). However, it is harder to understand why all of the "TPLF" stamps from Gloucestershire should also have a very distinctive and unique signature[112] (Plate 5.12) as this seems to be a duplication, unless this was indeed intended as decoration.

Chapter 2 commented on a limited number of signatures that were produced with a comb[113] rather than a finger. These are found on both tegulae and imbrices. Using the dating established in Chapter 4 it can now be seen that the practice of combed signatures on tegulae and combed ornamentation on imbrices started towards the end of the third century and is therefore a useful dating feature where the tegulae are non-diagnostic.

## 5.19 Tally marks

Seventeen tally marks were observed in the survey of Beauport Park tile: there were no marks on RIB 2481.75 (which had Group B and C cutaways) and none on the Group B variants of RIB 2481.102. However some of the Group C variants of RIB 2481.102 had tally marks and all of the tegulae with RIB 2481.7 where the bottom edge was intact had them. As RIB 2481.7 has both Group C and D variants this would be consistent with tally marking starting circa AD 200 and continuing thereafter. It might also suggest that RIB 2481.102 continued in use somewhat later than RIB 2481.75 because none of the Group C variants of this die had tally marks. A potential flaw in this analysis is that there is one tegula with Group B cutaways and stamp RIB 2481.43 that also has a tally mark of two diagonal lines, however the lines are more scratches than the normal indentations, so it is just possible that this marking was unintentional.

Tally marks are rare on other sites and extremely rare on civilian sites where only Leicester, in this survey, has produced a tally mark on a Group C tegula. Three tally marks were observed at Chester: two on tegulae stamped with RIB 2463.4, one with a Group A cutaway and the other a Group B and the third was on a non-diagnostic fragment stamped with RIB 2463.30 which elsewhere has been found on a Group B tegula. Slack has a tally mark on a Group A tegula stamped with RIB 2470.2 and York one on a Group B tegula stamped with RIB 2460.21. All of these military tally marks are earlier than those at Beauport Park.

Twenty five tally marks have been recorded from the

[112] See McWhirr & Viner 1978, 364, however there is one fragment Gloucester 646 that has the distinctive signature but does not have the stamp in its usual position above it (although it could be elsewhere on the missing part of the tegula).

[113] This combing, which was for decoration or identification, should be distinguished from that regularly found on box-flue tiles which was to provide a key for the plaster.

Silchester forum/basilica[114] and a group comprising seven different tally marks has come from the London amphitheatre.[115] Brodribb lists a further seven military sites which have produced marks and also a further seven civilian sites but these have only produced single examples.[116] It would seem that tegulae with tally marks are strongly correlated with buildings of an official or military nature. It may be significant that the only groups of tally marks found in a civil context have been at London where the amphitheatre was situated adjacent to the fort, and at Silchester where the suspicion of military involvement has persisted, but with little evidence to support it.

Brodribb speculated that tally marks could be used to sequence the order in which the tiles were fired: in effect that they were some form of numeric or pseudo-numeric system.[117] He recorded a total of 249 tally marks from all the fragments recovered, 151 being on tegulae and 98 on brick.[118] In total eighteen different types of tally mark were observed, of which thirteen occurred on tegulae and fourteen on brick. The top five marks on tegulae and brick each accounted for 80% of the total number of incidences. Three of these marks were common on both tegulae and brick and two appeared frequently on one type but not on the other. On this evidence, a numeric system seems unlikely as the range of the most popular "numbers" is so limited and it is not the same on tegulae as on brick. This is reinforced by the seventeen tally marks observed in this survey from Beauport Park where of the eight tally marks on RIB 2481.7, four comprise one design, three a second design and one a third.[119] Likewise there are three (different) designs on RIB 2481.102 tegulae but one is represented by three examples.

[114] Timby 2000, 116 however most, if not all of these marks were on brick.
[115] Betts forthcoming, 224.
[116] Brodribb 1987,132 but these may have been on brick.
[117] RIB II.5, 92 states that they are numerals.

[118] Brodribb 1979, 154.
[119] Brodribb's more extensive data (1979, 155) has seventeen examples of RIB 2481.7 with tally marks of which sixteen are represented by just two different marks.

# 6

# ROOF CONSTRUCTION

## 6.1 Introduction

This chapter explores how tegulae worked on the roof, why tegulae were designed in the way they were and how they interacted with the imbrices. In particular it considers how well tegulae meshed together and whether graduated size ranges were used.

It considers the pitch of roofs, whether tiles would slide and how they were held in position. It explores the use of nails and dowels to secure tegulae and how the practices differed between military and civilian roofs and through time.

It looks at how roofs were constructed, how they evolved and the implications of these changes on the design of tegulae. It analyses how the efficiency of tile design developed and investigates why Silchester and Caerwent have some anomalously inefficient tegulae.

Finally it brings the various strands together to propose a sequence of roof construction and tegula development.

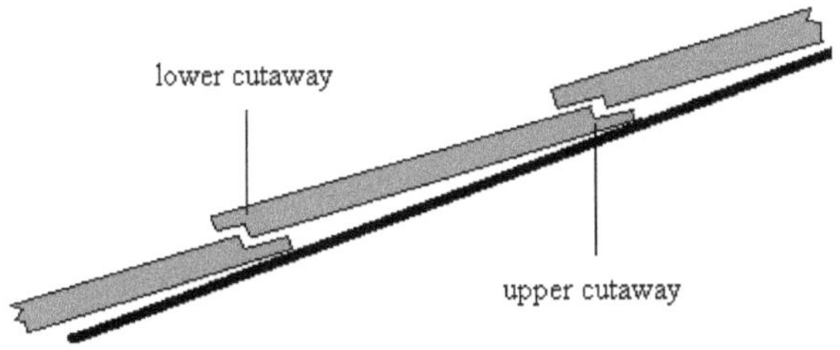


Figure 6.1: Double overlap of tegulae on the roof

## 6.2 Tegula design

Tegulae were designed to be placed in columns ascending the roof with each upper tegula overlapping the flanges of the tile beneath such that the water flowed over the tiles down the roof constrained by the, effectively continuous, flanges at the sides. Imbrices were used to cover the gap between the columns of tegulae and these in turn were overlapped with the upper imbrex sitting over the one beneath to ensure that there was no path for the water to penetrate. The effectiveness of the roof would have been determined by how tightly the tiles fitted together and, in particular, the way the tegulae meshed with each other. If the fits were poor then there would have been gaps through which the wind and rain could have penetrated, especially as these roofs were set at a shallow pitch.

The upper and lower cutaways on tegulae were designed to facilitate the overlapping of these tiles on the roof. Tegulae were designed to have a double overlap whereby the lower cutaway portion of the tile above covered the top end of the flange of the tile beneath and the upper cutaway portion of that tile lay beneath the base of the tile above as shown in Figure 6.1.

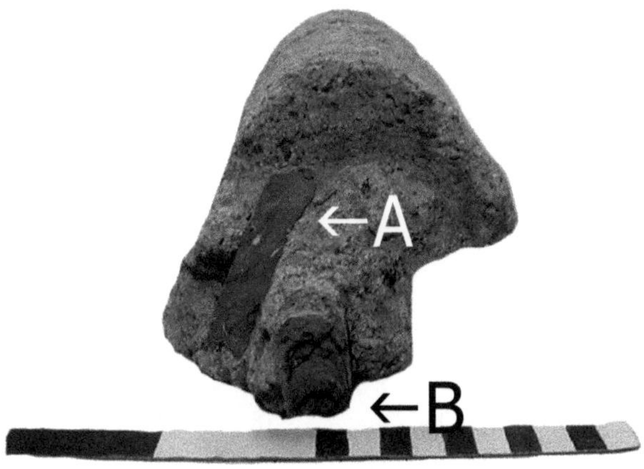


Plate 6.1a: Dorchester bathhouse roof fragment (upright)

Plate 6.1 shows an unusual survivor from the Wollaston bathhouse in Dorchester, Dorset that demonstrates this arrangement. Plate 6.1a shows the fragment upright with the mortar forming the profile of the upper imbrex and the diagonal element of tile showing in the section (letter A) being the end part of the lower imbrex that the upper imbrex was covering. At the bottom is the flange of one of the tegulae with its base broken away (letter B). Plate 6.1b shows the same fragment from the underside: the tile on the right (letter B) is the bit of flange with the base broken away seen on Plate 6.1a and on the left is the lower part of the tegula that would have overlapped with it (letter C). It can be seen that the end of the flange of the lower (right hand "B") tegula butts up against the end of the lower cutaway of the (left hand "C") tile above, and that the missing base of the lower tile would have run beneath the upper tile for the length of the upper cutaway. The depressions in the mortar on the lower part of the plate were left by the next (shadow) column of tegulae and it can be seen that the two adjacent columns of tegulae were some five centimetres vertically out of line (letter D) because the upper cutaway of the tegula in the shadow column ends in the deeper depression at the bottom left-hand corner of the plate.

Plate 6.1b: Bathhouse roof fragment (inverted)

| Cutaway group | Breadth taper % | Flange taper % |
|---|---|---|
| A | 1.7 | 31.4 |
| B | 1.9 | 21.6 |
| C | 4.3 | 31.0 |
| D | 3.1 | 26.0 |
| R | 4.2 | 4.7 |

Figure 6.2: Flange and breadth taper (% of wider end)

The meshing of the tegulae was facilitated by their overall trapezoidal shape such that the upper breadth was greater than the lower breadth. This breadth taper expanded the space at the upper end for the overlapping tile to fit into and reduced the breadth at the lower end such that less space had to be accommodated by the next tegula beneath. The flanges of the tegulae were also tapered being wider at the bottom than the top. The tapering of the flanges had two advantages: it increased the space between the flanges at the upper end thereby allowing more room for the overlapping tile to fit between. It also produced a greater bulk of clay at the lower end which allowed a larger lower cutaway to be carved out. This had the effect of reducing the lower breadth, thereby making it easier to fit between the flanges of the tile beneath. The extent of these tapers which are derived from the relatively small number of complete tegulae is shown in Figure 6.2. There is no time trend to the flange width taper whilst breadth taper appears to have increased in the second half of the production period. The flange taper of the regional forms is markedly less which reflects the different design philosophy of their cutaways.

The other major factor that determines the effectiveness of the meshing of the tegulae is the shape of the lower cutaway. This has been discussed in Section 3.4 and Figure 3.12 where it was shown that the both the ease of manufacture and the engineering effectiveness of the cutaway form improved through the cutaway sequence and hence through time.

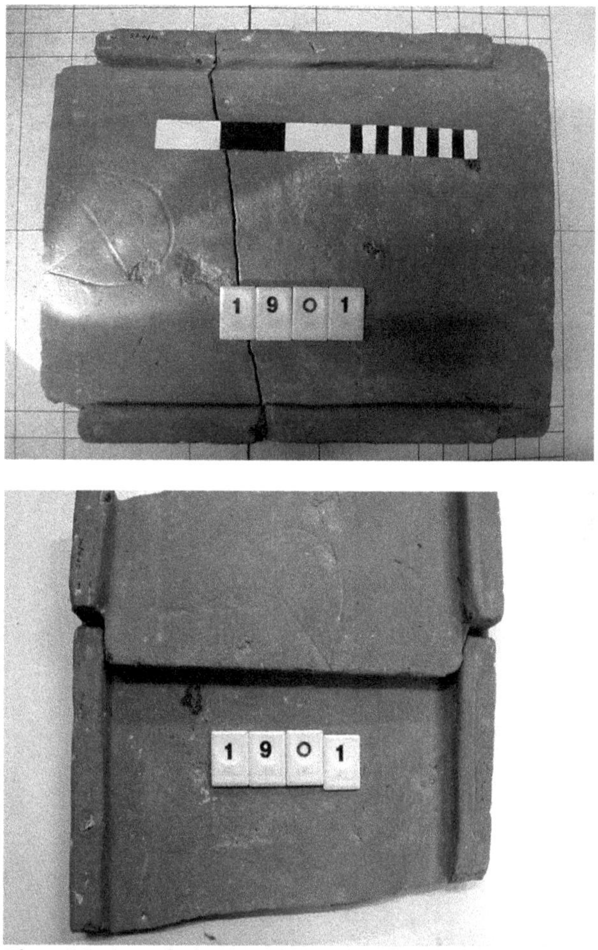


Plate 6.2: Sparsholt 1901 demonstrating tegula meshing

The regional cutaways produce a rather different story: Plate 6.2 shows a Type 7 tegula from Sparsholt firstly as a complete tile, and then with the lower half placed within the flanges of the upper half to show how tiles like this one would have meshed together on the roof. It can be seen in the lower plate that, whilst the fit is excellent, this arrangement does not present a continuous flange to the water flowing down the roof. There must be a probability that water would have penetrated through the junction, especially if the roof was flexing in high winds. If the imbrices were fully mortared into position over the gaps between adjacent columns of tegulae then the risk of water penetration would have been reduced. These tegulae would have been ideal for manufacture in an inverted box mould and, unlike other designs, would not have required any further work to finish the cutaways when removed from the mould.

Type 8 tegulae from South Shields (Plate 2.37 and 2.38) offer no engineering or manufacturing advantage. These tegulae would have had less than half the average 100mm overlap of the national forms because the sloping upper cutaway runs to the edge of the tile. Any gaps between overlapping tegulae would therefore be much more susceptible to weather penetration.

Type 9 tegulae from Corbridge (Plate 2.29 and 2.30) produced a double overlap but would have required considerably more effort to manufacture than any other design. The depth of engagement with the tile beneath was limited to about 5mm which would not have been as positive as the national forms: this could have made this roof design less stable.

Plate 6.3: Beauport Park 164 meshing on 165 and reverse

## 6.3 Tegula meshing results

Unique amongst British sites, Beauport Park has complete tegulae that must all have been on the roof at the same time prior to the building being buried by the slippage of an iron-working spoil heap:[1] this is therefore an ideal site to test how well tegulae meshed in practice. Pairs of complete tegulae, all with the same *Classis Britannica* stamp[2] (and therefore presumably made at broadly the same time), were selected and placed together. As is shown in Plates 6.3 and 6.4 most of these pairs of tegulae would only mesh effectively one way round but not when the tiles were reversed.

Plate 6.4: Beauport Park 164 meshing on 166 and reverse

This is a somewhat surprising result. It cannot be directly tested at other sites because there is not the same supply of complete tegulae that can be shown to have all come from the same roof at the same time, but it can be approached in another way. If a roof were comprised of identical tegulae then the bottom half of each tegula as they overlapped ascending the roof would be the same as the bottom half of the tegula on which it sat. The quality of the fit would be the same between all the pairs of tegulae and would also be the same as that which would be achieved if a single tegula were cut in half and its lower end sat within the flanges of the upper end of the same tile. This concept was shown visually on the example from Sparsholt in Plate 6.2. That tegula meshed with itself very well and may be termed a "self-meshing" tegula.

Adapting this approach to unbroken tegulae, the distance between the lower cutaways can be measured against the inside breadth between the upper flanges to see if a tile is self-meshing. In fact the assessment is slightly more complicated because it also depends upon whether the profile of the lower cutaway is sufficiently high to clear

[1] Brodribb & Cleere 1988.

[2] RIB 2481.102.

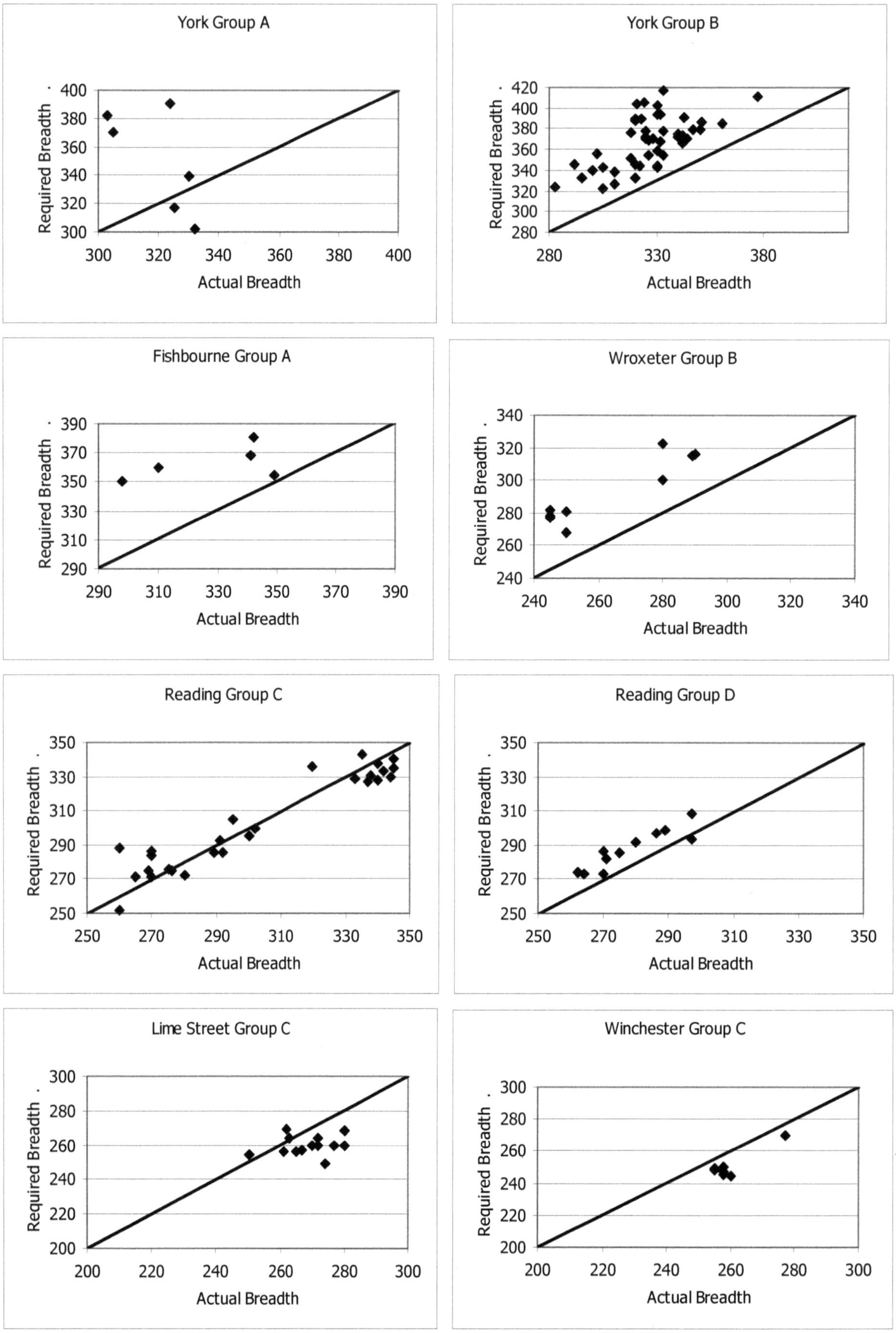


Figure 6.4: Meshing of tegulae from selected sites

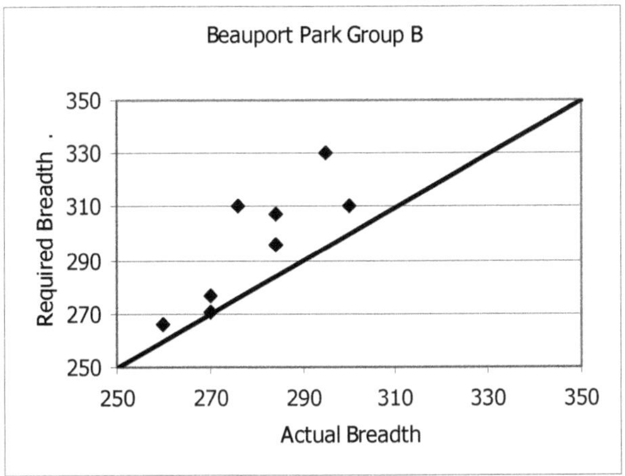


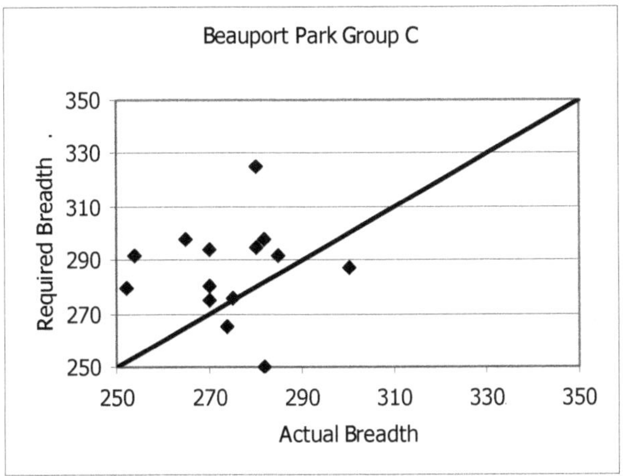


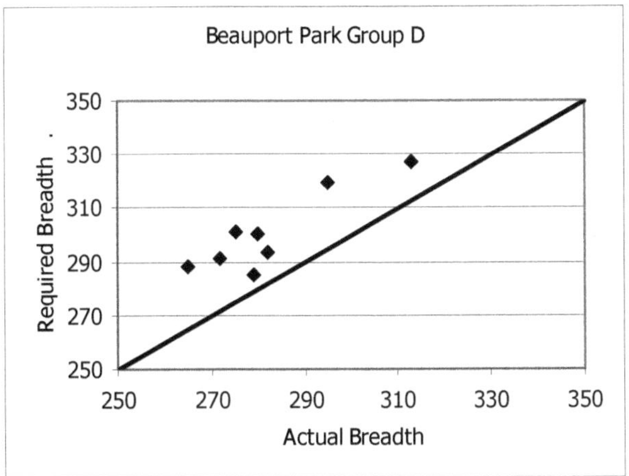


Fig 6.3: Meshing of Beauport Park tegulae

the top of the upper flange: this requires a little geometry which is shown in Appendix 5. The geometry allows the "required breadth" to be calculated and compared with the actual measured inside breadth. If the actual inside breadth is greater than the required breadth then the tegula will be self-meshing, but if it is smaller then the overlapping tegula will ride up on the flanges of the tile beneath creating a gap for the elements to penetrate and possibly to destabilise the roof as well. This calculation is plotted in Figure 6.3 for all the complete Beauport Park tegulae. Tegulae to the right of the diagonal line have a required breadth that is less than the actual measured inside breadth and will therefore be self-meshing, whilst those to the left of the line have an actual inside breadth that is less than the required breadth and will therefore not mesh.

Only three of the Beauport Park tegulae were self-meshing so it is worth examining other sites to establish whether Beauport Park is typical or unique. Figure 6.4 provides this data for six further sites. It can be seen that the Group A and B tegulae from York, Fishbourne and Wroxeter were similar to Beauport Park but that all the Group C and D tegulae were either self-meshing or very close to it. It may possibly be of significance that all the sites with poorly meshing tegulae were either military or had military antecedents (the military evidence for Fishbourne is discussed in Section 8.5).

## 6.4 Exploration of graduated sizes

It seems extremely unlikely that the Roman engineers were ignorant of the fact that some of their tegulae did not mesh or that they would persist in the significant labour of producing lower cutaways when these were actually ineffective. An alternative explanation is therefore required: an obvious possibility is that the sizes of tegulae were graduated as they ascended the roof in a similar manner to some modern tile designs. This would make meshing much easier because the overlapping tegula that had to fit between the flanges of the tile beneath would be smaller than the lower tegula and therefore more likely to fit.

A typical bathhouse might have rooms that were six metres wide, a barrack block eight metres wide and a villa ten metres wide.[3] Allowing for the double overlap of the cutaways shown in Figure 6.1, and using the range of average tegula sizes calculated in Chapter 3, yields a requirement for between nine and seventeen tegulae for each column on these roof slopes. If the tegulae were graduated then their breadth would have reduced as they went up the column and the gap between adjacent columns of tegulae would have grown as shown in Figure 6.5. The size of the gap would have had to have been limited to that which could be bridged by the imbrices that covered the gap. Imbrices are all considerably wider, and therefore use more clay, than is necessary to cover the gap between touching columns of tegulae: they must therefore have been designed to accommodate a gap between these columns.

[3] Bathhouse measurements taken from Mumrills, Castlecary, Balmuildy, Bearsden, Duntlocher and Camelon (Keppie 2004, 206/7), barrack blocks South Shields and Wallsend (Hodgson & Bidwell 2004, 124), and villas Sparsholt, Newton St Loe, Newport, North Leigh (Johnston 1994 and Wilson 2004, 79).

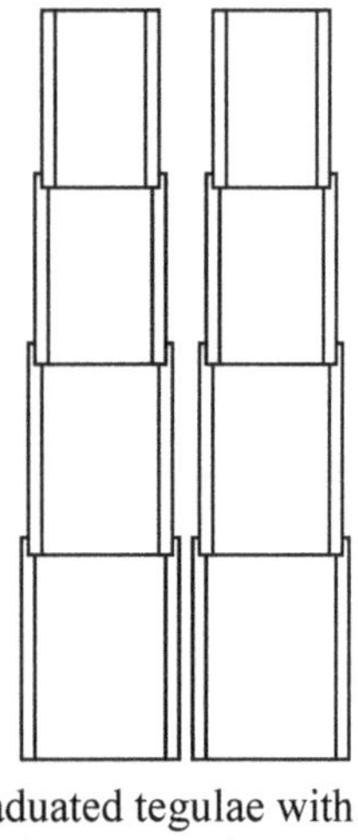

Figure 6.5: Graduated tegulae with expanding gap between columns

The imbrices overlapped in the opposite way to the tegulae such that the higher imbrex was required to be larger to fit over the imbrex beneath. To some extent this was achieved by the width of the imbrices tapering from the bottom to the top, but there could also have been graduated sizes of imbrices such that they increased in size as they went up the roof, with the imbrex with the largest width being at the top of the column.

The largest upper width of an imbrex in the survey was 193mm measured on a tile stamped by Legio VI at York; however a more typical upper width would be around 170mm. If the top imbrex had an upper width of 170mm then roughly 100mm would be lost in bridging over the flanges of the tegulae, as is shown in Figure 6.6 below. This would then leave 70mm as the maximum gap between adjacent columns of tegulae that could be accommodated by such an imbrex. Assuming that the bottom row of tegulae in each column were touching, then there would be half a centimetre reduction in breadth per tegula permitted for graduation on a typical roof made up of columns of fourteen tegulae.

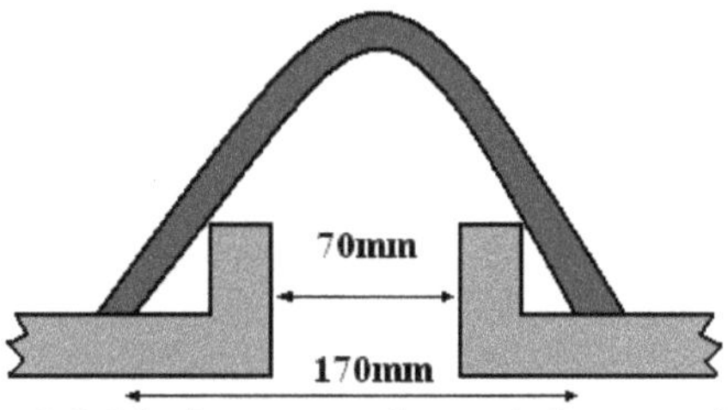


Figure 6.6: Maximum gap that an imbrex can bridge

Rook[4] has noted that many modern Italian tegula and imbrex roofs have a column of inverted tegulae instead of imbrices on roughly every fifth column. This gives an interesting texture to the roof and also allows corrections to be made to the line of the tegulae columns. If such a practice had been adopted in Roman Britain then this would have added some 300mm to the permissible gap every fifth column of tegulae or 60mm per column. This would broadly double the total allowable gap to 130mm and therefore facilitate a graduation of tegulae breadths of more than 10mm per tile.

In Chapter 3 it was noted that many of the clusters of complete tegulae had a spread either of breadths or of both lengths and breadths. These would be consistent with tegulae being produced in graduated size sets to facilitate the meshing on a roof. From Figure 3.3 it is known that the effect of differential shrinkage is unlikely to be greater than +/-2%, so tegulae from the same source showing size variations greater than this are likely to be deliberately different sizes and may possibly be graduated sizes.

The size of the complete tegulae from Beauport Park is plotted in Figure 6.7 using normalised measurements: that is actual measurements divided by the survey average for that cutaway group. No discernable pattern emerges from the Group C tegulae but the Group B and D tegulae offer the possibility of an alignment of breadths. In the case of the Group B data these breadths are separated by 8mm which is not inconsistent with the suggested gap calculated above. It is also notable that all three cutaway groups include representatives of the large (450mm length) Beauport tegulae as well as the smaller sizes.[5] As the Beauport Park bathhouse had rooms that were roughly half the size[6] of the "typical" bathhouse, they would only have needed half of the typical nine tegulae in each column and the range of sizes shown in Figure 6.7 should therefore have been sufficient. The distribution of Group D breadths suggests a more differentiated size range for these tegulae but it may be that some of the intermediate data points are missing.

Figure 6.8 (overleaf) gives the normalised size data for the same six sites shown in Figure 6.4. Some size variation should be expected to reflect the gradual reduction in tegula size that took place through time (as demonstrated in Chapter 3) and which should therefore be visible on any large site such as York. However York Group B shows a mass of tegulae with a far wider range of breadths than lengths and which may therefore represent a graduation of sizes. Fishbourne tegulae appear to fall into size groups but it is difficult to claim any pattern with the other sites. With the exception of Lime Street, and possibly the unprovenanced tegulae from Winchester, all the other tegulae come from a multiplicity of locations within towns. As a result they must represent the output of many tilemakers and episodes of production so it is perhaps not surprising that no pattern emerges. The Lime Street and Winchester tegulae were shown to be self-meshing in Figure 6.4 (as were the Reading tegulae) and would therefore not be

[4] Rook 1979, 298.

[5] The tegulae with tally marks are also represented in both small and large sized tegulae.

[6] Brodribb & Cleere 1988, 223.

required to be produced in graduated sizes. The absence of graduation may well be reflected in the comparatively tight size distribution of these tegulae shown in Figure 6.8.

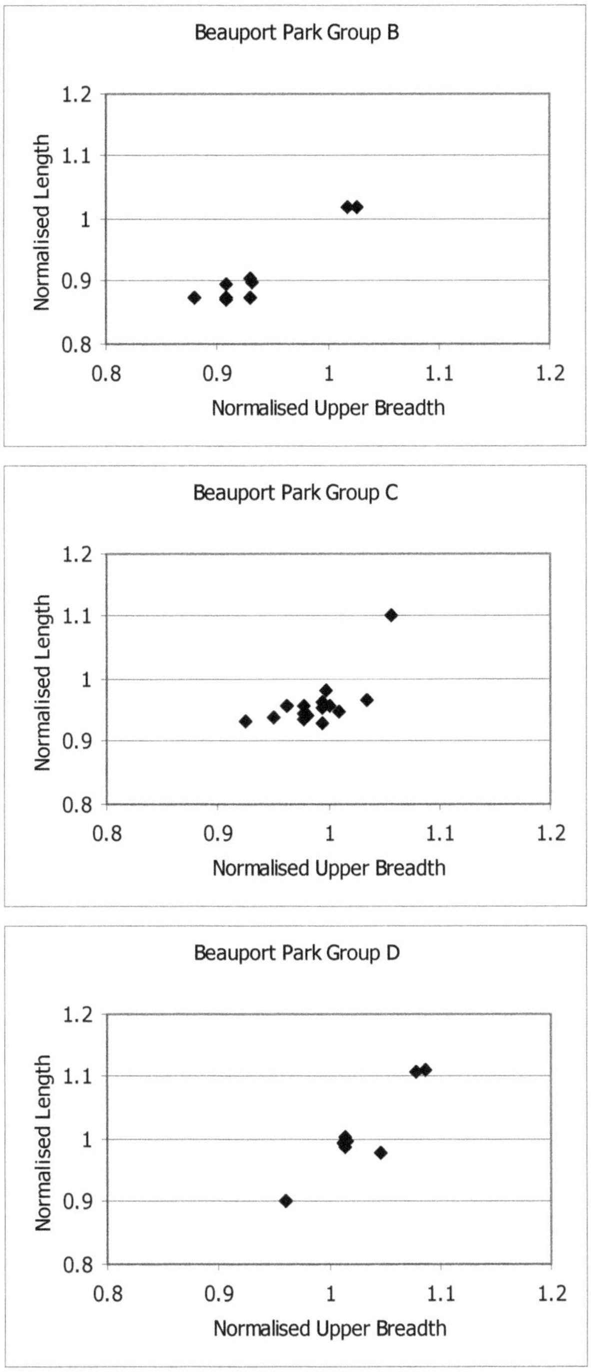


Figure 6.7: Beauport Park tegula size variation

The evidence is far from conclusive but some facts are clear: tegulae in general appear to have a greater range of breadth variation than length and far greater than could be accidental or the result of differential shrinkage; many tegulae would produce very unsatisfactory fits on the roof unless they were deployed in graduated sizes; and the tilemakers expended considerable effort in producing lower cutaways on their tegulae and presumably knew what they were doing.

It might also be noted that graduated tegulae would be ideal for tiling over an apse. On such a roof, instead of the gap between columns of tegulae being allowed to increase as they rose up the roof, the tapering of the roof would force the columns together but this would only work so far before the tapering of the roof required the number of columns to be reduced.

## 6.5 Graduation of imbrex sizes

Imbrices also show an interesting size pattern: the dimensions of the Beauport Park ones are shown in Figure 6.9. Imbrices with die types RIB 2481.7,[7] 8 and 9 all have a length around 390mm and those with dies RIB 2481.62, 63 and 72 have a length of 340mm.[8] Significantly, both these groupings have a similar 40mm range in lower widths. Although all of the Beauport Park imbrices appeared to have been made using an upright former,[9] this should not have resulted in any width variation because the clay had to be held tightly against the former over which it was draped to be smoothed and would also have needed support when being stamped. The width variation is therefore probably intentional.

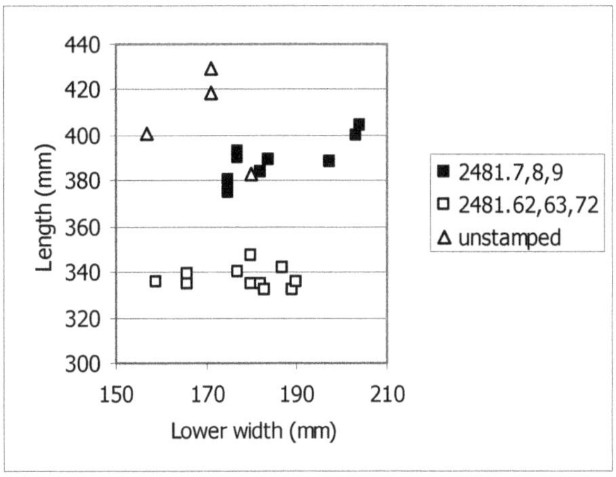


Fig 6.9: Beauport Park imbrices

[7] Only one imbrex had this stamp.
[8] It is notable that the unstamped imbrices form a separate cluster suggestive of a different manufacturing source to the stamped imbrices cf Section 5.8.
[9] See Section 2.8.

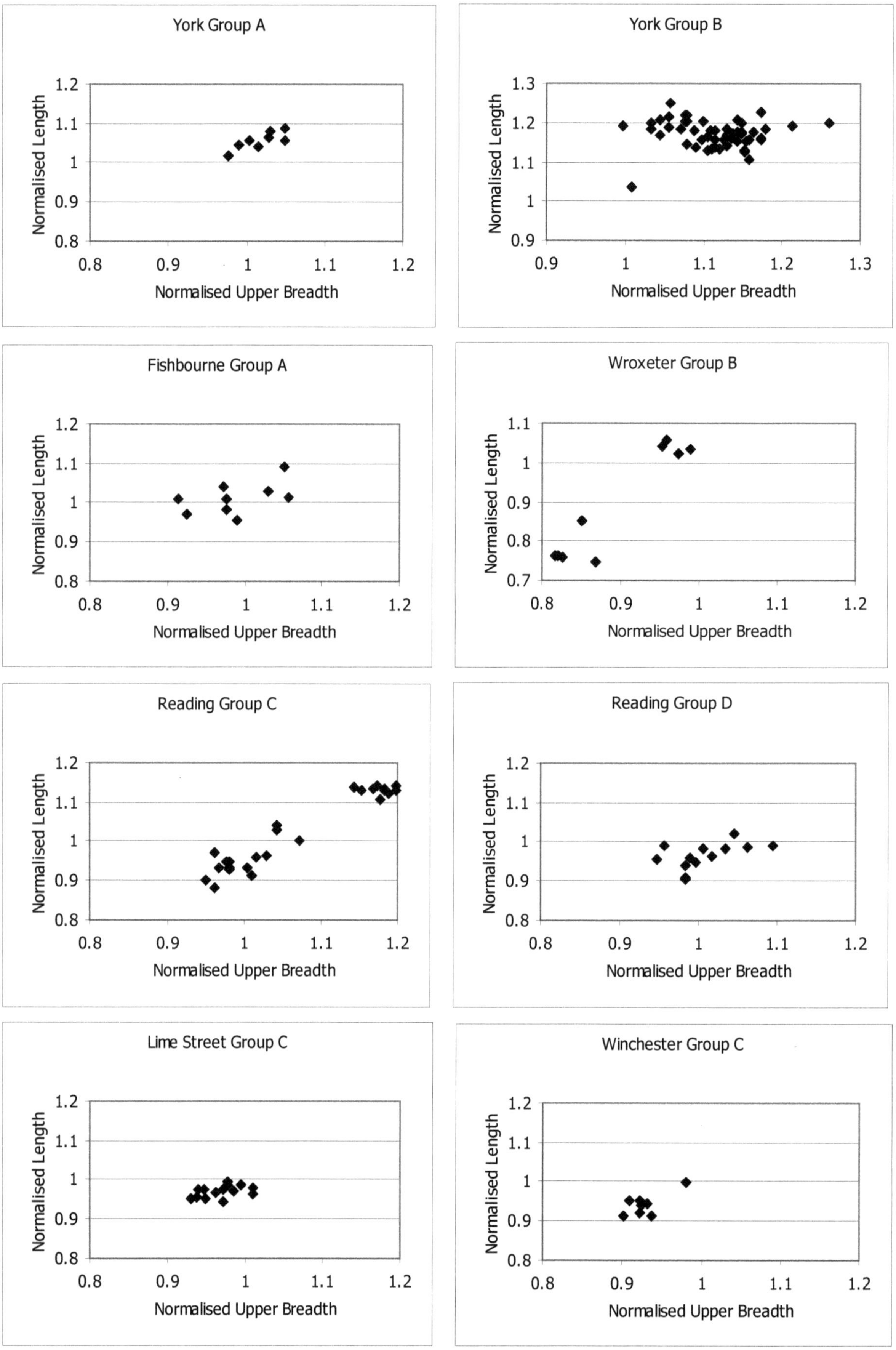


Figure 6.8: Tegula size variation from selected sites

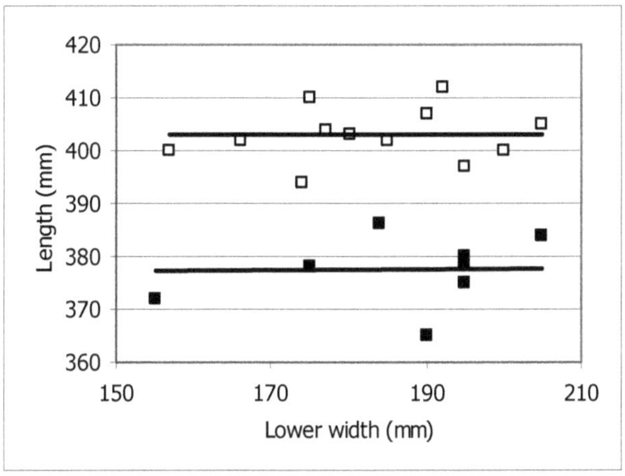


Fig 6.10: Reading Museum imbrices

At first sight the Reading imbrices (Figure 6.10) do not appear to have any particular distribution, but by comparison with Beauport Park chart, it is possible to see two lines of imbrices: one with a length centred on 403mm and the other with a length on 378mm.[10] Using these lines the width range is 60mm which more closely equates to the maximum 70mm variation anticipated in the previous section.

Such evidence as there is from the imbrices would appear to be supportive of the concept of graduated sizes. The imbrices with the narrower widths were placed at the bottom of the roof where the columns of tegulae were touching and those with the wider widths towards the top as the gap between the tegulae columns grew.

## 6.6 Use of mortar

Imbrices were designed with one end wider than the other such that they could be overlapped in a column but, as with the tegulae, when pairs from Beauport Park were actually tested together, the fit was less than satisfactory: either the lower width of the overlapping imbrex was too small or the height of the imbrex beneath was too great. For this reason imbrices could only satisfactorily have been mounted with the use of mortar.

Mortar is normally dissolved by acidic soils and surviving mortar from roofs is quite rare. By comparison mortar from walls, and on secondary tiles used in walls, is relatively common, presumably due to its more enclosed and protected environment. Nevertheless, as is shown in Plate 6.5 from Fishbourne, it does occasionally survive. Similar evidence has been observed in the survey at Silchester, Piddington, Caerwent and Gestingthorpe and in every instance the mortar is limited in extent suggesting that the practice was to insert just enough mortar to create a satisfactory seating for the overlapping imbrex. By contrast, the evidence from Dorchester in Plate 6.1 shows that the whole of the inside of that imbrex must have been filled with mortar before being inverted onto the flanges of the tegulae but this is unmatched elsewhere.

Plate 6.5: Fishbourne 2152 with mortar profile of imbrex

If imbrices were secured with a dab of mortar then the question must also arise as to whether the potential gap where the tegulae overlapped was also sealed with mortar as is shown in the reconstructed roof from Chester (Plate 6.6). There is no evidence in terms of mortar still attached or having left a mark on a tegula that survives, but this of itself does not disprove the proposal because almost all the mortar used on roofs has disappeared. Probably more telling is that mortar would not be very effective applied into a thin crack which would have been subjected to a fair amount of flexing from wind loading and thermal movement, as well as differential expansion of the tegula and the mortar. In all likelihood any mortar used in this way would have fallen out in a relatively short time. As a much larger amount of mortar was used to secure the imbrices it would seem much less likely to shift, although this would not prevent the overlapping imbrex losing its bond with the mortar and ultimately slipping off the roof as was suggested by the evidence in Section 5.17. If it was necessary to seal the gaps between the tegulae, then moss or some other organic material might have been a better option.

Plate 6.6: Reconstructed roof from Chester

[10] These two imbrex lengths may correspond to the two distinct groups of Reading Group C tegulae shown in Figure 6.8.

## 6.7 Roof pitch

Vitruvius stated that temples of the Tuscan order should have a roof pitch of one in three[11] which equates to an angle of just under twenty degrees.[12] This is consistent with modern Italian tegula and imbrex roofs[13] and is also consistent with the collapsed gable end from the Redlands Farm villa in Northamptonshire[14] which was measured at 22.5°. However the collapsed façade at Carsington, Derbyshire, where some near complete tegulae were found in the adjacent area,[15] had a slope of 40° if the gable end has been correctly identified.[16] This angle seems far closer to the Meonstoke collapsed façade where the slope was 47.5° but this was tiled in stone.[17] There must be some doubt as to whether the tegulae found at Carsington actually came from the roof of the collapsed façade[18] because, whilst under ideal conditions with well meshing tegulae and a firmly nailed lowest row, it would be possible to mount tegulae and imbrices at this angle, such conditions were rarely met. Moreover, using a 40° slope rather than 20° would require over twenty per cent more tiles.[19] There is a further collapsed gable end at Batten Hanger villa in Sussex where the roof pitch was over 60°, but this roof was most probably thatched.[20]

Assuming that the Roman builders aimed for a roof pitch of 20°, not all of the different roofs on even a mildly complex building could be at this angle without making it both difficult to build and unsightly to view.[21] A range of pitches from perhaps 15° to 30° may therefore have been adopted.

Experiments have been conducted which purport to show that tegulae will only begin to slide down the roof when the angle of slope exceeds forty degrees.[22] However the stability of a roof is somewhat more complex. If the roof was ideally constructed, only the bottom tegula would lie flat against the rafters and all the other tegulae would have their lower ends resting on the tegulae beneath whilst just the upper edge of the tegulae should rest on the rafters. As was shown in Plate 6.1b, the end of the lower cutaways butted up against the end of the flanges of the tile beneath (provided they meshed properly) and it was this that primarily prevented any of the upper tegulae slipping downwards.

Unless the pitch of the roof was extremely steep, perhaps the forty degrees mentioned above, a tile would be unlikely to slide under gravity alone because the braking force created by friction between the tegula and the rafter would probably be greater than the component of the gravitational force down the slope of the roof (the weight of the tile appears on both sides of this force equation and therefore the result will be independent of the weight of the tile). However, the real danger is not this "static" situation but the dynamic loading caused by the wind: a single tegula placed on a twenty degree roof would hold its position until a strong wind caused it to rock or lift slightly. Each time this happened it would settle back into a slightly lower position. Over months, especially as the rafters became wet and attracted algae, it would slowly slip down and ultimately off the roof.

If the tegulae do not mesh then the gap between overlapping pairs will allow the wind to enter and start rocking the upper tile. As this tegula will already be less securely abutted against the cutaway of the tile beneath, this rocking is likely to cause movement and, over time, the tegula can be expected to slip down the roof. Only frequent maintenance would prevent the ultimate catastrophic cascade of the whole roof.

## 6.8 Nail holes

The stability of the roof depended in particular on the security of the bottom row of tegulae to which all the downward thrust of the column of tegulae above was communicated through the lower cutaways abutting the end of the flanges beneath and through the mortared imbrices. The most obvious way to secure this row was with nails: Brodribb[23] quotes the *Lex Puteolana* as evidence for the nailing of the lowest course of tegulae, albeit this dates to 105 BC and the practice could have been different in Britain. In the probable absence of gutters, the bottom row of tegulae would have needed to project out over the wall to prevent the water flowing straight down the walls which would be an additional reason for securing that row with nails.

In this survey one in four tegulae was provided with a preformed nail hole and almost all the holes were placed in the top centre of the tegula which is entirely consistent with securing the bottom row or rows. The only notable exception to this were tegulae with holes adjacent to the flanges (Plate 2.14) which were presumably to secure the gable end tiles. Some holes were blind, failing to penetrate right through the tegulae. Some of these blind holes may have been manufacturing defects but at Norfolk Street, Leicester (Plate 2.15) they appear in such numbers that it is clear that they were deliberate: the

11 Vitruvius IV, 7, 5.
12 Contra Brodribb 1983, 20 who interpreted this as 30°.
13 Rook 1979, 295.
14 Keevill 1996, 51.
15 Ling 1992, 233-236.
16 It is possible that the rotational forces created as the gable end toppled would cause the upper end of the wall to expand by more than the lower end thereby exaggerating the angle of the gable, but this effect could not possibly account for the difference between 20° and 40°.
17 Frere 1990, 355.
18 Perring 2002, 120 lists Carsington as a slate roof.
19 Cos20°/Cos40°=1.23.
20 Pers comm: John Magilton.
21 Shirley 2000, 22.
22 Rook 1979, 295: in his experiments the angle of repose was 40° on planed boards and 35° when one tile was placed on another.

23 Brodribb 1983, 24.

intention being that the roofer only broke through those holes that were required when assembling the roof.

| Cutaway Group | Preformed holes | Total tegulae | Preformed % | Chipped holes |
|---|---|---|---|---|
| A | 3 | 34 | 9 | 1 |
| B | 13 | 93 | 14 | |
| C | 44 | 177 | 25 | 12 |
| D | 31 | 59 | 53 | |
| others[24] | 9 | 44 | 20 | 1 |
| Total | 100 | 407 | 25 | 14 |

Figure 6.11: Proportion of preformed holes by cutaway

Figure 6.11, which is based upon tegulae with complete upper breadths, shows that the proportion of tiles with preformed nail holes increased dramatically through time from just 9% for Group A cutaways to 53% for Group D. However this conceals some significant differences in the major contributory sites as shown in Figure 6.12.

| Site | Preform holes | Total tegulae | Preform % | Chipped holes |
|---|---|---|---|---|
| Ashtead | 1 | 9 | 11 | |
| Beauport Park | 8 | 42 | 19 | |
| Bignor | 6 | 12 | 50 | |
| Caerleon | 1 | 18 | 6 | 6 |
| Chester/Holt | 1 | 17 | 6 | 1 |
| Dorchester | 5 | 14 | 36 | |
| Fishbourne[25] | 3 | 13 | 23 | |
| Lime Street | 0 | 14 | 0 | |
| Mansell Street | 2 | 11 | 18 | 4 |
| Norwich St E | 1 | 9 | 11 | |
| Piddington | 8 | 11 | 73 | |
| Reading | 19 | 45 | 42 | 4 |
| South Shields | 8 | 8 | 100 | |
| Winchester | 6 | 9 | 67 | |
| Wroxeter[26] | 4 | 9 | 44 | |
| York | 2 | 72 | 3 | |
| Total | 75 | 313 | 24 | 15 |

Figure 6.12: Proportion of preformed holes by site

It can be seen that the legionary sites at Caerleon, Chester/Holt and York have contributed far less than the average with only four preformed holes between them: one of the holes (from Chester) was in a Group B tegula and the remainder were Group C. The two holes from York came from a tile coffin[27] using unstamped tegulae which were most probably not legionary. It would therefore appear that the legions did not generally use nails to secure their roofs, at least in the first two centuries. This raises the possibility that the legions laid their tegulae on a bed of mortar or clay.[28] The absence of mortar adhering to the underside of the tegulae should not rule this out because mortar is very rarely observed on imbrices and the flanges of tegulae despite clear evidence, as in Plates 6.1 and 6.5, that it was used. There were however six chipped holes on Group C tegulae from Caerleon which would suggest that circa AD 200 Legio II may have changed their roofing practice albeit without altering the tile production to preform the holes![29]

If the legionary Group A and B data, which account for a significant proportion of the complete tegulae are excluded, then Figure 6.11 is modified as shown in Figure 6.13.

| Cutaway Group | Preformed holes | Total tegulae | Preformed % |
|---|---|---|---|
| A | 3 | 19 | 16 |
| B | 12 | 35 | 34 |
| C | 44 | 177 | 25 |
| D | 31 | 59 | 53 |

Figure 6.13: Proportion of preformed holes excluding legionary Group A and B

Given the comparatively small sample for Groups A and B it may well be that non-legionary roofs averaged roughly one nail hole in four until a step change occurred with Group D.

| Cutaway group | Length % | Upper breadth % |
|---|---|---|
| A | 0.94 | 0.90 |
| B | 0.97 | 1.02 |
| C | 0.98 | 0.98 |
| D | 0.98 | 0.98 |

Figure 6.14: Size of tegulae with preformed holes compared to the overall population

If the tegulae were graduated in size then one would expect that the largest tiles would be on the bottom row which should be statistically detectable when the average size of tegulae with preformed holes is compared with the general population. However, even when the (larger) legionary tiles are excluded from the general population, the tegulae with preformed holes are actually marginally smaller than the population at large, as shown in Figure 6.14, although the Group A result is based upon only three holed tegulae and should therefore be discounted. This would either suggest that tegulae were not graduated in size, or that those sites that did use graduated sizes did not use nails.[30] In other words there were two roof

[24] Mainly tegulae where it was not possible to ascertain the lower cutaway form.

[25] Eleven of the Fishbourne tegulae were Group A and only one of these had a preformed hole whilst both the remaining two Group C tegulae had preformed holes.

[26] Three out of the four Wroxeter preformed holes were blind.

[27] This was a very different structure to the legionary tile tombs that have also been found in York.

[28] Brodribb 1983, 24 refers to this as the practice used by the Greeks and sometimes adopted in Italy.

[29] In fact there is one preformed hole on a Group C tegula from Caerleon so perhaps the tile making procedure was ultimately changed.

[30] The evidence from the Group C chipped holes points in the opposite direction: here the average length and breadth were 10% and 7% larger than the population as a whole. As these holes must have been chipped out at the time they were to be placed on the roof, this could be rather stronger evidence in favour of graduated sizes; however this may have been a transitional situation. Also apparently contradictory is Lime Street, London where there were no nail holes, but all the tegulae were self-meshing and there was no evidence for any graduation in size: these tiles were presumably a single manufacturing batch without holes rather

designs: one with graduated sizes that did not require nails and the other with self-meshing tegulae that did require nails.

The Group D tegulae introduce a step change in the frequency of preformed holes with over half the complete tegulae having such holes. Some of these tegulae also have much larger, perfectly cylindrical holes (Plate 2.16) which were clearly designed for a dowel or wooden peg rather than a nail, which is suggestive of a further development in roof structures. Although there is a southern bias in the observation of dowel holes, which is in keeping with the preponderance of Group D sites in the south,[31] they are nevertheless found in some numbers at South Shields. It is therefore probable that the additional nail holes on Group D tegulae are representative of a further general evolution in roof design: perhaps a steepening of the roof pitch.

## 6.9 Arrangement of tiles on the roof

A tegula and imbrex roof differs from most modern tiled roofs in that it can be laid in columns.[32] This would make it entirely practical to test the tegulae out in columns on the ground prior to mounting them on the roof. Likewise the imbrices, given their different widths, could also be tested out on the ground first. Ideally the length of the imbrices should match that of the tegulae if the neatness of the roof is to be preserved. An imbrex from Piddington (Plate 2.55) shows the line of mortar that was used to bed the imbrex above creating an 80mm overlap. Brodribb[33] has noted similar instances yielding overlaps between 50mm and 90mm. If the 80mm overlap is taken as typical, this would compare with the rather larger overlaps on the tegulae as calculated in Figure 6.15 below.

| Cutaway Group | Lower Cutaway[34] | Upper Cutaway | Overlap (mm) |
|---|---|---|---|
| A | 55 | 54 | 109 |
| B | 52 | 48 | 100 |
| C | 50 | 45 | 95 |
| D | 50 | 42 | 92 |
| R | 47 | 37 | 84 |

Figure 6.15: Overlap of tegulae by cutaway group

Applying this data to Beauport Park, where there appeared to be two main imbrex lengths of 340mm and 390mm (Figure 6.9) and two main tegula lengths of circa 390mm and 450mm, would suggest that the imbrices and tegulae could not remain in step going up the roof unless the overlap on the imbrices was closer to 40mm. At Reading it was suggested that the imbrices had two lengths of 380mm and 405mm (Figure 6.10) whilst the main tegula length groupings were 370-400mm and 470mm. These tegulae and imbrices could not have been kept in phase because the smaller imbrices are too long for the smaller tegulae whilst the larger imbrices are too short for the larger tegulae.

There is a significant variation in tegula sizes such that, if all the different sizes represent contemporaneous production, it would be difficult to maintain the symmetry between adjacent columns of tegulae rising up the roof. Plate 6.1b showed that the adjacent columns of tegulae were 50mm out of line at Dorchester. On this evidence one has to conclude that neither the tegulae nor the imbrices would have formed a uniform and matching pattern on the roof but instead they would have had a somewhat random and disorganised appearance. Adam[35] shows a plate of a reconstructed roof from Pompeii using original tegulae and imbrices which were neatly laid out in matching rows and columns, but closer inspection shows that many of the tegulae were barely touching let alone had overlapping flanges such that the rain would have poured through the roof. Likewise several of the imbrices failed to bridge the gap between the columns of imbrices.

The arrangement of tegulae in essentially independent columns on the roof facilitated repair and renovation, indeed Vitruvius[36] said that they could be changed with ease. For example: if part of the roof became damaged, only those columns of tegulae that were affected needed to be moved (together with the adjacent columns of imbrices). If a proportion of the tegulae needed to be replaced then new ones, possibly of a different size or cutaway design, could be inserted in self-contained columns without disrupting the older tegulae in the adjacent columns. If more extensive renovation, perhaps replacement of the roof timbers, was required then all the tegulae could be removed and reinstalled on the new timbers.

## 6.10 Roof structure

Vitruvius provides little information on how roofs should be formed and the translation of his technical terms is not always certain. Using Rowland and Noble Howe's interpretation of Vitruvius's description (which is shown in Figure 6.16), for large buildings Vitruvius says that the purlins should be above the principal rafters and that the common rafters should be above these with the tiles on top.[37] They translate "*asseres*" as common rafters which is consistent with Vitruvius's statement[38] that dentils imitate the exposed ends of the *asseres*/common rafters

than a random collection where some holes could have been expected.

[31] Winchester also has two dowel holes on Group C tegulae.

[32] On a modern roof each tile often overlaps with the two tiles beneath and therefore the tiled area expands diagonally up the roof as the tiles are laid.

[33] Brodribb 1983, 67/68.

[34] Using the average by sites data for consistency with the upper cutaway data.

[35] Adam 2005, 214.

[36] Vitruvius VI, 8, 8.

[37] Vitruvius IV, 2, 1.

[38] Vitruvius IV, 2, 5.

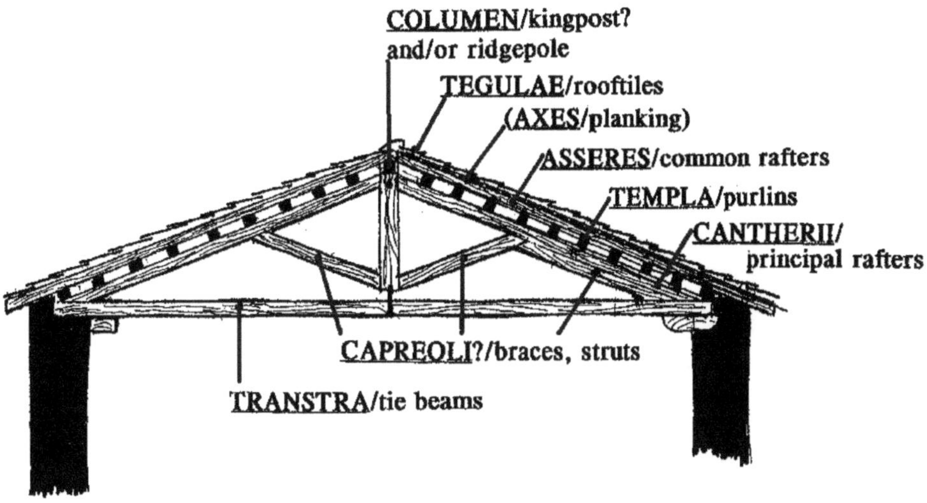


Figure 6.16: Roof structure (Rowland & Noble Howe 2002, 219)

although other scholars, for example Adam[39] as below, interpret the word differently.

For more modest buildings Vitruvius dispenses with the tie beams and braces. This is more consistent with Adams observations of ordinary houses in Pompeii which had purlins resting on the gable end walls with rafters placed over these,[40] albeit these houses do not include principal rafters as described by Vitruvius.[41] Perring[42] has observed that in Britain many gable end walls appear to be reinforced to carry purlins, as suggested by Adam, which may be the cause of the occasional collapse of these gable ends.

Adam provides a photograph of the surviving part of the eaves of a house from Pompeii[43] which shows rafters with a round cross-section spaced, somewhat irregularly, at roughly one per tegula. There are no principal rafters visible so it may be assumed that these were common rafters sitting on purlins. Above these common rafters was a layer some 8cm thick on top of which the tegulae were placed. Adam says the layer was comprised of laths, although later[44] he describes a process of covering the laths with layers of *asseres* at right angles to each other which would be more consistent with an overall thickness of 8cm. In the photograph the layer appears to be consolidated with mortar or daub which would surely have been necessary to provide a viable surface on which to mount the tegulae.

Adam also produces two further schemes[45] with tegulae placed directly on the common rafters (in which case the rafters would have needed to be regularly spaced) and onto a thin compact layer of laths set over the common rafters. Rook[46] suggests boards may have been placed over the rafters onto which the tegulae were laid. De la Bedoyere builds on the idea of boarding,[47] apparently following modern Italian practice, by inserting upright strips of timber (which he also calls *asseres*) between the columns of tegulae to wedge them into position. Boarding might work with modern machine-sawn planks and timbers where all the surfaces are flat and square but, even if the Roman builders had accepted the considerable expense of using hand sawn planks, they would have been very difficult to mount on the rafters to form a flat surface and any imperfections would have been magnified once the tegulae were laid. Moreover the retaining strips of timber to hold the tegulae in lines are manifestly not present in the elements of mortar in Plates 6.1 and 6.5, nor in any of the other pieces of mortar observed in this survey.

Adam's proposal of a layer of laths over common rafters could well be consistent with the first and second century legionary practice in Britain. Such a layer, assuming it was consolidated with mortar or daub as suggested by Adam's photograph, would be much more tolerant of the poorly meshing tegulae that the legions appear to have been producing at that time. Moreover, the tegulae would not have required any additional fixing beyond the mortar which would explain the absence of nail holes. It may also be that, because no great precision was required in laying the tegulae on such a bedding, it was easier to

[39] Adam 2005, 208.
[40] Adam 2005, 208.
[41] Adam goes on to describe other (larger) buildings that employed triangulated roof trusses with purlins above which would fit Vitruvius's description.
[42] Perring 2000, 119.
[43] Adam 2005, 207.
[44] Adam 2005, 208.
[45] Adam 2005, 209.
[46] Rook 1979, 298.
[47] De la Bedoyere 2001, 29.

use larger tegulae which would require fewer to cover a given area (but note the comments on efficiency in the next section). The normal absence of mortar on the undersides of these tegulae is entirely consistent with the general absence on imbrices where it was clearly used, although those tegulae which are found with mortar on the undersides may well result from use on such roofs, rather than from their secondary use in walls to which the mortar is normally attributed.

However, not all roofs in Britain could have been made in this way because most civilian and all later tegulae have a proportion of nail holes and nails would be ineffective on a mortar or daub substructure. Even in Rome not all roofs could have been made with this substructure because the literary references to peeping-toms peering in through the tegulae[48] and to attackers throwing tegulae down onto the occupants of the room beneath would have been impossible if there was 8cm of laths and mortar beneath the tegulae.

It seems unlikely that these later roofs utilising nailed tegulae would have had any battening placed across the rafters because, unless the tegulae were all spaced equally up the roof which the analysis in the previous section suggests was not the case, the battening would have made it more difficult to lay the tegulae, especially as only the top edge of each tile would have rested on the battens. The most likely arrangement would therefore seem to be simply a series of closely spaced common rafters onto which the tegulae were placed directly.

The use of wooden dowels with Group D tegulae suggests some improvement in timber technology because rather greater precision is required in their use than with an iron nail. Moreover, the fact that half the tegulae now have holes suggests that it would be more than the bottom row that was secured. This would require a rafter to be situated under the centre of each column of tegulae for the nails or dowels to be fixed into, but to prevent the tiles rocking there would presumably have been at least one other rafter beneath each column, and to preserve the symmetry probably two. As a result the roof was densely packed with common rafters which may have been a change from earlier arrangements. The greater use of nails or dowels on Group D roofs could have been necessitated by the rather smaller Group D tegulae reducing the stability of the roof, or possibly, an increase in roof pitch to better manage the less clement British climate.

## 6.11 Weight and efficiency

Potentially a larger tegula should be more efficient in terms of weight of clay in covering a given area than a smaller tegula because proportionately less tile is lost in the overlap. However the larger tegulae also tend to be thicker, partly because the clay technology of the Group A and B tegulae was not so advanced and partly because the thickness needs to increase as the tile gets larger to provide the same degree of structural strength. These tegulae also have longer cutaways and therefore larger overlaps, as was shown in Figure 6.15. To assess the relative effects of these variables, the average volumes of clay required to make the tegulae, the incremental area they would cover on the roof and therefore the cubic volume of clay required per unit of roof area is calculated in Figure 6.17.

| Cutaway Group | Tegula volume (cc) | Length covered (cm) | Area Covered (sq. cm) | Volume/ area |
|---|---|---|---|---|
| A | 5049 | 38.0 | 1406 | 3.6 |
| B | 4047 | 34.9 | 1180 | 3.4 |
| C | 3068 | 32.1 | 1000 | 3.1 |
| D | 3071 | 31.5 | 982 | 3.1 |
| R | 3247 | 31.0 | 954 | 3.4 |

Figure 6.17: Area and volume of tegulae by cutaway

This table shows that there was a marked improvement in the efficiency of the tegulae from Group C onwards, such that the tegulae on a Group D roof would have weighed 14% less than a Group A roof. This is a significant difference because not only would 14% less clay have had to be extracted and prepared, but the firing and transport costs would have reduced proportionately, whilst the roof timbers and, quite possibly, the walls would not have needed to be so thick to carry the lesser weight. This advantage would, however, have been reduced to the extent that the pitch of the roof was increased.

The anomaly in the regional result in part reflects the paucity of data, but is mainly due to South Shields production being rather crude with much thicker tegulae than would have been expected in this period. The Type 7 tegulae from the Hampshire area, taken on their own, would have yielded a ratio of 2.8, continuing the progression of efficiency improvement revealed by the rest of the data and delivering a 22% improvement over Group A.

## 6.12 Ridge Tiles

The ridge of the roof would have been exposed to the greatest buffeting by the elements and would therefore have needed to be very securely fixed or use very heavy components that could not be lifted by high winds. In an ideal situation the common rafters and the tegulae overlying them would all have met neatly in a line at the apex of the roof, but in reality it is clear that this would not have occurred with the tegulae (as shown in Section 6.9) and it is unlikely to have been true for the common rafters. To cope with these problems, ridge tiles tend to be heavy semi-cylindrical tiles with a large diameter. A large diameter would have been required to provide the ridge tile with sufficient height because, unlike the

[48] Brodribb 1983, 15.

imbrices which were effectively sitting on a flat surface, the ridge tiles were sitting over an obtuse angle of some 140° and thus the sides would have rested beneath the apex of the roof. Imbrices would clearly not be a suitable alternative to be used on the ridge except as ornamentation placed upon the top of a strongly mortared ridge (see Plate 7.4 from Bath).

## 6.13 Silchester and Caerwent public buildings

The extreme length of the lower cutaways from Silchester was noted in Chapter 3. Most of these tegulae arise from a few large deposits in Insula IX: the assemblages comprise predominately very homogeneous Group C tegulae with long lower cutaways together with a few Group B and rather more Group A tiles. The deposits had been carefully placed as levelling material and included relatively few imbrices. They did not arise from the collapse of adjacent buildings but appear to be tiles imported into the insula that had been stripped from just one or two buildings.

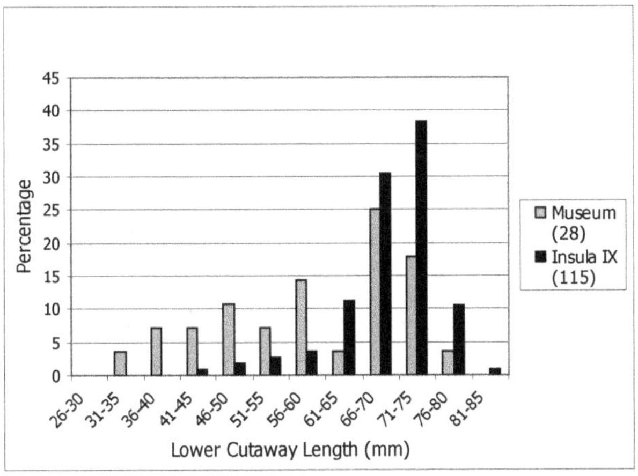

Figure 6.18: Distribution of Group C cutaway lengths

These Group C tegulae can be compared with the assemblage from the Victorian excavations of Silchester held in the Reading Museum. As can be seen in Figure 6.18, the Museum tegulae have a much broader cutaway length distribution than those from Insula IX (the number of measurements for each group is shown against the legend). Arguably the Museum tegulae have two modes: one around 56mm long which corresponds reasonably with the national average for Group C and the other around 70mm which closely matches those found in Insula IX.

All the Museum tegulae are complete and their sizes are plotted in Figure 6.19 below. It can be seen that the Group C tegulae form two clusters. The lower cutaway lengths of these two clusters of small and large tegulae are plotted in Figure 6.20.

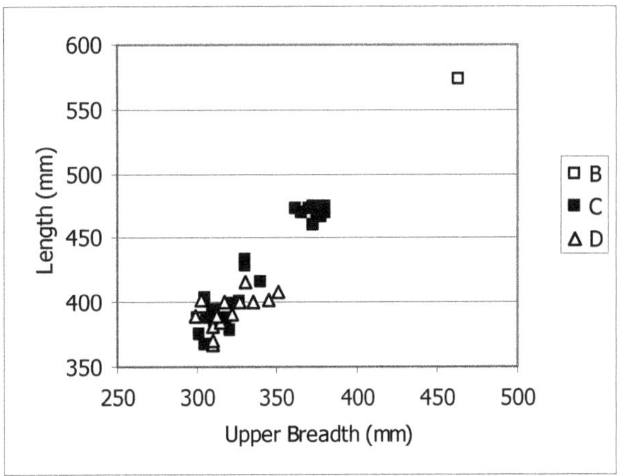

Figure 6.19: *Insula* IX and museum tegulae

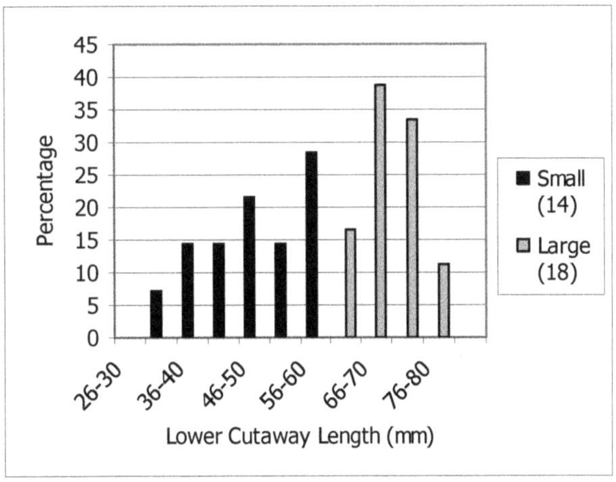

Figure 6.20: Distribution of Group C lower cutaway lengths based upon overall tegulae sizes

Figure 6.20 clearly shows that the long cutaway lengths can be associated with the cluster of larger Group C tegulae. From this it is possible to calculate the comparative efficiency of the smaller Group C tegulae versus the large, as discussed in Section 6.11. Unsurprisingly, because of the area lost in the longer cutaway overlap, the calculation shows that the larger tegulae require 15% more clay to cover the same area as the smaller ones. This is a huge inefficiency when augmented by additional transport and larger roof beams and one which the Roman engineers would surely have recognised. A public building seems the most likely destination for such expensive tegulae.

It is difficult to believe that there was any directly functional benefit from the extended cutaways but it is possible there was some aesthetic benefit. If the gaps on normal roofs were filled with moss, this might have appeared unsightly on an official building and the extended cutaways could have provided a sufficient overlap to make the use of moss unnecessary. It is also possible that on a tall building, such as the basilica, access to the roof may have been more problematic and the use of a longer overlap may have extended the time required between maintenance episodes.

The only other site that has produced a similar distribution of Group C lower cutaway lengths is the forum basilica at Caerwent. The cutaway lengths of the tegulae from the Caerwent forum basilica are compared with those from the Reading Museum in Figure 6.21 below. It can be seen that the distribution of the Caerwent tegulae is also very much bi-modal, the first mode matches the general Silchester production but the second mode of "public building" production is some 10mm greater than the Museum tegulae mode.

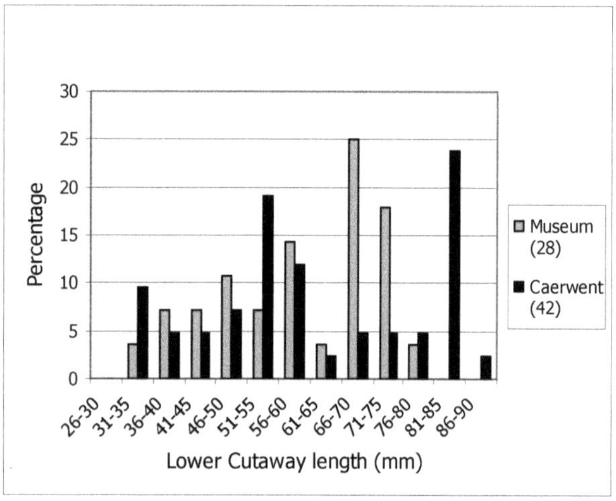


Figure 6.21: Cutaway length comparison between sites

There is a similarity between the date and design of the Caerwent and Silchester forum basilicas[49] and the above data would suggest that this similarity may also extend to the unusual cutaway length of the tegulae used on the roofs of other public buildings in these towns. However, an alternative explanation could be that the large inefficient Group C tegulae were considerably earlier production than the standard Group C tegulae. The relatively low proportion of Group B tegulae found on these two sites has already been noted in Section 4.5 which could be explained by a much earlier transition to Group C. If this were the case then these early Group C tegulae might be expected to be larger than normal and more akin to the normal size of Group B tegulae which were generally in use in that period.

## 6.14 Conclusions

The evidence appears to point towards three phases of roof design: **Phase I**, which was followed by the legions in the first and second centuries, was a roof built up with a layer of laths consolidated with mortar or daub, onto which the tegulae were placed. The tegulae normally had no nail holes and were held in place by the bedding of mortar. The tegulae were not self-meshing and may have been produced in graduated sizes to facilitate meshing. This practice was also adopted by sites that had military connections. The general absence of mortar which might have been expected to adhere to the undersides of the tegulae is entirely consistent with its absence on most imbrices and the flanges of tegulae despite the fact that mortar was clearly necessary to help seat the imbrices over each other and to secure them to the roof.

**Phase II**, which was adopted by private builders, perhaps from when the use of tiled roofs started in Britain, and by all builders from the late second century, involved placing the tegulae directly onto the common rafters and securing (probably) the bottom row with nails. These tegulae were self-meshing and were not produced in graduated sizes; their size however reduced in time in line with the overall gradual reduction of tegula sizes. This advance was made possible by the improvement in tegula technology whereby there was a step increase in the breadth taper and an improvement in cutaway design from Group C onwards. This enabled the tegulae to become self-meshing which gave the roof the additional stability that was necessary for direct mounting. The high incidence of (unscheduled) chipped nail holes on Group C tegulae suggests that not all manufacturers grasped all the relevant changes simultaneously.

**Phase III**, at some point in the third century, the roof pitch was increased and timber technology improved to the point where wooden pegs or dowels could be used instead of nails (albeit not universally). To provide security on the steeper pitch, every other tegula was secured and a common rafter was aligned with the centre of each column of tegulae to accept the pegs or nails.

There is also evidence for a putative **Phase IV**. Three sites within fifty kilometres of each other on the Hampshire/Sussex/Surrey borders – Crookhorn,[50] Batten Hanger[51] and Wykehurst Farm, Cranleigh[52] - have yielded flat ceramic tiles which were respectively heptagonal, pentagonal and pear shaped. These tiles had nail holes and were presumably intended for use on roofs although there is no evidence of their actual use on a roof. Crookhorn is a fourth century site and Batten Hanger may well have been active into the fifth century.[53] The design of these tiles was presumably intended to imitate similar products in stone slate.

On the continent there may also have been a further phase of evolution when imbrex and tegula roofs were replaced by roofs solely comprised of alternating rows of imbrices with one face up and the other face down which is noted by Adam[54] as occurring in Gaul in the later

[49] Fulford & Timby 2000, 573.

[50] Soffe et al 1989, 75: the tiles are in the Portsmouth Museum store.

[51] The tiles are in the Chichester Museum store.

[52] Brodribb 1987, 18: the tiles are in the Guildford Museum store.

[53] Pers Comm: James Kenny, Chichester District Archaeologist.

[54] Adam 2005, 214.

Roman period.[55] The archaeological evidence for this in terms of imbrices without tegulae would be very obvious and has not been noted in Britain, so this phase did not reach Britain before the end of the Romano-British period. The drivers behind these phases of roof evolution are considered further in Chapter 8.

A house was an important statement in the Roman world. It would be positioned to provide an impressive approach and maximum impact.[56] A tiled roof would have been an important aspect of this. Deliberate selection of clays to achieve particular colours[57] or the application of coloured slip would have enhanced the overall effect, perhaps producing patterns on the roof to increase the impact. It is therefore surprising that the relationship between the length of the tegulae and the length of the imbrices make it unlikely that they marched in step up the roof. Moreover, the evidence from Dorchester would suggest that even the adjacent columns of tegulae were mis-aligned.

[55] Brodribb 1987, 23 describes a similar system adopted by the Greeks in the seventh century BC.
[56] Martins 2003, 92.
[57] Tinson 2002, 25.

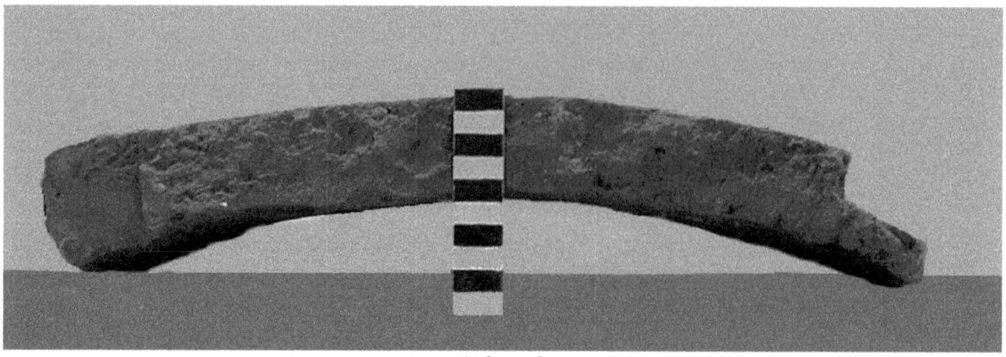

Ashtead

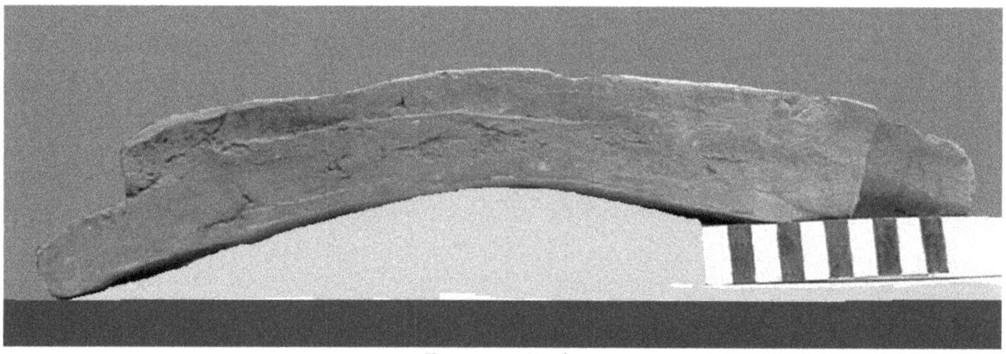

Beauport Park

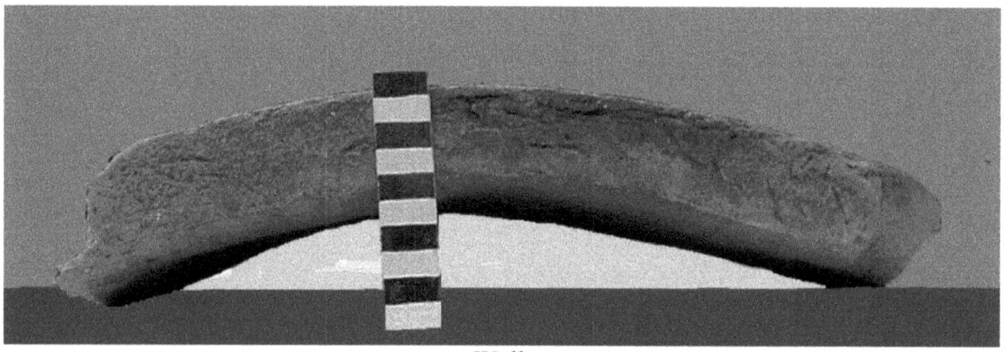

Wall

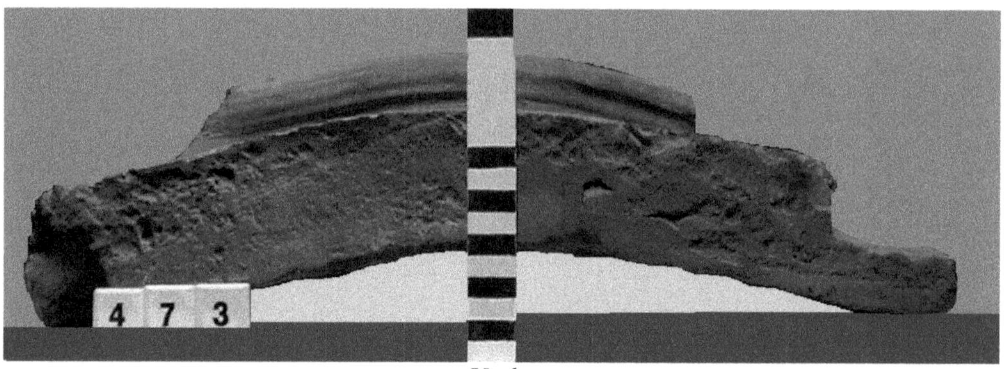


York

Plate 7.1: Examples of convex tegulae

# 7

# VAULTED ROOFS

## 7.1 Incidence of convex tegulae

In the survey, of the 480 tegulae with complete lengths that were examined, 22% were longitudinally convex[1] such that the tiles formed an arch shape as shown in Plate 7.1. This chapter demonstrates these were unlikely to be manufacturing defects and shows how they would have been used on roofs.

The degree of arching is referred to as "convexity": when the tegula is placed upon a flat surface it is the distance (in millimetres) between the bottom of the tile at approximately its mid-point and the surface below (Figure 7.1).

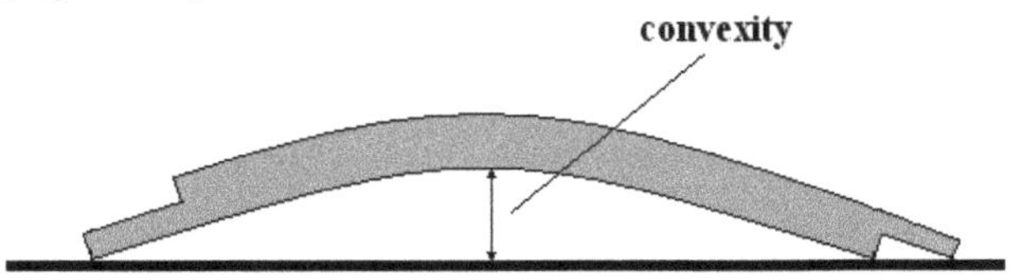


Figure 7.1: Measurement of convexity

Most tegulae have minor flaws, so in order to differentiate the genuinely convex tiles from those that were merely blemished, all tiles with a convexity of less than 10mm were regarded as being flat. Tegulae with a convexity between 11mm and 20mm were labelled as "mildly convex" and those with a convexity of 21mm or greater were identified as "severely convex". All the tegulae in Plate 7.1 were severely convex and their average convexity was 42mm.

Using this methodology, of the 480 tegulae with complete lengths that were surveyed, 78% were flat and 22% were convex, of which 15% were mildly convex and 7% severely convex.

Curved objects are less likely to survive in the ground unbroken than flat ones because there is a greater propensity for unsupported voids to be created, making it more likely that the object will fracture when a load is placed upon it. This can be seen with imbrices which have a much lower survival rate than tegulae and tend to break at their point of maximum curvature. It seems likely that convex tegulae will also suffer differentially when compared to flat tegulae and, as a consequence, the observed 22% of convex tegulae that survive may understate their true contribution to the original tegula population.

There appears to be no geographical bias in the incidence of convex tegulae, with examples being noted from Winchester to Corbridge and from Kent to Caerleon. The proportion of convex tegulae is between 20% and 30% for Groups A, B and C but is less than 10% for Group D. There were two mildly convex tegulae with regional cutaway forms.

## 7.2 Deliberate or flawed manufacture?

Webster[2] has commented upon the Roman tilemakers ability to create *bipedales* that were absolutely flat and square for use on bathhouse floors where no distortion could be tolerated. Convex tegulae are often described as wasters but it seems unlikely that almost one in four tegulae from across the country could be kiln rejects, especially when many of the same sites that have yielded convex tegulae fail to produce defective floor, wall or flue tiling. Moreover, if these "defects" were caused by shrinkage in the kiln then one might have expected the tiles to bend *upwards* as the greater mass of the flange contracted inwards, rather than *downwards* as observed. Indeed some convex tegulae demonstrate precisely this anticipated shrinkage (Plate 7.2) which has caused the flanges to rupture as they have tried to contract in opposition to the established shape of the tile. In contrast, concave tegulae, which ought to be the natural shape of deformation, have not been observed on any site. Buckled tegulae, whereby the tile bends to bring the two flanges closer together, are occasionally observed and almost certainly resulted from placing the tegula on

[1] In addition Boon 1984, 17 noted a group of 28 "somewhat arched" tegulae from the amphitheatre baths at Caerleon which sadly do not appear to have been retained and Betts 1990, 166 noted the curved shape of a number of tegulae at Dalton Parlours.

[2] Webster 1979,287.

its side before it was sufficiently dry to sustain the weight.

Plate 7.2: York 472 convex tegula with ruptured flange

Many of the convex tegulae had unusually smooth undersides. The undersides of most tegulae were rough, reflecting the process of throwing the clay into the mould and beating it to shape, but some undersides appear to have undergone a smoothing process. In some cases this may have been achieved by the tegula being rubbed against a former and in others it has been caused by a wire being passed between the underside of the tegula and the former. Plate 7.3 shows examples from Winchester which (as discussed in Section 2.3(a)) all appeared to have come from the same mould: the right hand tegula was flat and had a rough underside whilst the other two tegulae were convex and had smooth undersides. Clearly the two convex tegulae must have undergone a further manufacturing process that produced the smooth undersides, most probably being rubbed against a former.

Plate 7.3: Winchester smooth and rough undersides

Plate 2.25 shows a typical example from Beauport Park with clear striations caused by dragging a wire beneath the tegula whilst it was still relatively plastic. (This may be compared with marks left by the wire in the manufacturing experiment shown in Plate 2.42). This tile was severely convex yet, from the quality of the striations, it does not appear as if the tegula had bent after the wire had been used. The most likely reason for using the wire would have been to release a tegula that had become stuck to its baseboard. This would have been done by tensioning the wire against the baseboard but it would be almost impossible to let the wire ride up over the clay to scoop out a symmetrical segment (thereby creating a convex tegula) without either the wire going right through the tile or the carved out section being much greater on one side or the other. If on the other hand the tegulae were being deliberately manufactured to be convex then, after forming in the conventional way, the tiles could have been placed across an arch shaped former when they were in a semi-plastic state and either pressed against it (which could have led to them becoming stuck and the use of the wire to remove them) or being gently rubbed against the former until the shape was achieved (which could have led to a smoothing of the underside).

However it should be noted that not all convex tegulae had smooth undersides and that a minority of flat tegulae also had smooth undersides. There is also a tendency for the Group D and regional tegulae (neither of which were convex to any significant degree) to have smooth undersides which is a natural consequence of being made in inverted box moulds as discussed in Section 2.6(b).

The bathhouse at the Beauport Park iron-works near Hastings in Sussex was isolated from the rest of the site and was buried by the slippage of the slag heap probably in the fourth or fifth century.[3] As a result it is highly probable that all the tiles found at the bathhouse site were exclusively from the bathhouse and that they had all been on the roof together. The complete tegulae with stamps RIB 2481.7, 43 and 75 were flat[4] but thirteen of the tiles with RIB 2481.102 (roughly two thirds of the total with this die) were convex. That these convex tegulae were mounted on the roof is therefore virtually certain.

## 7.3 Possible roof structures

There are several possible forms that a bathhouse roof could take. Cunliffe's reconstruction of the roofing of the Great Bath of Sulis Minerva (Figure 7.2) nicely demonstrates the principal alternatives for covering a bathhouse. The Period I Great Bath employed a conventional tiled roof supported by a triangulated truss which was replaced in Period III by a large vault, a fragment of which was found in the bath itself.[5] The triangulated truss contained all the lateral forces within the timbers and thus exerted only a downward force on the walls thereby allowing a simpler construction of the building. The disadvantage was that the horizontal beams would have encouraged condensation to form and drip onto the bathers beneath whilst the timbers would have quickly rotted in the constant warm, wet atmosphere.[6]

[3] Brodribb & Cleere 1988, 243.
[4] There were two exceptions: Beauport Park 181 and 196 which had concavities of 15mm and 16mm respectively.
[5] Cunliffe 1969, 119.
[6] Vitruvius VII, 3 discusses suspended ceilings which could have taken a vaulted form and been inserted beneath the main roof which would

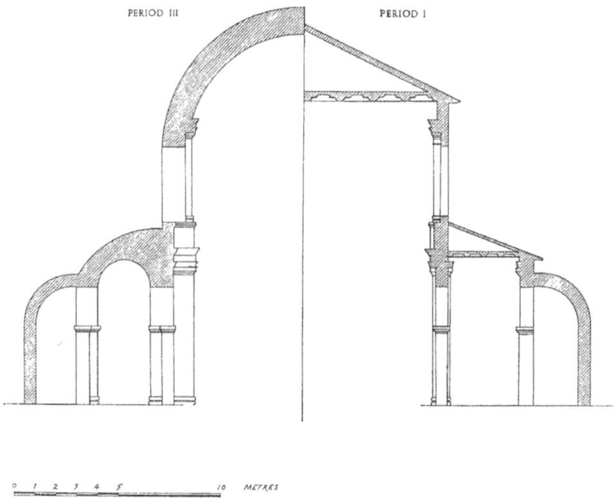


Figure 7.2: Reconstruction of the roofs of the Great Bath at Bath (Cunliffe 1969, 119)

In contrast, the vaulted roof in Period III was made entirely of tile and concrete which would not have rotted and the shape would have inhibited dripping. However, the vault would have exerted a lateral force on the walls which necessitated a stronger structure beneath. No roof tiles could be directly associated with this vaulting and Cunliffe therefore concluded that it was left bare, relying on a thick waterproof outer coating of concrete and imbrices,[7] as shown in Plate 7.4. This approach is rare in Britain but was not dissimilar to that adopted in the southern part of the Roman Empire, for example on the Hunting Baths at Leptis Magna,[8] where the vaults were covered with concrete but otherwise left bare. In contrast the presumed depiction of a bathhouse on the Simpelveldt sarcophagus now in the Leiden Museum[9] shows a pitched tiled roof on walls that were tall but without buttresses and therefore most likely rested on a triangulated truss similar to Period I at Bath.

Plate 7.4: Element of vaulted roof from the Great Bath

have avoided these problems but would have significantly reduced the internal height of the roof.

[7] Cunliffe 1969, 98.

[8] Ward Perkins and Toynbee 1949.

[9] Rook 1992, 32.

Bidwell's reconstruction[10] of the legionary fortress baths at Exeter has an alternative approach with a conventional tiled roof over a vaulted bathhouse but this was ingeniously achieved by resting the rafters on a wall[11] that effectively stood behind the one from which the vaulting sprang (Figure 7.3[12]). This device did not use a triangulated truss to contain the lateral thrust from the roof because the tie beams would have pierced the vault; instead it used the mass of masonry created by the secondary wall to absorb the lateral thrust from the roof.

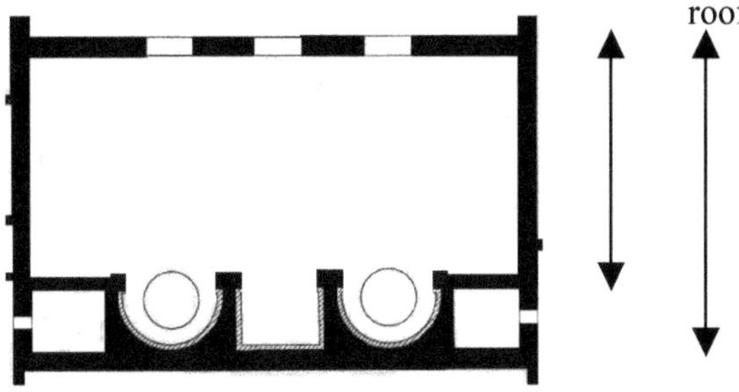


Figure 7.3: Exeter fortress baths *caldarium* (after Bidwell 1979, 52)

A further alternative is suggested by a fragment from the roof of Bewcastle bathhouse which had white concrete placed over the vaulting. This had been built up to form a flat angled base onto which sandstone roofing slates were laid to form a pitched outer skin to the roof.[13] A similar arrangement was adopted in the Villa dei Sette Bassi.[14] Bewcastle was one of five bathhouses adjacent to Hadrian's Wall that were all built to a common plan. On four of these bathhouses the otherwise unsupported wall of the first warm room was reinforced with buttresses independent of the topography of the individual sites.[15] These buttresses were presumably to resist the lateral forces exerted by the vaulted roof which suggests that the white concrete did not consolidate the roof to the extent that it could contain the lateral forces, but instead sat over the vault and would actually have added to the lateral thrust. Many of the bathhouses on the Antonine Wall also incorporate buttresses presumably reflecting the same consideration.[16]

[10] Bidwell 1979, 51-54.

[11] If the sole-plate for the pitched roof was at the same level as the base of the barrel vault then the slope of the roof would have been approximately 42°; however the pitch could have been reduced to a more acceptable 30° by the simple expedient of raising the height of the outer wall by 1.6m (to 9.5m).

[12] The building continued beyond the top of the figure with a, presumably, symmetrical counterpart such that the thrust of the two vaults and pitched roofs were balanced at the centre of the overall building.

[13] Gillam, Jobey & Welsby 1993, 14.

[14] Adam 2005, 208.

[15] Gillam, Jobey & Welsby 1993, 10: the other three bathhouses were at Chesters, Netherby and Carrawburgh; there appeared to be no buttresses at Benwell.

[16] For example Keppie 2004, 206-7.

The remains of the Chester *thermae* incorporated a mass of roof tile that was found overlying the *tubuli* and concrete that formed the vaulting for these baths.[17] These remains could be consistent with either the timber type of pitched roof used in the Exeter fortress baths or the solid version found at Bewcastle.

## 7.4 The Beauport Park roof

At Beauport Park, Brodribb and Cleere[18] discovered fragments of vaulting in Rooms 4 and 5 of the bathhouse (Figure 7.4), representing one of the later phases of its development. Each room had a separate vault with an east-west axis. Tegulae were also found associated with these rooms and they therefore surmised that a conventional tiled roof had been placed over these vaults although they did not speculate on how this was achieved.[19]

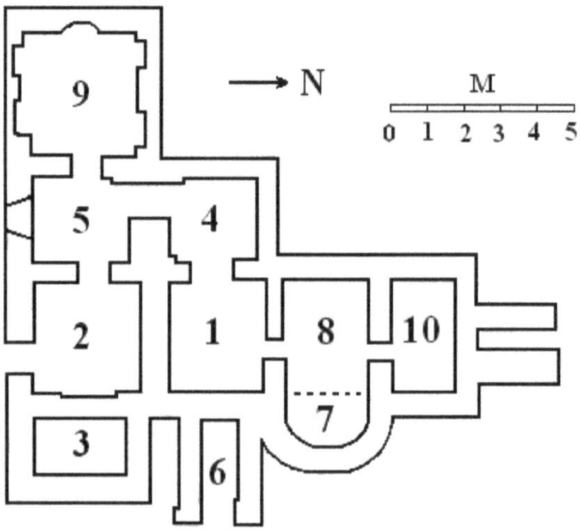


Figure 7.4: Plan of Beauport Park bathhouse (from Brodribb 1979, 140)

The most obvious roof structure would be one similar to the bathhouse at Bewcastle with the tiles placed over a built up layer of concrete to form a pitched roof. However, the fragments of vaulting (Plate 7.5) show no signs of any additional concrete overlying the extrados and one might doubt the ability of this relatively flimsy structure of alternating tile and stone courses to carry the additional load of both tiles and concrete. Moreover, unlike Bewcastle, the outer walls of Rooms 4 and 5 had no buttressing and the Room 5 wall was further weakened by a window which reduces the likelihood of this solution still more.

An alternative would be a timber superstructure to carry the tiles as at Exeter but this would have had an even greater lateral thrust without the benefit of the additional masonry that there was at Exeter to absorb it. This therefore was not the solution.

It seems unlikely that the Beauport Park vaults could have been left bare like the Period III roof at Bath because the fragments of surviving vaulting definitely did not have a protective outer coating and would have been quickly eroded by the effect of rain and frost. Moreover, in contrast to Bath, roof tiles were found that appeared to be associated with the rooms that were vaulted.

Plate 7.5: Fragment of vault from Beauport Park (Brodribb & Cleere 1988, Plate XI.B)

None of these alternative solutions seem to provide the answer for Beauport Park, but there is one further radical possibility which would neatly solve both the difficulty of covering the Beauport Park vaulting and explain the presence of convex tegulae. This is that convex tegulae were placed *directly* onto the curved extrados of the vaulting to provide the protective outer coating as shown in Figure 7.5. Flat tegulae would not work, but convex tegulae whose curvature matched that of the vaulting would produce a satisfactory effect.

The end of the bottom tegula would have rested on the wall and the meshing of the cutaways of the tegulae above would have transferred their weight down, as well as locking the bottom tegula in its near vertical position. The tegulae could not be fixed with nails but would need to be mounted on mortar like the Phase I military pitched roofs in Britain (see Chapter 6). Such an arrangement would have added only a minimal amount to the lateral thrust of the roof which may therefore explain the absence of buttresses.

[17] Mason 1990, 217.
[18] Brodribb & Cleere 1988, 226.
[19] Brodribb 1979, 142 and Brodribb & Cleere 1988, 235.

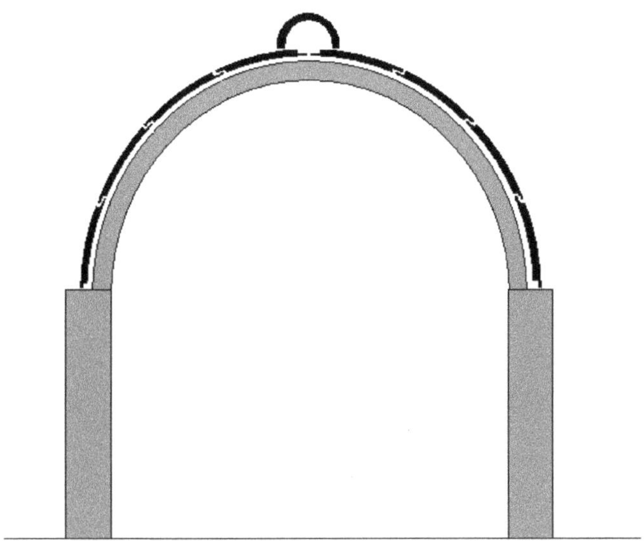

Fig 7.5: Vaulted roof tiled with convex tegulae

This hypothesis can be tested against the convex tegulae with RIB 2481.102 stamps. The average convexity of these tegulae was 13.5mm and their average length was 396mm.[20] These dimensions would produce a natural semi-circle of 2.86m diameter.[21] Allowing for the 250mm thickness of the vaulting,[22] this means that the tiles would ideally suit a vault spanning a room that was 2.36m wide. In fact Rooms 4 and 5 were 2.6m and 2.8m wide respectively, which matches the theoretical calculation within the sort of error that might be expected from the relatively small sample of convex tegulae that were available. This result therefore demonstrates the plausibility of this proposal.

Ideally, the imbrices should also be bent to match the curve of the tegulae, but an inspection of the Beauport Park imbrices shows that this was not the case. However, the dimensions of these imbrices were such that they would not have stacked over each other as required even on a flat roof without the use of mortar,[23] so presumably simply a little more mortar was used to seat them on a vaulted roof.[24]

Some convex tegulae show signs of mortar on their undersides but normally it is absent; however the general absence of mortar from the tegulae and imbrices is similar to that on ordinary pitched roofs where it must also have been used (especially on Phase I military roofs). The survival of the concrete vault at Bath must be due in part to its use of a much harder pozzolanic mortar which would have been faster setting and waterproof but perhaps also dangerously rigid. Such mortars (which would have been more expensive to produce) do not seem to have been widely used on roofs in Britain.[25]

If vaulted roofs were tiled as suggested then the last row of tiles would be virtually vertical and hence any antefixes would be both functionally and aesthetically redundant: the presence of antefixes would therefore be a signal of a conventionally pitched roof. Antefixes, even on military sites, are rare, so their absence does not signal a vaulted roof; nevertheless it is noteworthy that Beauport Park has produced no antefixes. By contrast antefixes were found at the Exeter fortress baths and the few tegulae remaining from the site appeared flat which reinforces Bidwell's reconstruction.

The drains under the hypocaust were composed of overlapping tegulae or imbrices and this is the likely form of the guttering on the roof. The dimensions of the walls in Rooms 4 and 5 were such that guttering composed of imbrices would fit conveniently on top of them and still leave space for the springing of the vaults, provided they were tiled as suggested. No external down-pipes were discovered and it is possible they may not have existed, with water simply running out of the gutter into a collection area at the base of the building. Room 1 had a vertical pipe in one corner composed of pairs of imbrices which had the appearance of a makeshift box flue; however, instead of being used for transporting hot gases upwards, it may have been piping for transporting the rainwater downwards into the drainage channels beneath the hypocausts. This would be similar to the down-pipe arrangements in the Baths of Caracalla; some of these were formed of tiles within the width of the walls which discharged into the drains beneath the building.[26]

Brodribb and Cleere[27] draw attention to the similarities between the Beauport Park bathhouse and (*inter alia*) the bathhouse at the Welsh fort of Gellygaer. It is therefore

[20] Calculated from all the tegulae with die RIB 2481.102.

[21] The radius R of the vault can be calculated from the diagram where

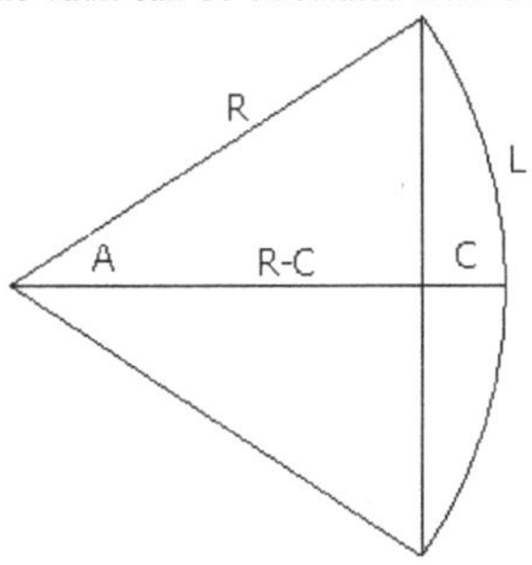


the length L of the tegula has been measured on its upper surface (which is effectively along the circumference) and 2A is the angle it subtends at the centre and C the convexity. Then $2A/360 = L/2\pi R$ and $\cos A = (R-C)/R$ which may be easily solved for R by iteration.

[22] Brodribb & Cleere 1988, 227.

[23] If a roof were comprised of identical tegulae and imbrices then, if the larger end were to fit over the smaller end of the next imbrex, the internal height and width of the larger end of each imbrex would need to be greater than the external height and width of the smaller end of the imbrex. At Beauport Park only two out of the twenty imbrices that could be measured met this criterion.

[24] Bending clay in two directions at the same time without buckling would in any event have been extremely problematic.

[25] The practice is much more common in Italy where volcanic ash could be used for making these pozzolanic mortars (Vitruvius II, 6, 1).

[26] DeLaine 1997, 153-154.

[27] Brodribb & Cleere 1988, 243-4.

interesting to note that Gellygaer has also yielded convex tegulae (albeit none that were complete).

## 7.5 Evidence from other roofs

A section of the roof incorporating a convex tegula from the Wollaston bathhouse at Dorchester, Dorset is shown in Plate 7.6.[28] It appears to come from a pair of additional heated rooms that were built against the bathhouse and re-roofed in the late third century.[29] Other rooms in the bathhouse were vaulted and there is evidence to suggest that these additional rooms were also vaulted. The width of these rooms was 6.5m which is almost twice the convexity of this tegula, albeit this may be a misestimate as the full length of the tegula did not remain. The underside of the tegula shows signs of mortar (Plate 7.6b) and, as it seems exceedingly unlikely that this bit of roof was ever reused elsewhere, this must imply that the tegula was placed on a bed of mortar on a vaulted roof.

Plate 7.6a: Dorchester 2470 part of roof over vault

Plate 7.6b: Dorchester 2470 mortar on underside

The end of the imbrex is filled with mortar rather than the impression of the imbrex beneath so this piece of roof (letter A) must have come from the bottom of the column of imbrices. As the bottom tegula would be nearly vertical against the vault, the imbrex would project horizontally outwards from the vault and potentially block any gutter placed on top of the wall to take the water off the roof. It is therefore interesting to see that this imbrex has had its end broken off and the mortar filling tapered such that gutter is not obstructed. This detail would not have been required on a normal pitched roof.

In Italy, the nave of the Basilica of Constantine in Rome was vaulted and a large fragment of the collapsed vaulting from the nave survived. The outer surface of the vaulting was made of concrete onto which tegulae had been laid whilst the concrete was still wet.[30] The two pieces of surviving tegulae, which were still in situ, were coloured red and yellow respectively; perhaps suggesting some decorative effect was intended in addition to weather-proofing.

One of the roofs of Trajan's market in Rome has a hemi-spherical dome which bears the clear impression of rows of tiles in the mortar substrate. These impressions were square with a side of 0.56m to 0.58m.[31] MacDonald interpreted these impressions as having being made by *bipedales* which had been used to protect the structural concrete from the weather. Bipedales would have been unsuitable for this task because uncovered gaps would have existed between them where rainwater would have been trapped. Convex tegulae with capping imbrices would, however, have been entirely appropriate. The impressions left by such tegulae might also be square: for example if one takes one of the larger tegulae measured by Adam at Pompeii[32] (0.52x0.66m) and assumes commensurate cutaways of 70mm[33] then the overlapping of the tegulae would result in a square impression with a side of 0.52m, almost exactly as MacDonald observed.

## 7.6 Other convex tegula assemblages

The largest accumulation of convex tegulae is from the Yorkshire tile tombs: one tomb[34] is comprised almost exclusively of severely convex tegulae and two others incorporate a number of convex tegulae. Betts[35] speculated whether convex tegulae could have been deliberately manufactured for use in such tombs but this seems unlikely because the severely convex tomb is not the product of a single manufacturing batch but comprises legionary stamped tiles bearing several different stamps suggestive of either different periods of

[28] This roof may have used a similar mortar to that at Bath.
[29] Pers Comm: John Magilton.

[30] Minoprio 1932, 7.
[31] MacDonald 1965, 83 and Plate 81.
[32] Adam 2005, 213.
[33] This compares with the average cutaway length for the smaller British tegulae of this period of 57mm.
[34] Excavated at Dringhouses in 1833 according to the Yorkshire Museum plaque.
[35] Betts 1990, 166.

production or that of different cohorts. The other tombs incorporating convex tegulae are also somewhat disparate and one contains tegulae from both *Legio IX* and *Legio VI*. However, the fact that the tomb builders had access to convex tegulae from production that seems to span at least a number of years suggests that either all such convex tegulae were set aside for use in tombs or, more likely, that they were readily accessible and therefore a reasonably widespread aspect of normal tile production.

As discussed in Section 6.13, the Silchester Group C tegulae in the Reading Museum divide into two size clusters and the larger tiles were probably used on a public building. In fact the larger tegulae are also distinguished by the majority being convex whilst the smaller tegulae are almost all flat. This would suggest that the public building from which the larger tegulae came was vaulted and, based on the average convexity of this group, the building had a span of around 4m. This measurement means that it was not the main basilica building whose span was a minimum of 12m[36] but may well have been the public baths.

## 7.7 Nail holes

If tegulae were to be placed directly onto the vaulting as suggested, then the best way of securing them would have been with mortar. Nails would not have been used as these would not have provided a satisfactory anchor into the mortar and masonry of the extrados (although it is possible that wooden dowels would have been more suitable). This conclusion is supported by the observation that none of the convex tegulae from Beauport Park has preformed nail holes whilst five of the flat tiles do.

| Convexity (mm) | No nail hole | Preform hole | Total tegulae | %age preform | Chip'd holes |
|---|---|---|---|---|---|
| up to 5 | 137 | 59 | 196 | 30 | 7 |
| 6-10 | 42 | 16 | 58 | 28 | 1 |
| 11-15 | 27 | 7 | 34 | 21 | 2 |
| 16-20 | 10 | 2 | 12 | 17 | 1 |
| 21-25 | 11 | 1 | 12 | 8 | 1 |
| 26+ | 16 | 0 | 16 | 0 | 1 |
| TOTAL | 243 | 85 | 328[37] | 26 | 13 |

Figure 7.6: Distribution of nail holes with convexity

If the hypothesis about vaulting is correct then the manufacturing specification of a production order for convex tegulae would not have included preformed nail holes, whereas a proportion of preformed nail holes would have been part of the manufacturing specification for flat tegulae destined for the Phase II or III pitched roofs discussed in Chapter 6. There were 328 tegulae in the survey where the presence or absence of a nail hole could be observed, and of these 85 had preformed nail holes.[38] The distribution of nail holes in tegulae with convexity is shown in Figure 7.6.

Only one of the 28 severely convex tegulae had a preformed hole and this was a dowel hole which does not necessarily conflict with the theory. Only two of the twelve tegulae in the 16 to 20mm convexity range had preformed holes and again one of these was a dowel hole. Rather more tegulae in the 11-15mm range have nail holes (although still less then for flat tiles), suggesting that many of these very mildly convex tiles were indeed manufacturing defects.

If the theory that convex tegulae should not have nail holes were wrong, and it was just a chance event that those tegulae that distorted into a convex shape through shrinkage just happened not to have preformed nail holes, then the probability of this occurring can be calculated. The probability of randomly selecting 28 tegulae with none or one hole from a population where one in four tegulae have a nail hole is 0.003.[39] In other words, the probability that the manufacture of the severely convex tegulae was accidental and that coincidentally none of these possessed a nail hole is three in a thousand. If one accepts that dowel holes do not conflict with theory then there would be no failures amongst the severely convex tegulae which moves the probability of the hypothesis being wrong to three in ten thousand.[40]

## 7.8 Conclusion

Convex tegulae were on the roof at Beauport Park. Nationally too many convex tegulae are found for these to be manufacturing errors which, in any event, might be expected to produce concave tegulae, of which none are found. The convex tegulae appear to be specially produced because they do not have nail holes and often have smoothed undersides whereas the reverse applies to flat tegulae. The convex tegulae provide a solution to the Beauport Park roof which is not provided by any of the alternative design proposals. The Wollaston bathhouse at Dorchester provides visual Romano-British evidence of how such a roof might have looked and the guttering detail would not have been required on an ordinary pitched roof. In Rome there are at least two buildings

[36] Fulford & Timby 2000, 575.

[37] This total differs from that in Figure 6.11 which included tegulae where just the upper half was complete and others that were complete but broken.

[38] Only preformed nail holes are counted because chipped holes represent a decision made after the tile had been manufactured. However preformed blind holes are included as it is clear that some manufacturers deliberately produced holes in this manner allowing the roofer to pierce the residual clay where required, but leaving the tile fully waterproof if a hole was not needed.

[39] The probability of x nail holes occurring in n tegulae when the average occurrence of nail holes is one in four or 0.25 is $0.25^{x}.(1-0.25)^{n-x}.n!/x!(n-x)!$ (Sokolnikoff & Sokolnikoff 1941, 502). Therefore the probability of none or one nail hole in 28 tegulae is $0.25.0.75^{27}.28!/1!.27! + 0.25^{0}.0.75^{28}.28!/0!.28! = 0.0033$.

[40] Probability is $0.75^{28} = 0.0003$.

where evidence of tegulae placed directly on vaulted roofs has survived.

However, whilst this makes a persuasive case for the use of convex tegulae, there remain too many mildly convex tegulae with preformed nail holes for this to be a complete explanation. In any event the use of convex tegulae appears to have ceased towards the end of the third century as only five convex Group D tegulae and two convex regional ones have been observed.

Nevertheless, if the theory is correct then, as there are only four times as many flat tegulae as convex ones in this survey, there should only be four times as many buildings with ordinary pitched tiled roofs as bathhouses with vaulted roofs covered by convex tegulae. As every villa will have had its own bathhouse a ratio of 4:1 for the roof areas is perfectly reasonable, but in an urban or military context the result is more surprising. It suggests that either tiled roofs were rarer and therefore more of a status symbol than previously appreciated or there were rather more private bathhouses in towns than might have been expected, most of them being tiled with convex tegulae.

# 8

# LOGISTICS, COSTS AND ECONOMICS

## 8.1 Introduction

The first part of this chapter examines the microeconomics of tile production. It investigates how tileries operated, the underlying logistics and organisation of such tileries, their costs of production, selling prices and area of supply, and probes how they responded to competition and demand shortages. Using the example of the *Classis Britannica,* it then considers how a military tilery might have operated and the output that it could have produced.

The second part of the chapter looks at the wider political and economic conclusions that can be drawn from the tile evidence. It examines the causation that lay behind the sequence of cutaway changes and uniform reduction in tegula size. It then explores the insights that tegulae can provide into technological innovation and the state of the economy and how tegulae mirrored trends in other sectors of industry.

## 8.2 Logistics of tile production

There are two graffiti from Italy stated to be on tegulae that refer to two men making 440 tiles in a day[1] and a list of four men each making 220 in a day.[2] A further less complete graffito comes from Regensburg in Germany "....coli produced 110 tiles"[3] which was presumably half a day's work at the *standard* rate of 220/day. The only British graffito referring to production is one from Silchester:[4] "by hand 199". This graffito is stated as being on a brick, but bricks ought to be considerably quicker to produce than tegulae. It could be that the British tilemakers were slower, or that the other graffiti have been misattributed to tegulae, but the most likely explanation is that the Silchester graffito is actually on a tegula which in fragmentary form is indistinguishable from many bricks.

[1] CIL III, 11383.
[2] CIL III, 11381.
[3] von Elbe 1977, 325.
[4] RIB 2491.14.

There are also a number of graffiti consisting solely of numerals, the most relevant of these are from Hartfield in Sussex and Ahrweiler in Germany. The Hartfield one[5] is on a tegula and consists of the numbers CCXV and CCXIIII whilst the Ahrweiler graffito[6] is on an imbrex with the numerals ...CCX, ...XXX..., and ...CX... . Both sets of numbers are in columns and mirror the list in CIL III 11381 noted above. The Ahrweiler graffito is damaged and could perhaps have listed the numbers 210, 230 and 220. The interpretation of these numbers as starting with a two hundred is supported by another Ahrweiler graffito which reads CCX... . Imbrices should be quicker to manufacture than tegulae but may have taken longer to store because of the more cumbersome formers. In any event, the tegula production would have needed to keep in step with the imbrex production and a daily quota for each would appear to have been around 220. If the tegulae were Group A then this is equivalent to 1.1 cubic metres of fired clay per day. By way of comparison a nineteenth century brick-maker would have been expected to produce 1000 bricks per day which equated to some 1.3 cubic metres of fired clay.[7] This was some 20% more clay than the Romans were producing per day, but would seem to confirm the interpretation of the graffiti.

There is no standard Romano-British kiln size, but a firing chamber measuring 3m by 2m would not be untypical.[8] There is little surviving evidence for the height of a tile kiln but again 2m would not be unreasonable.[9] Later tegulae were very roughly 400mm long, 300mm wide, 20mm thick, with a flange height of 50mm and a flange width of 25mm. As shown in Figure 8.1, if these tiles were stacked into the kiln on their ends in pairs, with the flanges interlaced but with a 15mm clearance all round, then a pair of tegulae would occupy a floor area of 100mm by 345mm. A column of twenty interlaced pairs of tegulae would extend to exactly 2m. Between eight and nine such columns would fit within a

[5] RIB 2491.43.
[6] Redknapp forthcoming.
[7] DeLaine 2001, 262.
[8] The arithmetic average size of the 54 tile kilns listed in McWhirr 1979, Table 6.1 is 3.1m by 2.7m.
[9] McWhirr 1979, 99 implies 2m as a sensible height.

length of 3m. Four such layers would fit comfortably within an overall height of 2m giving a grand total of roughly 1440 tiles that could ideally be placed in this typical kiln.[10] In broad terms, at a rate of 220 tegulae per day, this equates to one week's output of one tilemaker.

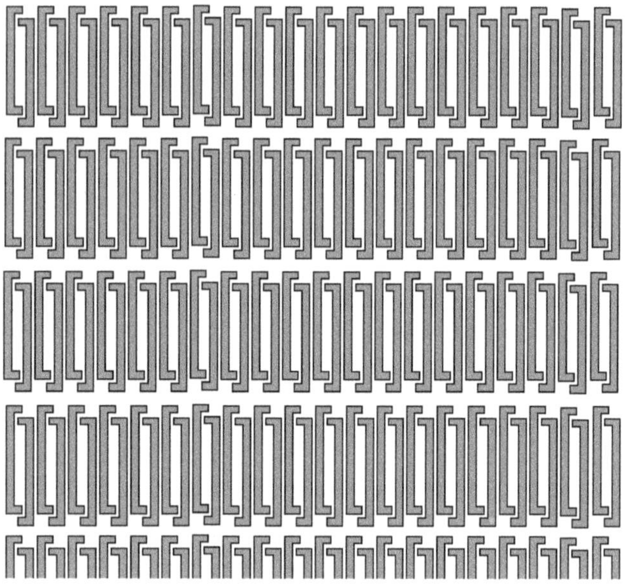

Five more rows

Figure 8.1: Plan view of a possible kiln load

The firing cycle of a kiln involves loading the kiln, sealing the superstructure, heating up the kiln, firing the tiles, cooling the kiln, dismantling the superstructure and unloading the tiles. On a modern tile kiln this whole process takes a little short of two weeks.[11] A Roman kiln superstructure might take longer to seal but could probably be removed earlier after firing which would allow the tiles to cool faster. Delaine suggests 60 hours firing time and three to five days cooling albeit for a much larger kiln, but notes that firing times are roughly proportional to the size of the kiln.[12] In contrast Soffe ventures that the firing and cooling cycle would take several weeks.[13] Balancing these conflicting estimates, for this illustrative analysis, a fortnight is adopted. This implies that two "typically" sized kilns would need to work in tandem to keep a single tilemaker fully employed just making tiles. Although there are some larger sites such as Minety, the majority of civilian sites have just a single kiln.[14] As the size of the kiln would have been a considered decision, this implies that at most commercial production sites it was the intention that the tilemaker should also be involved in other activities: most likely in loading and unloading the kiln.

The kiln load above equates to 168 square metres of tegulae (but slightly less for imbrices), all of which would need to be laid flat to dry for as much as a week prior to being turned onto their sides to finish hardening. If this were to be done under cover then it would require a building at least 10m by 15m in size if the tiles were set out as a single layer. In the last phase of development at Piddington,[15] circa AD 300, a large building (10m by 5m) had a stoke hole for a tile kiln inserted at one end which was fed from within the building. If this building was also used for tile making and tile drying then the heat from the kiln would have fed back into the building and caused the tiles to dry more quickly, perhaps halving the time. Arbury[16] and Crookhorn[17] kilns also have structures over or adjacent to the kiln and both of these sites were also probably in production in the fourth century, which could suggest that an integral drying area was by then the norm.[18]

If tile manufacture was not continuous (as suggested by the relationship with kiln sizes) and some sort of racking was installed, then it becomes possible that all of the production activities could have been contained within the one building.[19] The tilemaker would have spent the first ten days forming tiles when the kiln was operational and heat into the workshop therefore at its maximum. Imbrex formers would be less easily stacked than those of tegulae, so a large batch of these would be formed first whilst the workshop was still clear, such that they would be ready to be turned off their formers as the workshop filled up. During the latter part of the second week the tilemaker would unload the kiln and then reload with the previous ten day's production which by that time should have dried sufficiently to be fired. The tilemaker would also have been involved in moving the formers and stacking the tiles once they were sufficiently dry to be turned. As a result he would not have been able to match the 220 tegulae per day of a dedicated tradesman,[20] but this would have meant that his output was consistent with the capacity of the kiln. This strategy would maximise overall production for a given size of building, and explains why most of the kilns were single examples rather than pairs.

Once fired the tegulae, assuming they were not for

[10] In reality it is unlikely that a load would have consisted solely of tegulae.

[11] Pers comm: Peter Minter of Bulmer Brick and Tile Co Ltd.

[12] DeLaine 2001, 262-3.

[13] Soffe et al 1989, 99.

[14] McWhirr 1979, Table 6.1 lists the 36 then known tile kiln sites in Britain and 25 of these sites consisted of just a single kiln.

[15] Friendship-Taylor 2002, 8.

[16] McWhirr 1979, 175.

[17] Soffe et al 1989, 101.

[18] Green 1979 interprets a structure at Itchingfield as a tilemakers' workshop but this differs from Piddington in that it is not connected to the kiln and has an external drying area. The possible early second century drying shed at Hartfield was also not connected to the kiln (Rudling 1986, 227).

[19] If Plate 2.51 is indicative of the space consumed in drying imbrices then it is doubtful that the space would have been sufficient for these as well.

[20] It is evident that the graffito listing four workers who each produced 220 tegulae per day was akin to a tile-making factory where individuals would specialise rather than this typical Romano-British model. Section 8.7 considers the *Classis Britannica* operations which would have equated much more closely to factory production.

domestic use, would need to be despatched to their destination. In the absence of river or sea transport, carts would have to be used for which the load was limited by law to about 500kg.[21] This limit would mean that each load would comprise a maximum of eighty tegulae[22] and that some sixteen trips would be necessary to move a week's output of one tilemaker. However, if as suggested above, the tilemaker only made tiles every other week in order to keep pace with the kiln, then this would spread the sixteen trips over fourteen days or broadly one per day. If all the customers were less than 10km from the kiln then it might just be possible to do this with a single ox-cart, but it seems more likely that at least two would have been required. These could have been owned and run by the tilery or been independent contractors.

## 8.3 Costing commercial production

DeLaine has produced an estimated breakdown of labour costs for producing bricks principally based upon ratios derived from Diocletian's Price Edict and nineteenth century building surveyor's handbooks.[23] These handbooks predate mechanisation and may therefore be a valid proxy for the brick-making industry which may have changed little since Roman times. The costs, expressed in man-days per cubic metre of brick, are shown in Figure 8.2 below.[24]

| Activity | Man-days/m$^3$ |
|---|---|
| Quarrying clay | 0.22 |
| Loading and carrying clay 250m | 0.91 |
| Preparing clay and forming bricks | 2.15 |
| Carrying and loading kiln | 0.55 |
| Obtaining fuel for kiln[25] | 3.96 |
| Firing | 0.20 |
| Unloading | 0.12 |
| Total | 8.11 |

Figure 8.2: Initial cost estimate

DeLaine's figures are based upon a kiln capacity of 100m$^3$ which is almost ten times the typical Romano-British kiln size. Each time a kiln was fired, fuel was required to heat up the surrounding structure as well as the tiles themselves. Manifestly the more firings that were required for a given quantity of tiles, the less efficient the process and the more fuel that would be required. Using plausible assumptions, of the order of twice as much fuel would be required to fire the same amount of tiles in a 10m$^3$ kiln compared to a 100m$^3$ kiln.[26] Gathering this additional fuel would therefore add roughly 50% to the overall cost of production shown in Figure 8.2.

The data in Figure 8.2 is expressed in terms of man-days per cubic metre of fired clay produced. Using the data in Figure 6.17, one cubic metre of fired clay equates to approximately 200 Group A tegulae or 250 Group B tegulae. If the production rate graffiti referred to 220 Group A tegulae, which is a demanding but not impossible[27] rate, then the man-day figures in Figure 8.2 broadly equate to the number of people that would need to be employed to service one skilled craftsman dedicated to making the tegulae.

As the labour content in forming bricks is minimal one might expect that most of the labour attributed to this category in Figure 8.2 was spent in clay preparation and laying out the bricks to dry. It is therefore appropriate to add one skilled man-day to that value when considering the production of tegulae, but all the other figures should be unaltered. Amending Figure 8.2 for the smaller kiln size and pro-rating the quantities to be sufficient to produce 220 Group A tegulae per day, then the manpower required could be as in Figure 8.3 below.

| Activity | Men employed |
|---|---|
| Quarrying and moving clay | 1 |
| Preparing clay and moving tegulae | 3.5 |
| Forming tegulae | 1 |
| Obtaining fuel | 8.5 |
| Total | 14 |

Figure 8.3: Manpower costing for 220 tegulae per day

Wood gathering is best done in the winter months when the leaves are off the trees and, in Britain, tile-making is best done in the summer months when the weather is suitable for drying. Only five tiles have been found with graffito dates in Britain; one is for mid-July, three for August and one for 26th September[28] which supports the proposition of summer tile-making. It is therefore possible that instead of the fourteen men identified in the costing model in Figure 8.3 being employed in *seasonal* manufacture for four months, in fact there were just five men working *throughout* the year. For eight months they would be gathering fuel and preparing clay and for the four summer months they would be running the kiln and

[21] Boon 1984, 51.

[22] Brodribb 1987, 11 gives the average weight of the Beauport Park tegulae as 13.6lb.

[23] DeLaine 2001.

[24] Derived from DeLaine 2001, 262 Table 11.B3 adding in the labour equivalent for fuel and converting into unskilled man-days per cubic metre.

[25] DeLaine 2001, 234.

[26] If one assumes that the external dimensions of the 100m$^3$ kiln were 5m then the surface area exposed would be 125m$^2$. The smaller Romano-British kilns could have had an external dimension of 3m and thus an exposed surface area of 45m$^2$. It is arbitrarily assumed that the fuel required to fire a kiln is 50% dependent on the volume of tiles to be fired and 50% on the surface area that has to be heated and through which heat will be lost. Then the ratio of the heat required to fire 100m$^3$ of tile in a large kiln compared to the same quantity in ten firings of a small kiln will be (100m$^3$+125m$^2$) : 10x(10m$^3$ + 45m$^2$) = 225:550. On these assumptions, very roughly twice as much heat would be required to fire the same quantity of tile in a small 10m$^3$ kiln compared to a large 100m$^3$ kiln.

[27] Pers comm: Tony Minter of Bulmer Brick and Tile Co Ltd.

[28] RIB 2491.9-14. Boon 1974, 279 believed 26th September was the latest such date on any tile in the Western Empire. RIB lists one further graffito as a possible date (2491.15) but this consists simply of an "X" and a "D" so the interpretation as a date is highly speculative.

making the tiles. This would make more sense than fourteen men just working in the summer because what little agricultural work that was available in the winter months would presumably have been done by those already employed on farms on a permanent basis. In many tileries farm animals must have had access to the area of the drying sheds but this is not inconsistent with a landlord leasing a kiln to a tilemaker whilst running a farming business alongside and need not imply that the tilemakers were also engaged in agriculture.

The capital involved in tile production included the land on which the tilery was sited and from which the clay and fuel were taken. A building would be required for the tile making and drying, or possibly low standing covered "hacks" (Plate 2.50) were used for drying instead. The kiln would have been made from clay where the investment would have been mainly labour, likewise the arrangements for pugging the clay need not have been expensive. If the tilery also undertook delivery then oxen and carts would have represented a considerable investment together with a byre for the winter. Infrastructure in terms of a yard, roads and water supply may also have been significant. Moulds and palettes for forming and moving the tiles, picks and shovels for extracting and breaking up the clay and axes and saws for gathering the fuel would also have been required.

A study of brick-stamps in Italy has shown that the landowners (*domini*) and tile producers (*officinatores*) were separate and unconnected people.[29] Finley[30] has highlighted the papyri from Oxyrhynchus in which potteries were let on two-year leases. In these contracts the landlords undertook to provide raw materials and equipment and the tenants committed to produce a set number of jars for which they were paid a fixed price. It is unclear whether the jars (which numbered between 15,000 and 24,000 annually) were the total output of the pottery which the landlords intended to sell on themselves or whether they were just part of the output which was to be sold to them at a discounted price in lieu of rent. Either way, the evidence suggests that the landlord and tenant had a contractual relationship for which the landlord expected recompense. This payment could be very considerable, indeed fortunes could be made from it,[31] and in Italy (and perhaps Britain) it gradually fell into the hands of the emperor and the senatorial class.[32]

Applying this to the model of a Romano-British tile-works, one would expect the landowner to have supplied the buildings and perhaps the equipment that was used in tile-making. In the case of the manufactory at Minety in Wiltshire, all of the facilities might have been owned by a single landlord who divided them into parts and leased each part to a different manufacturer. The amount the landowner charged for these facilities is not known, but in view of the Italian evidence, it would not be an unreasonable starting point to assume that his charge for the facilities, the clay and the fuel was of the same order as the cost of converting it into finished product. In other words, half of the selling price of the tegulae went to recompense the labour and half to the landlord for the facilities and raw materials. Applying this assumption to the data in Figure 8.3 would produce a basic price for 220 tegulae equivalent to the daily wage of 28 labourers or 4.6 *denarii* per tegula.[33]

However the tile producer could not expect to always be able to fill his capacity with orders and it would be more realistic to price on the basis of, say, 80% utilisation. As the rent would probably have had to be paid independent of utilisation, and as his labour costs would have been fixed in the short term, it would be sensible to add 20% to the overall costs to produce a selling price of 5.5 *denarii* per tegula. This would also hopefully have left some profit for the producer.

Unfortunately the price of tegulae is illegible in Diocletian's Edict so the validity of the foregoing analysis cannot be directly tested. However the price for a *pedalis* survives which is given as four *denarii*.[34] Almost all of the production cost in Figure 8.3 is proportional to the amount of clay consumed. The amount of clay in a *pedalis* is 3632cc[35] whilst that in a Group A tegula is 5049cc[36] so the cost of a tegula should be broadly 40% more than that for a *pedalis* or 5.6 *denarii*. This is remarkably close to the 5.5 *denarii* produced by the costing model and, whilst the precision is accidental, it suggests that the overall result is correct, albeit it may have been produced by some of the larger assumptions having off-setting errors.[37] Moreover it supports the internal consistency of the prices within the Edict because the tegula cost build up only relied on the labour rate from the Edict whilst the tegula price only used the *pedalis* price from the Edict.

## 8.4 Area of supply

To understand the overall economics of the tile-making industry, an estimate of the maximum distribution distance and therefore the total potential area of supply is required. The best evidence for this comes from the distribution of tile stamps, particularly military ones, but

[29] Helen 1975, 108.
[30] Finley 1985, 190-191.
[31] Helen 1975, 109.
[32] Finley 1985, 190.

[33] Daily pay of a labourer in Diocletian's Price Edict was 36 *denarii* (DeLaine 2001, 234).
[34] Reynolds 1989, 280: Price Edict Chapter XV, Column 3, Line 90.
[35] Using the dimensions from Brodribb 1987, 142.
[36] Figure 6.17.
[37] Arguably the comparison should be performed between *pedales* and Group D tegulae which would be contemporaneous with Diocletian, but as it was the quantity of clay that dominated the calculation, the cost per cc of Group A and D tegula are comparable: thus the cost build up for a Group D tegula would produce a price of 3.3 denarii and that calculated from a *pedalis* a price of 3.4 denarii per Group D tegula.

this section also examines other potential sources to help determine the answer.

Generally the legions did not establish new tileries away from their bases except when the whole unit, or a significant vexillation, was on detachment such as *Legio* II and *Legio* XX at Carlisle, or in the case of *Legio* VI possibly at Corbridge and Carpow. As a result the distribution of stamped tegulae around these bases gives a good indication of the distances that the military were prepared to transport tile. Ignoring dubious outliers, all of the legions have stamped tiles found at least 60km as the crow flies from their kiln sites and for *Legio* XX this rises to 100km.[38]

A similar picture emerges with the Gloucestershire private and municipal tile stamps which have also been found up to 80km from their assumed source.[39] It is possible that some common distribution system operated for the Gloucestershire and Minety tileries because both "ARVERI"[40] and "TCM"[41] tiles have been found in Warwickshire 80km away from their main distribution in the Cirencester area whilst "LHS"[42] and the municipal "RPG"[43] tiles have been found in the same town in Herefordshire 80km away from their main Gloucester/Cirencester distribution.

It was not just the Gloucestershire tilemakers whose tiles or stamps have travelled: "MCV"[44] stamps in Dorset and Somerset have been found 35km apart and the "IVCDIGNI"[45] stamps in Wiltshire and Berkshire have been found 70km apart. The location of the kilns are not known for these tiles so it is possible that these distances represent the extremes with the production centre in the middle.

Darvill has suggested that the "TCM" dies,[46] which have been found in clusters in Gloucestershire and Warwickshire up to 80 km apart, were not the work of a single factory but of an itinerant tile producer. He studied nineteen examples of these stamps, split between three dies, and identified six different fabrics which closely correlated with the find spots but cut across the dies.[47] Although the match was not always perfect he concluded that these must be the work of a peripatetic producer. He also identified that one of the "TPF" dies[48] appeared in two different fabrics, the second of which he was only able to identify at Hucclecote[49] where the local clays were not dissimilar. He therefore argued that the stamp must have been moved to Hucclecote rather than originating from Minety where a different fabric prevailed. Whilst this might be true, his argument was somewhat undermined by the existence of another "TPF" tile at Hucclecote that was in his Minety fabric. It may therefore be another example of the difficulty of drawing conclusions from apparent differences in fabrics.

Betts & Foot[50] have argued that calcareous tile found at sites accessible from the south coast from London as far as Exeter may have all been supplied by sea from a single tilery or group of tileries. They have dated the tiles between the mid-late second century and the end of the third century.[51] The argument is based upon the similarity of the fabric and form of the tegulae. In respect of the form they note, *inter alia*, that most of these tegulae have only two types of cutaway, that the flanges tend to have a rounded top surface, there tend to be no finger grooves running down the side of the flanges and that their sizes are relatively uniform.[52] Their two cutaway forms were Group B and C and these results are what might be expected of any small group of such Group B and C tegulae based upon the typology developed in Chapter 3. The sizes they quote are slightly smaller than the overall average in the survey but urban and rural tegulae tend to be slightly smaller than average, as does production from the south of the country.[53] Calcareous fabric is encountered throughout the country but tends to be associated with third century sites. Thus, whilst their hypothesis may be correct, it also closely matches the national typology for the period and their observations may simply be a microcosm of this and not represent any distinct grouping or distribution system.

Betts has studied the probable kiln sites supplying London[54] and observed that the city seems to have received supplies from all the main directions, with the exception of the west. These were typically some 25km distant, but in the third century some product was coming from Harrold in Bedfordshire which was 84km from London.

Castleford makes an interesting case study. The fort was established circa AD 70 and a vicus sprang up at much the same time. The fort was abandoned around AD 100 but the vicus continued, ultimately absorbing the fort in the later third century.[55] Bricks stamped by *Legio* IX

[38] Maximum distances as the crow flies for *Legio* II: Caerleon to Loughor is 80km; *Legio* VI: York to Catterick is 60km; *Legio* XX: Holt to Caernarvon is 100km; *Legio* IX has stamps at Scalesceugh in Cumbria which is almost 200km, but there must be doubt as to whether the Cumbrian *Legio* IX tiles originated in York.

[39] RIB 2489.4 (ARVERI): Minety to Radford Semele (on the Fosse Way just outside Leamington Spa) 80km; RIB 2489.21 (LHS): Minety to Kenchester (Herefordshire) 80km, Minety to Silchester 70km.

[40] RIB 2489.4A (xxxiii).

[41] RIB 2489.40A (i, xviii and xix).

[42] RIB 2489.21H.

[43] RIB 2486.5 (iv-xvi).

[44] RIB 2489.28: Stoke-sub-Hamdon to Dorchester.

[45] RIB 2489.18: Studley to Silchester.

[46] RIB 2489.40.

[47] Darvill 1980, 52.

[48] RIB 2489.44F.

[49] Darvill 1979, 319.

[50] Betts & Foot 1994.

[51] Betts & Foot 1994, 32.

[52] Betts & Foot 1994, 23 mentions lengths of 354-360mm and 383mm and lower breadths of 258-265mm.

[53] See Section 8.9.

[54] Betts 1987, 28.

[55] Abramson et al 1999, 19-20.

were found in the military bathhouse, a fragment of tegula stamped by *Cohors IIII Breucorum* was found in the vicus and two fragments of tegulae stamped by *Cohors IIII Gallorum* have been found in the general area of the fort and vicus. Betts believed that the *Legio* IX bricks originated at York and were delivered to Castleford by river transport.[56] The *Cohors IIII Gallorum* tegulae have already been considered in Section 4.8 which concluded that they originated from a tilery that was serving both Castleford and Templebrough (some 35 km distant as the crow flies).

However the most interesting tile is the one from *Cohors IIII Breucorum* found in the vicus which must surely have come from that unit's tilery at Grimescar, some 40km from Castleford. Betts confirmed that the fabrics were the same.[57] *Cohors IIII Breucorum* was based at Slack which was established at broadly the same time as Castleford and the fort was renovated circa AD 104[58] when the stamped tegulae were first used. The fort went out of use circa AD 125.[59] With the exception of a single stamped tegula found at Castleshaw[60] (about 10km west of Slack) all the other finds of *Cohors IIII Breucorum* tile have been at Slack and Grimescar. Although possible, it seems unlikely that *Cohors IIII Breucorum* were originally based at Castleford before moving to Slack and, even if they were, it would still not explain why they should have sent tile back to Castleford unless vexillations of the unit were responsible for both forts. Moreover the Castleford fragment came from the vicus rather than the fort although, unfortunately, it was unstratified.

The most likely explanation would seem to be that tiles were gathered from the derelict site at Slack, perhaps in the third century, for reuse in Castleford which was some 40km distant. This implies that distance was not an obstacle even when it came to the acquisition of secondary material.[61]

The movement of relief-patterned (normally) box-flue tiles may also provide a window into tegula distribution because most of the tileries that have produced relief-patterned tiles have normally produced a full range of bricks and tiles including tegulae.[62] The distinctive nature of the relief patterns allows their distribution to be plotted with considerable confidence and Betts, Black & Gower have produced distribution maps for each of the main dies.[63] Six of their dies[64] had distributions with extremes more than 100km apart and in the case of Die 9 they were 250km apart.[65] They suggest that one possible explanation for the wide distribution is that some dies may have had two distribution centres. For example, in the case of Dies 1 and 4 they suggest a primary centre in Surrey and a secondary one near St Albans.[66] This seems to be what must have happened with Die 32 which is found in a different fabric at Sandy, Bedfordshire than at the other sites.[67] However, although *some* dies may have been moved,[68] the fabric analysis supports a wide primary distribution with the distinctive Fabrics 1 and 7 both having a range in excess of 100km.[69]

All this evidence suggests that tegulae were regularly transported up to 40km and occasionally as far as 80km. Box-flue tiles being a more specialist product may have travelled further. Manifestly the distance that these products were able to travel was affected by the transport opportunities available: the *Classis Britannica* tiles, which also have a wide distribution, were presumably transported by sea where distance became almost immaterial. Other manufacturers may have used river routes, but it seems clear that some of these distances were travelled by road with ox-carts.

## 8.5 The economics of tile-making

To understand the economics of the tile-making industry it is necessary to understand the market for the products and how these were supplied. Darvill & McWhirr make the distinction between "light-sector" pottery producers who needed middlemen to break their output down into smaller quantities to fit the multiplicity of individual customers' requirements and "heavy-sector" tile producers who would have been supplying large batches direct to the end customer in response to specific orders.[70]

Tilemakers may have limited themselves to simply manufacturing the product or they may have also delivered to site. Some may have gone further and involved themselves in the design and roof erection process. In the case of roller-stamped box-flue tiles and bathhouses it is evident that there was a degree of specialisation and that some manufacturers were prepared to travel and produce their product locally. The strong inference from this must have been that they were involved in the design and erection of these bathhouses and indeed at Shakenoak in Oxfordshire the roller stamp is not just found on the box-flue tiles but also on bricks.[71]

56 Betts 1998, 232.
57 Betts 1998, 226.
58 Richmond 1925, 47.
59 Richmond 1925, 46-47.
60 RIB 2470.1 (i).
61 Cunliffe 1971, 187 notes that tegulae from Fishbourne were recovered and carted away following the third century fire that ended its life although it is not known where they were taken.
62 Betts, Black & Gower 1997, 52.
63 Betts, Black & Gower 1997, 26-39.
64 Dies 1,4, 5A, 9, 12 and 32.

65 The distribution of Mancetter mortaria, which were also a heavy product, was even more extensive (Fulford 1989, 184 Fig 2).
66 Betts, Black & Gower 1997, 29.
67 Betts, Black & Gower 1997, 33.
68 The organisation may have been more complex because at Beddingham in Sussex Die 5A is found in two different fabrics (Betts, Black & Gower 1997, 52).
69 Betts, Black & Gower 1997, 20 and 24.
70 Darvill & McWhirr 1984, 240-241.
71 Brodribb, Hands & Walker 1968, 39.

The limited amount of military box-flue production has already been noted and the possibility that the military subcontracted some technical aspects of bathhouse design and construction should also be considered. It appears that from the second century onwards the military increasingly used subcontractors and this is perhaps exemplified at Castleford and Templebrough where a single supplier appears to have been producing tiles both for the military fort at Templebrough and the civilian vicus at Castleford.[72]

However the roles may have initially been reversed. At Silchester the existence of tiles with a Neronian imperial stamp[73] and tally marks on twenty-five tiles from the Flavian forum-basilica[74] (but none from later periods) which were a strong indicator of military production,[75] suggest that there was an official involvement in the romanisation of the town in the first century. A similar position may have existed with the formation of Caerwent where the Group A tegulae, although unstamped, are similar to Group A tegulae from Caerleon with Boon's first group of Ai stamps.[76] An aborted foundation trench for Fishbourne palace[77] had tegulae with diagonal upper cutaways, which are almost only ever found on military sites, that had the same overall dimensions as tegulae of broadly the same date from the legionary fortress baths at Exeter. Frere[78] noted similar evidence of military involvement at Verulamium although this is disputed by Millett.[79]

Transport costs would have restricted the potential size of the market but there would have been a trade-off between the inefficiency of having lots of inadequately utilised tile factories that were close to their markets and fewer fully utilised producers but with higher transport costs. This trade off would have changed through time: in the building boom of the second century every town would have needed at least one tilery and maybe more, but by the fourth century there may have been insufficient demand to keep so many of producers busy, such that their number would have shrunk and their areas of supply increased.

Jones & Mattingly[80] assert that large-scale tile manufacture was associated with civitas capitals and that rural production was smaller scale, but the only specific evidence they produce is the proximity of the Minety kilns to Cirencester and their distribution map[81] is (unavoidably) inconclusive. In contrast Peacock[82] says that the majority of known kilns were more than 20km from the nearest civitas capital and were not therefore linked with the towns but were probably estate production. The truth probably lies in between: civitas capitals originally had local suppliers, but as demand waned and raw materials became scarcer, production moved out to the countryside. Many towns do have associated kilns[83] whilst others do not, or they have not yet been found.[84]

In a theoretical economic model the price of tegulae would be determined by the amount a local producer needed to cover his production costs and pay for the lease to the landowner. A distant producer who had surplus capacity would be prepared to bear some additional transport costs if he could secure orders in a neighbouring market in order to make some money to put towards the rent. In effect, particularly if his labour costs were fixed in the short term, then any price would be better than leaving the kiln idle, so long as it was greater than the cost of transport. In principle he would be prepared to travel further and further to secure orders until at the extreme his transport costs approached the local selling price in the most distant market.

The point at which transport costs equate to local production costs therefore defines the maximum area of supply for any tilery. Of course one would not expect many tiles to have travelled this far because it would only make business sense to cope with occasional shortages of orders. To do it on a regular basis would bankrupt the business because insufficient money would be earned to pay the labour. This is mirrored in the "frequency fall off" curves produced by Darvill & McWhirr[85] which show that the number of stamped finds diminishes markedly with the distance from the source.

Diocletian's Price Edict allows an estimate of transport costs as a ratio of a labourer's daily pay rate to be made as shown in Figure 8.4 overleaf.[86]

[72] See Section 4.8.
[73] RIB 2482.2-5.
[74] Timby 2000, 116.
[75] Section 5.19.
[76] However Greene 1977, 125 notes that the early pottery from Caerwent differed from that made at Caerleon.
[77] Pers comm: David Rudkin.
[78] Frere 1999, 196.
[79] Millett 1990, 69-72.
[80] Jones & Mattingly 1990, 217.
[81] Jones & Mattingly 1990, Map 6.36.
[82] Peacock 1979, 8.
[83] For example: Cirencester, Colchester and St Albans.
[84] For example: Caerwent, Leicester and Wroxeter. Using a crude analysis based on half the land area of *Britannia* (65,000 sq km) being suitable for tile-making and that a tilery had a normal supply range of 15km, but that two kilns would have been required in each area to span the whole of the Roman period, yields a total requirement for 186 kiln sites. McWhirr 1979, Table 6.1 list 36 sites which would suggest that 20% of the total have so far been discovered.
[85] Darvill & McWhirr 1984, 250 and 253.
[86] Taken from DeLaine 2001, 234. Betts 1998, 232 quotes broadly similar relativities. However Laurence 1999, 98-99 advises caution in using the Price Edict data because loads and forms of transport may not be directly comparable.

| Mode of transport | Man-days per tonne-mile |
|---|---|
| Ox-cart | 1.44 |
| River upstream | 0.26 |
| River downstream | 0.13 |
| Sea | 0.035 |

Figure 8.4: Cost of transport

There were roughly 165 tegulae to the tonne,[87] so the cost of transporting one tegula one mile by ox-cart was 0.3 *denarii.*[88] The price of tegulae was calculated as 5.5 *denarii* in Section 8.3, therefore the theoretical maximum distance of supply by road would be 18 miles (just under 30km); or a range of up to 60km with the kiln at its centre. However the local producer would also have incurred some transport costs. If it is assumed that in addition to the 5.5 *denarii*, his pricing allowed for distribution up to 15km, equivalent to a day's journey for an ox-cart, then the overall theoretical range becomes 75km. This seems consistent with the 40-80km actual distribution observed in Section 8.4 and provides a third "verification" of the Edict data, this time relating transport costs to both the pedalis price and labour rates.[89]

If the tile producer also owned the oxen then transport would have become a fixed cost because the oxen would have had to be fed even if there were no tiles for them to deliver. As a result it would make economic sense to deliver even further afield in order to keep the kiln and the oxen busy.

Of course there may have been times when there was no demand for tiles even within this extended area, and whilst the tilemaker could then have stock-piled production in the hope of future orders, this could only continue for a limited period before all the spare space or cash was consumed. The tilery would then have had to be moth-balled, the staff laid off and probably the contract with the land owner terminated. In a modern economy demand for most products tends to be continuous, especially as governments strive to avoid economic depressions, but in the less well managed Roman economy where the number of houses requiring tiled roofs will have been limited, one might anticipate that such tilery closures would not have been uncommon. The archaeological evidence for kilns being restarted after a period of disuse may be difficult to detect although there are many reports of kilns which supposedly only operated for a short period.[90]

Two alternative explanations for the wide distribution of tile stamps should also be considered. The first is that if two-year contracts, as at Oxyrhynchus, were in any way typical then the displacement of incumbent producers must have been a regular occurrence. It is therefore possible that the wide distribution of some tegula stamps, particularly that of some of the Gloucestershire tilemakers, was the result of relocation. The second possibility is that the landlord took payment in tiles which he used on buildings in a distant estate that he also owned. Economic considerations would not necessarily have applied to the transport of such tiles.

Superficially, the high cost of transport would make it attractive to produce tiles on site using clamp firing, especially where the demand was episodic and there was no local supplier. However, whilst clamp firing may have been an easy alternative to investing in kiln construction, the investment in moulds and palettes and the extraction and preparation of the clay would have meant that this was a task that could not be lightly undertaken. More significantly, clamps would have had a much smaller capacity and a much greater heat loss than a conventional kiln. They would have required many more firings to achieve the same output and as a result fuel consumption would have been greater, perhaps many times greater. The differential would have been compounded by a higher reject rate which would have further expanded the penalty. Figure 8.3 showed that fuel was over half the production cost of production, so one might anticipate that clamp production would be the exception rather than the rule.

The dominance of transport costs in all aspects of the process from the excavation of clay to the procurement of fuel and delivery of the finished product, suggests that the cost of production would have been dominated by the selection of the optimal geographic location for the tile works. The size of the kiln would also have had a significant impact upon costs and therefore efficiency, as was shown in Section 8.3, but not to the extent that it would negate the benefits of transportation. This suggests that unless river or sea transport was available, for example as at Holt, tileries are likely to have serviced relatively limited geographical markets in order to limit the impact of transport costs. However, by the end of the second century there is growing evidence to suggest that the availability of fuel was becoming a dominant factor to the extent that towns, such as Silchester, were forced to resort to the use of stone despite the abundance of local clay.[91]

[87] Calculated from Brodribb 1983, 26.

[88] Daily pay of a labourer in Diocletian's Price Edict was 36 *denarii* (DeLaine 2001, 234).

[89] Independent support for the Edict transport relativities can be deduced from Fulford & Hodder 1975, Figure 3 which shows two frequency fall off lines for the distribution of Oxford pottery: the first is for pottery that travelled by river transport and the second for road transport. Projecting these lines shows that distribution by river would have extended 6.5 times as far as by ox-cart. Assuming that the potters were prepared to spend the same amount on transport independent of the method then this implies than ox-carts were 6.5 times as expensive as river transport. This is very close to 7.4 multiple of ox-cart transport to the average cost of river transport in Figure 8.4 above.

[90] For example Tarbock: Cowell & Philpott 2000.

[91] Fulford 1989, 188.

## 8.6 Competition

When the potential market size was large it was easier for new suppliers to enter the market because they could be successful by taking a relatively small amount of sales from a rather larger number of competitors. However if the potential market size was small, in effect the supply area was limited by transport costs, then competition might be more difficult. A new entrant would have had a very direct impact upon the one or two existing suppliers to that limited marketplace, who would therefore have had to drop their prices to retain business. However, because price reductions were unlikely in the short term to increase demand for tiles, the only consequence would have been that the viability of existing players and the new entrant would have been significantly compromised. So whilst excess competition will surely have occurred, it seems unlikely to have been a major feature of the marketplace.

Instead of trying to compete on price, which might have been self-defeating, a more effective way would have been in terms of quality and branding. Private tile stamps may therefore have been brand marking whereby competing manufacturers sought to increase their sales by associating their brand with quality tiles. Diocletian's Edict refers to "*formae primae*"[92]in relation to tegulae implying that different grades of quality for the product were recognised. However, whilst it is true that there were few instances of poorly made stamped tile, it is not evident that stamped tiles generally were superior to the majority of unstamped tiles (but then the same could be said by an objective observer about branded products today).

Plate 8.1: Stamp with *tria nomina* from Piddington

In Italy, tile stamps may well have been an acknowledgement of the legal contract between the *domini* and *officinatores.*[93] This was clearly not the case in Britain where almost all private stamps took the form of initials or a single name which is assumed to identify the producer or tilery owner. Piddington was exceptional in having a stamp with a complete *tria nomina*[94] (Plate 8.1). Gloucestershire has yielded one set of stamps with sequential initials: "TPFA", "TPFB", "TPFC" and "TPFP". It is difficult to see why a tile producer would wish to identify his product in this manner (unless the initials denoted different quality grades), but much easier to understand if "TPF" was the landlord's initials[95] and "A", "B", "C" and "P" were used to denote successive tenants of his tilery at Minety.

It may well be no coincidence that the Gloucestershire/Wiltshire area was the home of the best known manufactory site and that it was is also the area where private tile stamps were most prevalent. There must have been several independent producers all with similar production costs at Minety, all making tiles to sell in the same market area. It is inevitable that there would have been competition between them and the best way to do this was through branding rather than destructive price competition.

Competition can be studied in individual buildings: where supplies appear to have been taken from a single manufacturing source this might suggest little competition, whilst multiple sources would suggest that there was a competitive supply market. Unfortunately relatively few examples of any stamp are found on civilian sites such that sites with more than one type of stamp are inevitably rare, although Greyhound Yard in Dorchester[96] had tile stamps from two different makers. An exception is Hucclecote villa in Gloucestershire which has produced an extraordinary proliferation of stamps. There were seven different dies from the municipal tile works and four different private tilemakers' stamps (assuming that different sets of initials represented different tilemakers), two of which were represented by three different dies.[97] Two of the reasonably prolific Gloucestershire stamps were not present. This would seem to suggest a competitive market with tiles initially being bought from the municipal tilery and subsequent supplies from the private tilemakers probably based at Minety in Wiltshire. This position was mirrored in Italy where buildings normally yield stamps from several different producers.[98]

Stamping of tiles dies out to a significant extent in the third century, however this trend is not just confined to tiles, but occurs in all manufactured goods.[99] Crookhorn has yielded two examples of a second century stamp[100] but none from the fourth century tilery. This could be an indication of a tilery choosing to stop stamping tiles, but the archaeological evidence suggests that the tile production on the site did not begin until the third/fourth century,[101] so the stamped tiles may simply be imports from another tile-works.

[92] Reynolds 1989, 280; Price Edict Chapter XV, Column 3, Line 88a.
[93] DeLaine 2000, 134.
[94] Tiberius Claudius Severus.

[95] Or perhaps the "TPF" could be read as *Tegula Prima Facto est* by A, by B etc.
[96] Bellamy 1993, 173.
[97] Analysis from RIB II.5: the dies were 2486.2, 14, 15; 2487.4, 5, 18; 2488.1, 4; 2489.40A; 2489.44C, E, F; 2489.45B, D, E; 2489.48C.
[98] Helen 1975, 20.
[99] Fulford 1994, 317.
[100] Crookhorn 2178: RIB 2489.42.
[101] Soffe et al 1989, 108.

## 8.7 *Classis Britannica* tilery structure

So far this chapter has largely dealt with commercial tile manufacture. Military tileries seem to have been larger and the way they operated may well have been different. Many military dies were used on all types of tile but there were some dies, represented by statistically significant numbers of examples, that were only found on tegulae and others only on imbrices (there were no similar instances exclusively on brick). This was particularly the case with the *Classis Britannica* stamps from Beauport Park where separate dies were used for tegulae and imbrices. As both tegulae and imbrices must have been mounted on the roof at the same time, it would not be credible to assume that the imbrices were all made at a different period to the tegulae. It therefore seems probable that either a different set of imbrex dies existed in parallel with those for tegulae, or some process caused different dies to be applied to the imbrices.

*Classis Britannica* stamped tile appears to have been introduced circa AD 120[102] and, assuming consistency with most other military production, probably ceased circa AD 260. To date 119 different dies have been found[103] but 48 of these were singleton examples, suggesting that there was a considerable tail of further dies that have not so far been recovered.[104] So, ignoring the nine French dies, there must still have been the equivalent of at least one British *Classis Britannica* die for each of the 140 years of stamped tile production (although this need not imply that they were issued annually).

Figure 5.19 showed that, although there were two dies being produced across the Group B to C cutaway transition, they were being used by two different groups of tilemakers. Moreover, although at Beauport Park the top six dies were almost exclusively found on either tegulae or imbrices, this was not always the case at Dover. For example the prolific RIB 2481.7 was almost exclusively found on tegulae at Beauport Park whilst at Dover it was almost always on brick;[105] RIB 2481.92 was similar.

A possible explanation for this could be that the *Classis Britannica* tilemakers worked at two separate benches, each probably consisting of three men consistent with the three signatures on RIB 2481.102 (Figure 5.19). Each bench had its own stamp which the men shared and applied to the tile themselves.[106] At any one time, one bench was given over to making one sort of tile (for example tegulae) whilst the other was making a complementary product (for example imbrices). Every week the work was rotated such that the first group was then put on imbrices and the second perhaps on floor tiles, but in the process of rotation each group held onto their stamp. The graffito that named four workers each of whom produced 220 tegulae[107] could be an example of just such a bench arrangement.

Brodribb[108] has calculated that 1400 tegulae would have been used on the Beauport Park roof. This quantity could easily have been produced by the three men on one bench inside a week. One might therefore reasonably expect that all the tegulae for one phase of the building would have the same stamp, but would carry the signatures of the three tilemakers on that bench. The imbrices would need to be made at the same time and would therefore have been made on another bench and used a different stamp.[109] When the work was rotated the first bench could have been producing bricks for Dover but still using their same stamp.

There may also have been a third bench where the apprentices worked. The apprentices did not have a stamp and were assigned the simplest of tasks which would be making the small flat *bessales*: none of these were stamped at Beauport Park.[110]

All this suggests there were two dies in use concurrently which would both need to be changed every other year to match the number of different dies that have been found. Although it is possible that they simply wore out, there may have been some administrative reason for the change. For example if a new factory overseer (*vilicus*) were appointed it is quite possible that new stamps would be issued such that the new official's output could be identified and he be held to account for any subsequent quality problems.

The existence of two different dies on tegulae that both spanned the Group B to C cutaway transition is now explained by the model because these tiles were presumably made after the benches had been rotated at the end of the week.[111] Moreover the decision to switch

[102] Williams 1981, 127.

[103] RIB 2481.

[104] At Beauport Park there were probably around 44 dies that were not recovered (see Chapter 5 footnote 105), but some of these could have been dies that have already been found at other sites.

[105] Williams 1981,124 categorised fragments, inter alia, into tegulae, imbrices and "flat". It is possible that the "flat" category included some tegulae, but even if a significant number of the 256 "flat" fragments were actually tegulae (compared to the 24 fragments identified as tegulae), it would still not compare with Beauport Park.

[106] It was suggested in Section 2.3(d) that at Beauport Park the tilemakers applied the stamp to the tile rather than the stamp being applied by an official at a later point as seems to have occurred on most other sites.

[107] CIL III, 11381.

[108] Brodribb 1979, 142.

[109] Imbrex length was strongly correlated with stamp dies (Figure 6.9) which supports this model, although other explanations are also possible.

[110] Brodribb 1987, 124.

[111] If the benches regularly rotated then one might expect the same stamp to appear on both tegulae and imbrices rather more frequently than it does at Beauport Park. However, as all the tegulae for the roof could have been made in less than a week, and as presumably the imbrices were made at the same time but on the other bench, then the

from Group B to Group C cutaways could not be implemented instantaneously, but was no doubt done over a period of months or years as each of the hundreds of moulds or palettes came round for renewal or maintenance. As a result both Group B and C tegulae would have been produced throughout this period.[112]

Superficially this model appears to address all the facts but there is a problem with the die RIB 2481.7. This die was found (on brick) in a stratified context at the Dover forts which was unlikely to date much later than AD 210.[113] At Beauport Park it appears on tegulae with Group C and D cutaways, so either the Dover dating needs to be stretched by at least twenty years or this die had a much longer life and so was not used in the manner described above.

All of the tegulae with RIB 2481.7 had tally marks and all of these, with one exception, fell into two forms. Each tally mark form is associated with its own group of signatures so it is tempting to suggest that the tally mark therefore identified the bench which made the tile. It is unclear why this would be necessary if the stamp was also being used to identify the bench as proposed above. However, given the difficulty with Dover and the fact that there were arguably four different tilemakers using this stamp,[114] it is possible that the practice may have changed in some subtle way in the third century.

## 8.8 *Classis Britannica* tilery output

On the basis of two work benches each with three men forming tiles, Figure 8.3 would suggest that there could have been 30 men employed at the tile-works and a further 50 engaged in gathering and delivering fuel. However, as discussed in Section 8.3, wood gathering is best done during the winter period and, in Britain, tile-making is best done in the (shorter) summer period, so it is possible that only 30 men were involved alternating between the two tasks with the seasons. Although it is often assumed that tile-making alternated with agriculture, it seems more likely, at least in a military context, that it was an all-year round activity.

In the four summer months, working an average of six days a week, such a team could have produced 135,000 tegulae if working on these exclusively. This compares with the 1400 tegulae that Brodribb estimated were on the bathhouse roof[115]. However, the same tile-works was also producing imbrices, bricks and box-flue tiles. Using Brodribb's average sizes,[116] 3.0 *bipedales* would have been required per square metre of floor area but 11.5 tegulae[117] would have been required per square metre of floor space covered if the roof pitch was 30°. Each *bipedales* would have had its corners resting on a *pila* comprised of two *pedales* and, say, ten *bessales.*[118] In principle there would have been the same number of *pilae* as *bipedales* except for the edges of the room. For a typical room this would have meant that there would have been 15% more *pilae* than *bipedales.*[119] This enables the number of other tiles relative to the tegulae on the roof, and the clay that these would have required, to be calculated as shown in Figure 8.5 below.

| | Volume (cc) | Number/ square metre | Number per tegula | Clay used relative to tegula |
|---|---|---|---|---|
| Tegula | 3068[120] | 11.5 | 1 | 1 |
| *Bipedalis* | 19976 | 3.0 | 0.26 | 1.70 |
| *Pedalis* | 3632 | 6.9 | 0.60 | 0.71 |
| *Bessalis* | 1685 | 34.5 | 3.00 | 1.65 |

Figure 8.5: Ratio of clay used in bathhouse

It can be seen that for each tegula used on the roof the equivalent of roughly four tegulae would have been required in the hypocaust beneath.[121] Estimates of the number of box-flue tiles would depend upon the height of the walls and the extent to which these walls were lined. If all the walls were fully lined then there would have been many more box-flue tiles than tegulae, but at Beauport Park this was not the case and a reasonable guess could be that the number of each was similar. Box-flue tiles would take more time to manufacture than tegulae but the clay used, which was the most important determinant of cost, would have been broadly similar.

Roughly the same number of imbrices and ridge tiles combined would be required as tegulae. There would also have been an unknown number of wall tiles which will be assumed to equate to the number of tegulae. Adding all of these tiles together produces a total requirement that equates to eight times the number of tegulae used on the roof or 11,200 tegula equivalents (which would have cost some 60,000 denarii if procured commercially). This represents less that 10% of the estimated seasonal capacity of the tile-works. It is impossible to know what the total *Classis Britannica* demand for tiles in Britain was, but the nine known establishments in the south-east were all supplied from the same source.[122] Most of these sites were considerably

same stamp would only be expected on tegulae and imbrices to the extent that the production happened to fall across the bench changeover or if supplemental orders were placed on the tilery.

112 How roofs could accommodate two different cutaway forms is discussed in Section 6.9.

113 Williams 1981, 126.

114 Figure 5.19 suggests that it was only two by relying on the number of curved forms in the signature but it is possible that the variations in the shape of these forms represent further tilemakers.

115 Brodribb 1979, 142.

116 Brodribb 1987, 142.

117 Floor area/Cos30.

118 Great Witcombe pilae have nine bessales and Bath ten.

119 Taking typical bathhouse room sizes as per Section 6.4 would suggest 16x12 bipedales were required which would rest upon 17x13 pilae.

120 Group C per Figure 6.17.

121 This might be a slight overstatement because the roof covers the wall area as well as the floors.

122 Peacock 1977.

larger than the Beauport Park bathhouse, but even with all of them being regularly renovated, it would seem that the postulated tile-works would have had more than ample capacity to have dealt with the demand.

This volume of production would have required a considerable drying area. A week's output of tegulae or equivalent tiles would require 1000m$^2$ to lay them out all abutting each other and perhaps half as much again if they were to be accessible. This is equivalent to a building of 40m by 40m, which seems improbable, so hacks or other arrangements must have been deployed. Animal impressions are virtually absent from the Beauport Park tiles suggesting that they were excluded from the drying area. Tegulae with die RIB 2481.102 all had stepdowns but none occurred on the other dies. This suggests that the bench using this die either had special racking for drying their tegulae or they laid them out in an overlapping fashion resulting in "skink" marks.

## 8.9 Drivers for change

The seriation data and other evidence show that the same sequence of cutaway forms was followed throughout the province and it is possible that the transitions between these cutaway forms all took place at broadly the same times. Certainly the data for the transition from Group A to Group B cutaways show the change should not have taken more than the twenty years AD 100-120. Indeed the data are entirely consistent with the switch happening in a single year AD 110 but such a rapid transition is exceedingly improbable. The evidence for the range of subsequent transition dates is less clear cut, although it is manifest that the sequence is followed throughout the country on both military and civilian production and, on very limited evidence, also appears to have been adopted in the rest of the Roman Empire. To produce this degree of uniformity there must either have been some central diktat or very powerful engineering, commercial or cultural forces driving the change.

The reduction in the average size of tegulae was also extremely uniform through the Roman period with almost all of the dimensions moving in proportion. The Prices Edict shows that tegulae were priced by the unit rather than by the area they covered. In such circumstances, given that the cost of producing tegulae was almost directly proportional to their size, it is unlikely that a tilemaker would ever knowingly use an oversize mould, but if the mould was fractionally smaller than the prevailing standard then it may have been left uncorrected. Thus, even without any deliberate deception, it is likely that the size of commercially produced tegulae would reduce steadily through time and perfectly reasonable that all the dimensions of the tegulae might reduce in proportion through the same process.

The one dimension that reduced more rapidly than the others was the length of the upper cutaway. This length determined the amount by which the tegulae overlapped and thus the number of tegulae required to cover the roof. The reduction in upper cutaway length may therefore have been introduced as an explicit cost saving measure.

These arguments go some way to explaining why commercially produced tiles might have steadily reduced in size, but they are inadequate as an explanation for the similar change in military production. Figure 8.6 shows the overall length of tegulae in the survey segmented by the nature of the site. It can be seen that military sites produce the largest tiles and rural (mainly villa) sites have the smallest tegulae, although the analysis is not perfect. Chapter 3 showed that the flange measurements mirrored those of the overall tile but had the advantage of a much larger data set. Following this reasoning, Figure 8.7 shows the lower flange height segmented by type of site and, as can be seen, a much more definite pattern is produced with a clear gradation *for each cutaway group* from military, through urban down to rural which produces the smallest tegulae.

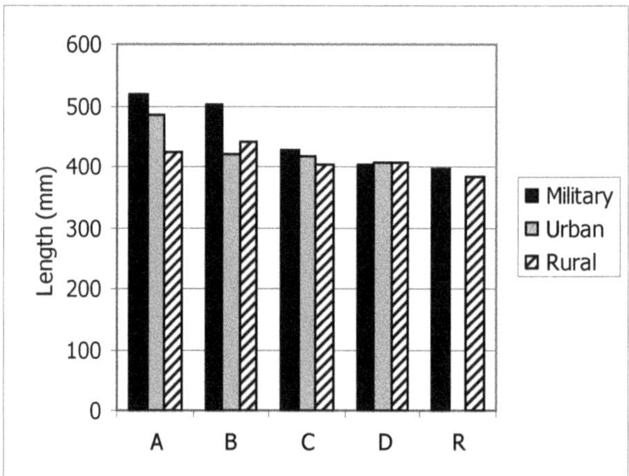


Figure 8.6: Comparison of length by site type (521)

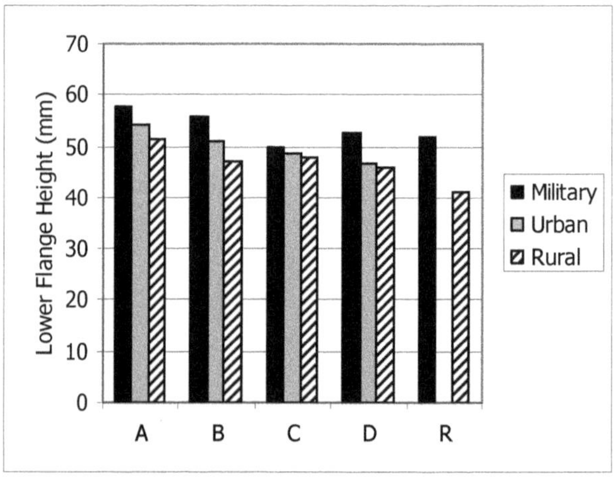


Figure 8.7: Comparison of flange height by site (1656)

This is a significant result. Manifestly this gradual evolution in tegula sizes on all types of site could not have been driven by commercial considerations alone.

Fashion is a more credible possibility with rural manufacturers leading the evolution towards smaller sizes followed by urban producers with the military lagging behind. However, it would be surprising if the villas were the driver of change and an alternative argument could be that larger town buildings were more amenable to the use of large tegulae. But whilst this might explain why urban tegulae were larger than rural ones, it would not explain why military ones were the largest of all; nor why they all gradually reduced in size at the same rate. Section 5.8 showed that the military gradually subcontracted their manufacture to civilian tilemakers who would presumably have used the same moulds for both their military and civilian customers. This may have had an influence on the residue of military production which also reduced in size even when it continued to be stamped (as was the case with *Legio* II production).

Perhaps a more plausible driver of this change could have been the evolution in roof styles postulated in Chapter 6. These different roof phases required different tegula characteristics and could therefore have been the principal force behind the changes. It is much easier to understand how such a major and visible element as a roof could drive the consequential changes in components like tegulae, whereas the reverse direction is more difficult to accept.

In Tarentum in 89 BC the qualification for membership of the town council was expressed as a minimum number of roof tiles.[123] If the tegulae were a little bit smaller, then more roof tiles would have been required thereby making it easier for the householder to achieve membership of the council, albeit without owning the actual requisite size of house. In fact property ownership was the primary qualification for membership of the *ordo* throughout the Empire and this will have required some method of assessing relative values. Cato calculated the total building costs of a villa in terms of a set cost per roof tile[124] so it is possible that Cato's methodology based on roof tiles was generally adopted for valuing the properties.

If the same area of roof that had been covered with Group A tegulae in the first century were re-covered with smaller Group C tegulae in the third century then 40% more tegulae would have been required.[125] If the pitch of the roof were also increased then even more tegulae would have been needed. Such an increase in tegula numbers would surely have required some regulatory response such as rebasing the property qualification for the *ordo*. However, the possibility that the regulatory response extended to defining the size of tegulae as well should not be discounted.

Diocletian's Price Edict shows the desire of the central authority to control economic activity throughout the empire and the existence of significant imperial and senatorial interests in tile production may well have reinforced this. It is also perhaps significant that the Edict, whilst fixing the maximum price of tegulae and identifying that this was for "*formae primae*", does not then specify any specific tegula size: one inference must be that this was already thought to be standardised, quite possibly through regulation.

## 8.10 Insights into technology transfer and investment

Finley has charged the Romans with being unsupportive of technological development, reluctance to invest and failing to properly understand economics and accountancy.[126] These views are not supported by the evidence of the Romano-British tile industry which grew from nothing to full commercial operation within fifty years,[127] with tiles for military bathhouses almost certainly being produced by civilian contractors before AD 100. No doubt, just as the indigenous mortaria industry was established by immigrant potters, perhaps sponsored or paid for by senatorial landowners, so too will have been the Romano-British tile-making industry. The military may also have played a part in setting up the early tileries. The logistics of tile manufacture were somewhat greater than for pottery production so it was a feat of organisation and training to build the industry so quickly. It was mirrored by the way the black burnished ware producers were able to hugely ramp up their production and distribution over the space of a few years in response to a shift in military procurement under Hadrian.[128]

Once established, tegula design evolved: new cutaway forms were sequentially introduced and tiles were reduced in size. It has been shown that these changes made the tegulae easier to manufacture, improved the integrity of the roof, reduced its weight and therefore the cost of the materials. The improvements to tegula design, particularly the reduction in the thickness of the tiles, could not have been achieved without improvements in the preparation of the clay and fabric technology. Likewise the increased use of nails for securing the tegulae to the roof and in particular the use of wooden dowels implies an improvement in wood technology.

[123] Brodribb 1983, 17.
[124] Cato *On Agriculture* 14.4-5.
[125] Figure 6.17.

[126] Finley 1965.
[127] Frere 1999, 37 records the first fired bricks as being from a hearth in Canterbury shortly before the conquest.
[128] Greene 1977, 124.

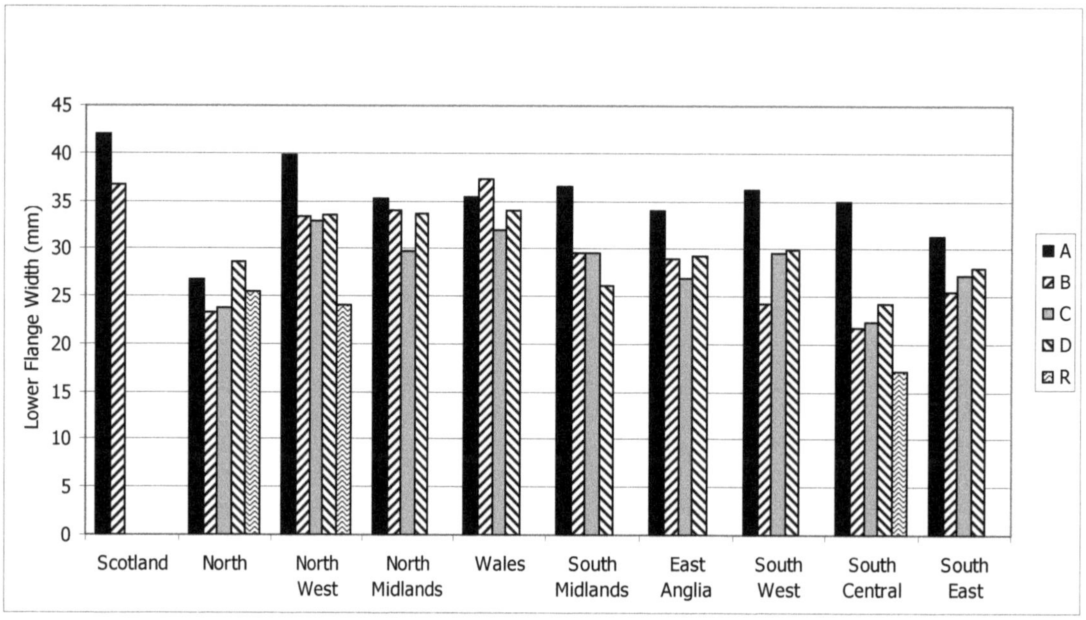


Figure 8.8: Regional development of lower flange width (1699)

A second wave of investment occurred in the latter part of the third century when many tileries converted to inverted box moulds for their tegulae and inverted formers for their imbrices and, perhaps, erected new buildings as combined workshops and drying areas.[129] The significant improvement in fabric quality and concomitant reduction in tile thickness raises the possibility that mechanical pugging machines were introduced although there is no direct evidence for this. These third century changes may have reflected a deeper change in the organisation of the tile industry or mirrored the release of creative energy in society more generally.

The evidence, particularly of competition in the Gloucestershire area, points to Minety being a manufactory site and the strong similarities in the form and dating of the stamps between Minety and Crookhorn in Hampshire[130] suggest the possibility of common ownership. It is conceivable the extent of common ownership, or perhaps ownership within a close group of senatorial families, was the cause of the uniform and contemporaneous adoption of the sequential cutaway groups across Britain. These families may well have had close ties with the military and therefore have either led or followed the same changes in military production of tegulae. Greene[131] has commented upon how these interests manipulated the markets they controlled to their own advantage and thus it is possible that they may have deliberately reduced the size of the tegulae for commercial gain.

The introduction of regional forms in the later third century, and changes in the method of manufacture and finishing of tegulae and imbrices, may therefore have been a consequence of a shift in ownership of the tileries, with perhaps new Romano-British owners bringing new ideas to an old industry. Such a change may have been brought about by a change in the economic viability as demand shifted from towns to villas and as local sources of fuel were exhausted and competition from stone tiles grew. However, it may also have represented a clawing back of provincial investment by Roman landowners as the anxiety of the times caused them to move their wealth back closer to Rome.

Far from Finley's ossified vision of Roman commerce and innovation, it is clear that there was considerable advance in tile making and roof technology. What distinguishes it from industries such as Samian production is that the industry was highly fragmented with a minimum of one hundred separate tileries servicing the British market, yet all of these appear to have experienced the same phases of tile and roof development at much the same time. If, as seems likely, these technological developments were also replicated broadly contemporaneously in the rest of the Empire then a very much more vibrant picture of technology transfer and development emerges than the one painted by Finley.

It is possible that a regional analysis of tegula size might give some indication of the rate of change across the country. At present the quantity of data is insufficient, but by way of a demonstration of what might be achieved with considerably more data, the result is shown in Figure 8.8. This chart suggests that there was a gradual decline in tegula size as one moved south, although the north of England, on very limited data, bucked that trend and produced almost the smallest tiles of all. This result

[129] It is notable that the inverted box technology, albeit without the upper cutaway insert, existed in the early second century at Leucarum and was probably subsequently used by the same unit at Carlisle, but apart from some isolated examples at Caerwent, the technology appears to have been lost until being rediscovered in the later third century.
[130] Section 5.14
[131] Greene 2000, 44.

would imply that the change started in the south (perhaps coming from across the channel) and gradually worked its way northward.

Finley also questioned the economic awareness and accountancy skills of the Romans and provincials. The cost build-up of a tegula shown in Section 8.3 yielded a price of 5.5 denarii/tegula and relied only upon the labour rate per day provided by Diocletian's Edict. This compared to a price of 5.6 denarii/tegula derived directly from the Edict. Whilst undue weight should not be placed upon the apparent precision of the numbers given the large assumptions made in the costing model, it nevertheless suggests that the Romans had a good understanding of costing. The distribution of tegulae discussed in Section 8.6 exactly aligned with that predicted by economic theory using the transport costs derived from the Edict and therefore further supports the internal consistency of the Edict and the costing methodology of the Romans. It also shows that the tilemakers had an excellent awareness of the principles of marginal costing: a point neatly demonstrated in Fulford & Hodders' analysis of the longer distances travelled by the fine (and more valuable) Oxfordshire wares compared to the humbler New Forest wares.[132] The evidence of branding as a response to competitive pressures amongst the Gloucestershire tilemakers is a further demonstration of economic awareness because the natural reaction of cutting prices would simply have damaged all the tilemakers and produced no short term increase in demand.

The idea of pricing by reference to the number of tiles on the roof as reported by Cato, and its use as a property qualification, shows pragmatic practicality. The steady improvement in the ratio of clay volume to area covered per tegula demonstrates a commercial approach to business improvement. The evidence from Beauport Park of the roof maintenance being conducted by a team that had accumulated a stock of tiles with a multiplicity of different stamps, rather than ordering a new set of tiles for each job, is a further demonstration of rational economic behaviour. The regular reuse of redundant tegulae in other structures, often with the flanges removed, shows careful husbandry. However, the Romano-British were not always so clever: the 12% of nail holes that are chipped rather than preformed demonstrates that one time in eight they had either failed to order the correct tegulae or to deliver the required tegulae in a timely manner.

## 8.11 Insights into the overall economy

Peaks and troughs of pottery consumption have been used to infer the condition of the economy as a whole and the evidence of tegula production might also be used in a similar way. Figure 8.9 below shows the data from the survey. The height of the bars is simply based on the number of tegulae recorded which will overstate the importance of Group C and D production if their periods of production have been correctly estimated at 100 and 140 years compared to 80 years for the other cutaway forms. However Group D suffered from competition with the regional forms and also stone roof tiles whilst the extension to a 140 year production period was based solely on the evidence of Maiden Castle temple. In contrast the prominence of Group C could be further overstated if, in fact, Group C was introduced earlier than AD 160 in some urban areas as suggested in Section 6.13. It is also distorted by an exceptionally large contribution of Group C tegulae from Silchester. As a consequence, the validity of the survey is inevitably limited, especially when broken down into type of site and form of cutaway, so conclusions drawn from it should be treated more as illustrations of what might be possible with a much larger survey rather than as statements of fact.

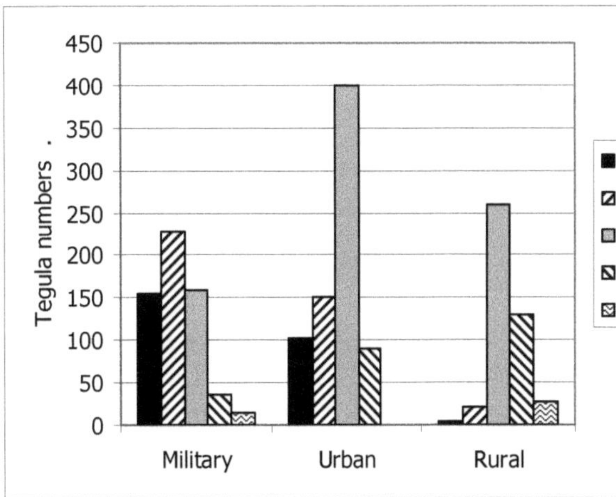


Figure 8.9: Relative building activity through time

With this caveat, it can be seen that military activity peaks with Group B in the second century but withers away from the mid-third century onwards. Urban activity peaks with Group C circa AD 200 and falls back in the later third century, but is not represented at all in the later regional forms. Rural building activity starts much more slowly but also peaks with Group C; it then reduces more gradually and has a continuing presence with regional cutaway forms in the fourth century. It is tempting to comment on the relative quantities of military, urban and rural tegulae but this would be extending the data too far and such a comparison might also have been influenced by differential survival rates on the different types of site.

The urban and rural peaks in Group C tegulae, on this very limited data set, may be contrasted with the apparent depression in the first half of the third century deduced from other sources. Faulkner[133] shows this as a chart of rural construction activity with peaks in AD 150 and AD 300 with the depression in AD 225. Fulford evidences it

[132] Fulford & Hodder 1975, 30 Figure 4.

[133] Faulkner 200, 71.

by the reduction in monetization, reduction in pottery consumption and perhaps also in the British garrison, thereby reducing the income flowing into the economy.[134] Going[135] described the situation as "quickly worsening depression turns into slump which becomes so severe that between c AD 210-240 it is extremely difficult to distinguish assemblages". This may actually be the explanation for the apparent contradiction because, as Fulford observes,[136] an absence of new dating evidence can cause subsequent events to be dated alongside events that were actually earlier. Alternatively, if Group C commenced somewhat earlier in urban areas then the peak of building activity could shift into the second century and be more consistent with the pottery evidence. Nevertheless, rebuilding activity at Verulamium after the fire of AD 155 was delayed[137] until around the turn of the century and in London the wall and new waterfront both date to the first part of the third century:[138] all consistent with the tegula data. This surge of building activity in London was accompanied by the appearance of new clay fabrics.[139]

The later third to mid-fourth century was a time of resurgence, particularly marked by the growth of villas.[140] The tegula evidence shows a continuation of activity in the towns and countryside but not as strong as for Group C. Pottery designs flourish and this liberation was also reflected in tegulae and imbrices as the hitherto rigid uniformity in tegula production and design began to break down. *Some* tilemakers (across the country) switched from the old four-sided mould to the inverted box mould for tegulae and from an upright to an inverted former for imbrices. At Stanton Low a mould with retractable sides was introduced.[141] Combed patterns started to appear on the gable ends of imbrices and on the south coast some tilemakers started using a comb for their signatures. Tiles were painted with coloured slip.[142] Roof pitches appear to increase. At the start of the fourth century regional varieties of lower cutaway forms emerged on rural sites in Hampshire and at military sites in the north. Probably later still, flat polygonal ceramic tiles were produced in the Hampshire/Surrey/Sussex area.

In parallel with the overall economic development of the province, tile-making experienced a gradual shift away from the towns as fuel sources were depleted[143] and the economics of production changed. In the southwest stone became more popular than ceramic tile from the mid second century onwards; this fashion extended eastwards to Hampshire and has been noted in Lincoln in the fourth century when ceramic was replaced with slate.[144] It was less common in the southeast where ceramic remained common into the later roman period.

Despite this, tile production continued well into the fourth century. McWhirr lists eight kilns with probable fourth century dates of which Crookhorn is the most secure with an archiomagnetic date of AD 330 +/– 20.[145] The latest tegulae in the survey from a secure context are those from Maiden Castle temple in Dorset which Wheeler placed as no earlier than AD 379.[146] It is possible that the tegulae from Batten Hanger are as late, but the dating is less certain.[147] The latest tegula from a northern site may be a Group D with upper cutaway inserts from the subsumed fort at Castleford, but this could be dated anytime after AD 250.[148] The latest northern tegulae with reasonably precise dating are those with a regional cutaway from the fort at South Shields which date to the early fourth century.[149]

[134] Fulford 1989, 191-3.
[135] Going 1992, 100.
[136] Fulford 1989, 193.
[137] Frere 1999, 236.
[138] Fulford 1989, 193.
[139] Betts, 1987, 28.
[140] Fulford 1989, 195.
[141] The tiles were stamped on the outside of the flange (Plate 2.4).

[142] For example Frilford, Stanton Low and Piddington, but Drury 1988, 79 has reported white slip on tegulae from the Chelmsford Mansio which were probably first or early second century with similar information on Stonea.
[143] Fulford 1989, 188.
[144] Perring 2002, 120.
[145] McWhirr 1979, 97-190. The kilns were Crookhorn, Hants; Scotland Farm, Hants; Park Street, Herts; Messingham, Humberside; Heckington Fen, Lincs; South Shields, Tyne & Wear; Arbury, Warwickshire; and Kenilworth, Warwickshire.
[146] Wheeler 1943, 74-75.
[147] Pers comm: James Kenny, Chichester District Archaeologist.
[148] Abramson et al 1999, 19-20.
[149] Bidwell & Speak 1994, 156.

# 9

# SUMMARY AND CONCLUSIONS

## 9.1 Introduction

This chapter summarises the main findings of this study. For clarity the summary omits the qualifications made in the main text but these caveats remain and the footnotes provide the references back to the sections where the full arguments are provided.

## 9.2 Production

Tegulae were made using wet clay that could not be rolled but was instead shaped with a tensioned wire.[1] All tegulae were made in four-sided moulds[2] until around AD 250 when some tile-makers switched to an inverted box mould.[3] The ends of the four-sided mould acted as a template to guide the wire.[4] The upper surface of the tile was smoothed with a straight edged piece of wood and the flanges were smoothed by hand, often resulting in the finger channels running down their sides.[5]

Production utilising the inverted box mould only became practical with the advent of Group D and Type 7 lower cutaway forms. It would have been a faster production method and the thinner bases would have required less clay, however the moulds would have been more expensive. It would appear that the same tilemakers who used the inverted box for tegulae also adopted an inverted former for imbrex production at around the same time.[6] The inverted former offered only a small efficiency advantage by eliminating the need to smooth the outer surface of the imbrex on the former, but also made it easier to decorate the gable end of the tile with patterns.

Once the tegula was formed, the mould was removed and the tile was taken on its palette to the drying area. Depending upon the weather and the drying arrangements, it might be left for up to a week before being turned off the palette, any finishing cuts completed and then rested on its side to complete the drying process.[7] There would have been a considerable investment in palettes, even a small commercial tilery might have required at least a thousand. In the case of the four-sided mould each palette would have had the lower cutaway inserts nailed into position. [8] Any change to the cutaway form would have entailed considerable work and would most likely have been gradually introduced as the palettes were modified in stages.

It is possible that the drying area was simply an open yard where the tiles were placed and covered with straw to shield them from the sun and rain. However many tiles show the impressions of animal feet, including that of birds, which never appear distorted by any intervening layer of straw.[9] It is therefore more likely that they were placed in covered hacks to shield them from the elements, although this did not always safeguard them from the ingress of animals. Stepdowns on the ends of many military tegulae, and some civilian ones from the south, may indicate that some sort of racking was used into which the palettes were slotted in order to optimise space utilisation.[10] In the later third century a number of integral buildings incorporating tile-making, tile-drying and a stoke-hole for the kiln appeared,[11] perhaps reflecting the technological leap seen with the inverted box moulds.

Tile production could only take place during the summer months when the weather was warm enough for the tiles to dry and it has therefore been assumed to be a seasonal activity. However, the workforce requirements for gathering fuel were such that it would have made more sense to gather most of the wood during the winter when

[1] Section 2.5.
[2] One exception to this are tegulae with triangular flanges that must have been made with an inverted box mould in the second century. These have been found at Leucarum in Wales and Carlisle and may possibly have been made by the same military unit.
[3] Section 2.6.
[4] Section 2.6.
[5] Section 2.6.
[6] Section 2.8.
[7] Section 2.6.
[8] Section 2.3.
[9] Section 2.3.
[10] Section 2.7.
[11] Section 8.2.

| Cutaway Group | A | B | C | D | R |
|---|---|---|---|---|---|
| Length (mm) | 488 | 449 | 416 | 407 | 394 |
| Upper breadth (mm) | 374 | 341 | 317 | 315 | 305 |
| Lower breadth (mm) | 366 | 335 | 306 | 308 | 312 |
| | | | | | |
| Lower flange height (mm) | 56 | 53 | 49 | 47 | 45 |
| Lower flange width (mm) | 36 | 32 | 29 | 28 | 20 |
| Lower tile thickness (mm) | 26 | 25 | 22 | 22 | 21 |
| Lower cutaway length (mm) | 57 | 54 | 52 | 49 | 45 |
| Upper flange height (mm) | 52 | 50 | 46 | 47 | 46 |
| Upper flange width (mm) | 24 | 23 | 18 | 19 | 21 |
| Upper tile thickness (mm) | 23 | 23 | 20 | 20 | 25 |
| Upper cutaway length (mm) | 54 | 48 | 45 | 41 | 37 |
| | | | | | |
| Finger channels (number) | 0.45 | 0.47 | 0.57 | 0.68 | 0.63 |
| Preformed nail holes (%) | 9 | 14 | 25 | 53 | 0 |
| | | | | | |
| Breadth taper (%) | 1.7 | 1.9 | 4.3 | 3.1 | 4.2 |
| Aspect ratio | 0.75 | 0.76 | 0.78 | 0.77 | 0.79 |
| Flange height taper (%) | -2.5 | 2.4 | 2.4 | 1.6 | -1.3 |
| Flange width taper (%) | 31 | 22 | 31 | 26 | 5 |
| Ratio of lower flange height to length | 0.112 | 0.114 | 0.114 | 0.117 | 0.125 |
| Ratio of upper to lower cutaway | 0.98 | 0.93 | 0.90 | 0.83 | 0.72 |
| | | | | | |
| Overlap on roof (mm) | 109 | 100 | 95 | 92 | 84 |
| Volume (cc) | 5049 | 4047 | 3068 | 3071 | 3247 |
| Area covered on roof (sq cm) | 1406 | 1180 | 1000 | 982 | 954 |
| Ratio of volume to area | 3.6 | 3.4 | 3.1 | 3.1 | 3.4 |

Figure 9.1: Summary data by cutaway group

the trees were bare. This would have resulted in a much more stable workforce.[12] Kiln size shows that commercial tileries were relatively small undertakings, possibly only employing five people.

The cost of fuel was the most important factor in the economics of tile production and it will have determined the location of the kiln.[13] Whilst it is apparent that the private tilemakers moved once they had exhausted the local supplies of fuel there is less evidence for this in military kilns despite their much higher output and therefore much larger consumption of timber. Clamp-firing was inherently inefficient and the high cost of fuel would have meant that these kilns would only have been used for very limited production where fuel was immediately available.[14]

Military production, as represented by Holt and the *Classis Britannica,* was on a much larger scale. In the case of *Classis Britannica* at least six tilemakers were working full time at two separate benches with a potential overall workforce of eighty (or thirty if the work was spread throughout the year).[15] Such a workforce could have produced all of the tiles required for the Beauport Park bathhouse in a fraction of the season.[16]

The number of tilemakers determined the number of different signatures: in a commercial tilery a variable number of semi-circles was more than adequate to identify individual makers, but in a military works the larger number of producers necessitated a wider variety of signatures.[17] In legionary tile-works where soldiers would be allocated to the task, the variety of signatures was greatest.

Signatures were made by drawing the fingers across the wet clay and always placed at the bottom of the tile as this was the end that was nearest the tilemaker in the manufacturing process.[18] Some manufacturers in the south of the country started to use combs to draw signatures[19] in the mid-third century which may have been a natural consequence of employing combs for decorating the gable ends of the imbrices. Tegula signatures would have been visible on the roof, however they were not for decoration but primarily to identify the tilemaker, although some identified the tilery owner.[20] There was a Gloucestershire signature that must have been for decorative purposes[21] and signatures on imbrices, which were also at the lower visible end, also seem more likely to have been for decoration.

[12] Section 8.3.
[13] Section 8.5.
[14] Section 8.5.
[15] Section 8.8.
[16] Section 8.8.
[17] Sections 5.18 and 8.7.
[18] Section 2.3.
[19] Section 2.3.
[20] Piddington and contractors "D" and "V" to *Legio* XX (Section 5.18).
[21] Section 5.18.

| Cutaway Group | | A | B | C | D | R |
|---|---|---|---|---|---|---|
| Length (mm) | Military | 518 | 500 | 428 | 403 | 395 |
| | Urban | 486 | 419 | 417 | 408 | |
| | Rural | 423 | 440 | 402 | 407 | 383 |
| Upper Breadth (mm) | Military | 388 | 371 | 325 | 328 | |
| | Urban | 366 | 316 | 320 | 316 | |
| | Rural | 370 | 293 | 307 | 302 | 303 |
| Lower Flange Height (mm) | Military | 58 | 56 | 50 | 53 | 52 |
| | Urban | 54 | 51 | 49 | 47 | |
| | Rural | 51 | 47 | 48 | 46 | 41 |
| Lower Flange Width (mm) | Military | 37 | 34 | 30 | 29 | 25 |
| | Urban | 35 | 30 | 30 | 28 | |
| | Rural | 33 | 26 | 28 | 28 | 17 |
| Lower Thickness (mm) | Military | 27 | 26 | 23 | 23 | 24 |
| | Urban | 25 | 23 | 23 | 21 | |
| | Rural | 24 | 23 | 22 | 22 | 20 |

Figure 9.2: Site data by cutaway group

## 9.3 Typology

The manufacturing efficiency and engineering improvements[22] argue strongly for the typological sequence shown in Figure 1.3. The seriation evidence from sites and tile stamps both provide strong, statistically significant, support for the typology.[23] The resulting dimensional evolution of the tegulae confirms the correctness of the sequence and the dating evidence reinforces this.[24] Equally significantly, all the subsequent analysis on stamps, roofs and industry economics is neatly explained by the use of the typology which, in many ways, is a still more convincing demonstration of its validity. The evolution of the cutaway forms follows a logical stylistic development[25] which can be seen on many sites: in particular the development of the Type 7 regional form from the Type 1 form is apparent at Crookhorn and Grateley.[26]

The proposed dating for the cutaway groups is:[27]

| | |
|---|---|
| Group A | 40-120 |
| Group B | 100-180 |
| Group C | 160-260 |
| Group D | 240-380 |
| Regional | 300 onwards |

Whilst the dating of the first and last groups is reasonably secure, there is greater uncertainty about the dates of the intermediate forms, particularly of Group C. The evidence from Silchester and Caerwent could point to a much abbreviated period of Group B production and an earlier introduction of Group C in these towns.[28] Inevitably the earlier dating evidence is heavily biased by military data and, as it appears the military were using a different roof structure until the later second century,[29] the possibility that they also lagged civilian practice on cutaway forms at this time should therefore not be discounted.

The average sizes of the dimensions of the tegulae in the survey are given in Figure 9.1.[30] All the measured dimensions are the average of the site data with the exception of the lower flange dimensions which take advantage of the larger dataset by using the average of all tiles. The aspect and taper ratios have been calculated on individual tegulae and then averaged by site and then these sites averaged to produce the recorded result. All the other derived data has been based on site average dimensions which have themselves then been averaged.

It is notable not just how the individual dimensions and features move in line with the typology but how their relative sizes remain in proportion as shown by the ratio of lower flange height to length. The regional cutaway data is less consistent, mainly due to the relatively small dataset and the consequential influence of the poorly manufactured *Cohors V Gallorum* tegulae from South Shields. Behind these averages there is, however, a considerable dispersion of actual values reflecting the diversity of different tilemakers and the length of time during which each cutaway group was being made.[31]

The segmental analysis of dimensions is less reliable due to the sparcity of the data but is nevertheless provided in Figure 9.2. This shows that the gradual reduction in size was apparent on each type of site and that military tegulae tend to be larger than civilian production throughout the Roman period.

The cutaway sequence appears to have evolved across Britain and, on very fragmentary evidence, across the

22 Section 3.4.
23 Section 3.6.
24 Sections 3.7 and 4.6.
25 Section 3.4.
26 Section 3.9.
27 Section 4.5.
28 Section 6.13.

29 Section 6.14.
30 The Silchester Group C lower cutaway length data has been excluded from the average as discussed in Section 3.7.
31 Section 3.10.

continent[32] at virtually the same time. The evolution of cutaways complements the improvements in weight, efficiency, roof design and wood technology. These improvements might have driven the change in the cutaway sequence but the possibility of some official intervention cannot be dismissed.[33]

## 9.4 Stamps

The legions started stamping tiles around AD 90 and stopped around AD 250, although the last *Legio* XX stamped tile is one with the *Antoniniana* cognomen that was probably made by a contractor rather then the legion itself.[34] *Legio* VI and *Classis Britannica* stamped every tegula and it seems probable that all the other legions, and possibly all military units, did likewise.[35] The life of legionary dies was around twenty years which indicated that many of these dies were in concurrent use.[36] In the case of *Legio* II each cohort must have had its own die;[37] *Legio* VI and *Legio* XX also probably had separate dies for each cohort.

All the legions contracted out the manufacture of box flue tiles for their bathhouses.[38] *Legio* XX always made extensive use of contractors for everything other than tegulae, but contractors were producing tegulae for them at Tarbock by AD 126, and within a few years all of *Legio* XX's output was produced by contractors.[39] Some of these contractors were recognised by the letters "D" and "V" (previously regarded as cognomina) and by the stamps with extended scripts.[40] *Legio* VI was using contractors in the second century,[41] but there is no evidence for contractors with *Legio* II until the third century.[42]

Ansate frames were almost only ever used on military production and, within that, almost all were *Legio* XX.[43] Military stamps were almost always in relief but half the civilian and official production used incuse dies.[44] Private tilemakers did not employ stamps until the second century and their stamps had a shorter life than military ones.[45] Tile stamping reduced significantly by AD 250, with the last civilian stamp being that from Stanton Low on the outside of the flange which probably dates to around AD 300.[46] A similar date can be given to the last military stamp of *Cohors V Gallorum* at South Shields.[47]

The proliferation of civilian stamps in the Gloucestershire area is attributable to branding which was a response to competition, in part originating from the multiple producers at the Minety manufactory.[48] The fixed cost of the rent for private tile works meant that if the kiln was underutilised it became economic to distribute tiles at least 40 kilometres away and occasionally at twice that distance, as is evidenced by the distribution of tile stamps.[49] Military stamped tiles have an even wider distribution but this may have been driven less by economics than simple military logistics. The cost structure of tegula manufacture is broadly consistent with the imputed price of 5.6 denarii per tegula derived from Diocletian's Price Edict.[50] The cost structure is also consistent with the evidence of competition and distribution ranges.[51]

## 9.5 Roofs

All tegulae until the advent of the regional forms were designed to have a double overlap when placed upon the roof and to present a continuous flange face to water running down the roof.[52] The earlier forms of tegulae, especially military tegulae, were non-meshing and would have been produced in graduated sizes in order to be successfully fitted onto a roof.[53] Imbrices were also produced in graduated widths.[54] The non-meshing tegulae had no preformed nail holes and were placed on a bed of mortar and daub which itself was placed over a thick layer of laths.[55] The military roofs from the end of the second century, and civilian roofs from rather earlier, used self-meshing tegulae with preformed nail holes on every fourth tile which were fixed directly onto the common rafters without the need for a bedding of mortar or a substrate of laths.[56] In order to become self-meshing, the breadth of the tegulae was tapered (as was the width of the flanges).

In the mid-third century a further stage in roof design emerged when the number of preformed nail holes was doubled and wooden dowels rather than nails were used on some roofs.[57] It is probable that the pitch of the roofs was increased. Concomitant with the evolution in roof design, the cost of tegulae per square metre of roof was reduced by shortening the overlap and the thickness of tiles was also decreased.[58] Production of tegulae

[32] Section 4.10.
[33] Section 8.9.
[34] Section 5.3-5.7.
[35] Section 5.8.
[36] Section 5.2.
[37] Section 5.3.
[38] Section 5.9.
[39] Section 5.8.
[40] Section 5.7.
[41] Figure 5.12.
[42] Section 5.8.
[43] Section 5.16.
[44] Section 5.15.
[45] Sections 5.2 and 5.14.
[46] Section 5.14.
[47] Section 5.11.
[48] Section 8.6.
[49] Section 8.4.
[50] Section 8.3.
[51] Section 8.5.
[52] Section 6.2.
[53] Sections 6.3 and 6.4.
[54] Section 6.5.
[55] Sections 6.8 and 6.10.
[56] Sections 6.8 and 6.10.
[57] Section 6.10.
[58] Section 6.11.

continued into the fourth century, possibly until the end of the century.[59] During the fourth century more radical roof designs were tested using flat polygonal tiles.[60]

The survey found that 22% of tegulae were convex which probably underestimated the number of convex tegulae in the original population.[61] The tegulae were always bent inwards (opposite to the natural direction). These tegulae appear to have been specially produced because few of them had preformed nail holes (which would have been part of the standard manufacturing specification).[62] Moreover the underside of many of these tegulae was smooth in contrast to the rougher surface of flat tegulae.[63] These convex tegulae were not wasters but had been used to match the extrados of vaulted roofs.[64] Examples of this exist in Rome and a fragmentary example survives from Dorchester.[65] Convex tegulae were found at Beauport Park where they must have been on the bathhouse roof and for which alternative roof designs were not possible.[66] The significant number of convex tegulae suggests that there were only four times as many buildings with normal pitched tiled roofs as there were bathhouses with vaulted roofs tiled using convex tegulae.[67]

## 9.6 Applications

This study has shown how a study of tegulae can assist in the interpretation of other archaeological evidence. It has helped fix the date of Viducius's production for *Legio* XX to AD 126[68] and has provided supporting evidence for the building of the Carpow fortress circa AD 180[69] with resultant adjustment to the post-Antonine history of Scotland. It has provided dates for many of the legionary and other military stamps.[70]

Several examples of statistical analysis based upon dimensional data have been deployed thereby demonstrating the potential of this approach. Scattergrams of pairs of dimensions have proved a useful tool in establishing linkages between individual tegula fragments, most fundamentally in demonstrating that the legions stamped all of the tegulae that they themselves produced.[71] At Chester and Silchester the modes of cutaway length distributions were compared with those from other sources to identify the origin of supply.[72] A different statistical approach was taken to establish the likely range of differential shrinkage between tegulae.[73]

A comparison of the quantities of cutaway groups generated by site types has provided an insight into military, urban and rural construction activity.[74] An analysis of the stamps from the Beauport Park bathhouse not only identified three phases of construction but also provided information about how the maintenance was undertaken.[75] The composition of the tegulae from successive seasons of the Silchester *Insula* IX excavation showed how the site developed and the intensity of activity at different periods.[76]

This work has highlighted a number of diagnostic features that indicate the production date or source of tegulae. Most of these features could be identified in the field by any competent archaeologist with minimal training. These features are encapsulated in Figure 9.3 overleaf.

## 9.7 Further work

A database of 2300 tegulae, whilst adequate for overall analysis, is too small to produce reliable segmental breakdowns. The accumulation of a much more diverse and extensive data set would not only allow the dates of the main cutaway sequence to be refined but might also enable dates to be applied to the individual cutaway types within these main groupings. It would also permit a more meaningful regional analysis which would help establish the rate at which the cutaway changes were transmitted across the country and thereby generate a better idea of the drivers behind these changes. More data might also make dating by dimensional analysis a more realistic possibility although this will always require a statistically valid sample from the site under investigation.

One of the more surprising conclusions is how the cutaway sequence appeared to have evolved across Britain at broadly the same time. Continental data will be essential to determine whether Britain fits into an overall pattern of evolution both in terms of cutaway forms and tegula dimensions. Assuming that the fragmentary evidence noted in Chapter 4 is representative, then this would make the transmission mechanism for these changes even more significant.

The analysis of Carpow has shown how diagnostic tegulae, and tile stamps in general, can aid the interpretation of other evidence. The ability to provide broad date ranges for all stamped diagnostic tegulae fragments should help illuminate the history of a number of the military units based in Britain.

[59] Section 8.11.
[60] Section 6.14.
[61] Section 7.1.
[62] Section 7.7.
[63] Section 7.2.
[64] Section 7.4.
[65] Section 7.5.
[66] Section 7.4.
[67] Section 7.8.
[68] Section 4.6.
[69] Section 4.7.
[70] Chapter 5.
[71] Section 5.8.
[72] Sections 4.7 and 6.13.
[73] Section 5.2.
[74] Section 8.11.
[75] Section 5.17.
[76] Section 4.11.

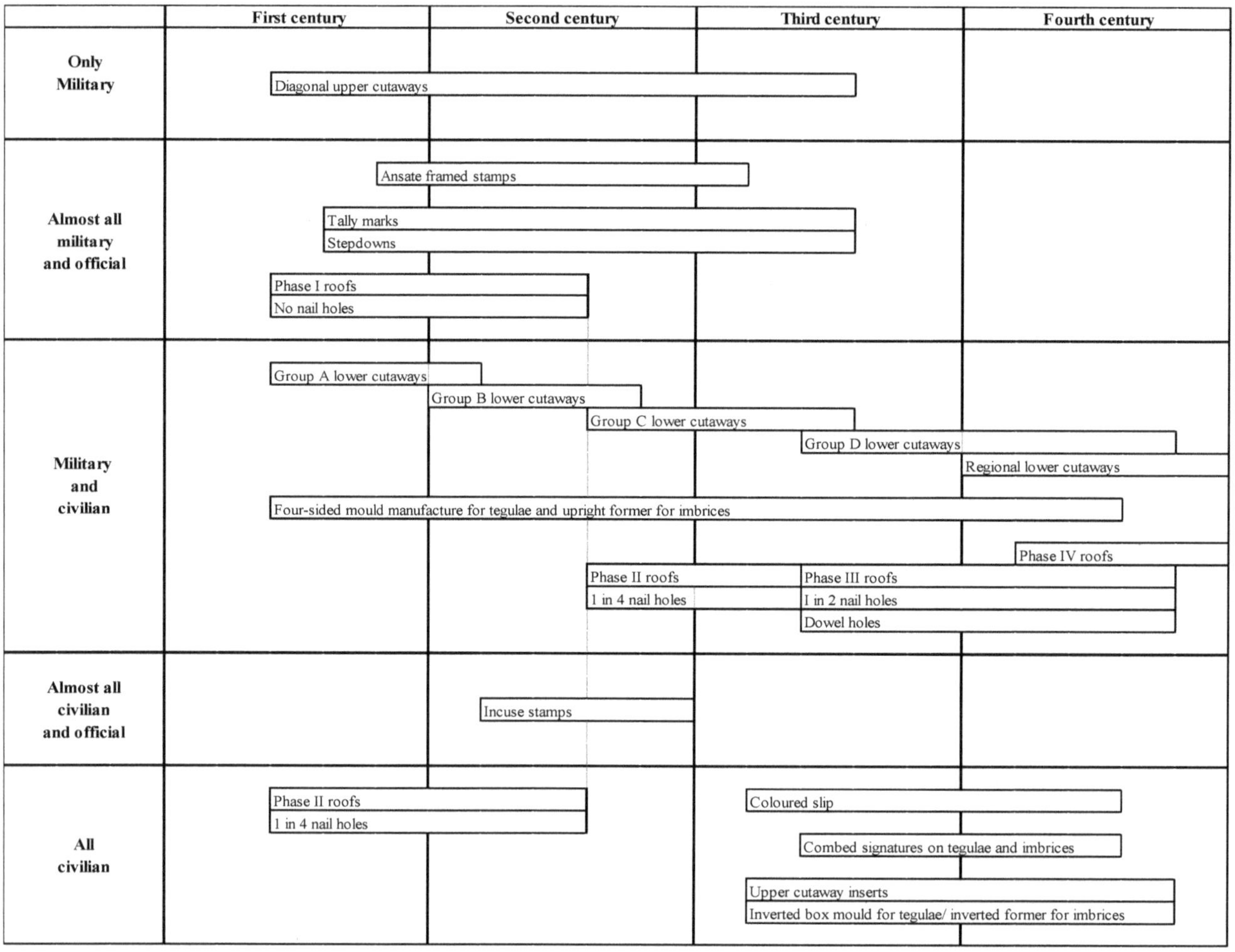


Figure 9.3: Summary of diagnostic features

It appears that the military originally provided the tegulae (or at least the expertise) to the towns and hence played a material role in the Romanization of the province,[77] but that by the second century the position had reversed with contractors taking an increasing share of military production.[78] Tile technology was then relatively stable until the mid-third century when new manufacturing methods, new tegula cutaway designs, new types of signature and slip-coloured tiles were introduced.[79] These economic and cultural shifts should be explored further.

Some problems remain: for example the explanation of stepdowns being produced by the racking for drying is not entirely satisfactory.[80] Tally marks were clearly a military and official practice but their precise purpose remains obscure.[81] The analysis of the *Classis Britannica* tile-works goes some way towards showing how it operated but a study of the Dover stamped tile[82] would surely improve this.[83] More data may show that some of the conclusions were statistical aberrations or suggest alternative solutions.

## 9.8 Conclusion

This work has hopefully demonstrated the wealth of artefactual, constructional and cultural information that an analytical and statistical approach to tegulae can provide. The typology established in Chapters 3 and 4 is amply vindicated by the way it has illuminated the analysis in all the subsequent chapters. Those sites where the largest number of tegulae have been examined[84] have generally generated the most interesting results: the potential for more extensive study of other sites is therefore significant. Much more data are necessary to test and refine the conclusions and to extend the potential of this study, in particular, continental comparators are now essential.

Gerald Brodribb regarded his pioneering work as a starting point for others,[85] and whilst this study has extended that knowledge, it still remains very much as a starting point.

[77] This would appear to contrast will Millett's (1990) suggestion of a less interventionist approach.
[78] Section 8.5.
[79] Section 8.11.
[80] Section 2.7.
[81] Section 5.19.
[82] The stamped tile from the Dover forts has not yet been released for public study.
[83] Section 5.17.

[84] Silchester, Caerleon/Caerwent and Beauport Park.
[85] Brodribb 1983, 4.

# Appendix 1: List of all sites entered on database

Alchester
Ashtead
Batten Hanger
Beauport Park
Benwell
Bignor
Buckinghamshire
Caerleon
Caerwent
Caerwent House
Canterbury Longmarket
Canterbury Whitefriars
Carlisle
Carpow
Carrawburgh
Castleford
Chedworth
Chelmsford
Chelmsford Mansio
Cheltenham
Chester
Chichester
Chignall
Chilgrove 2
Cirencester
Colchester Buc Church
Colchester CF
Colchester GBS
Colchester Rural
Colchester St Mary's
Colchester Urban
Cookham
Corbridge
Crookhorn
Dominant House
Dorchester (Dorset)
Dorchester (Oxon)
Dorchester Hospital
Dover
Drayton
Eccles
Exeter
Fishbourne
Frilford
Gatehampton
Gelligaer
Gestingthorpe
Gloucester
Gosbecks
Great Holt
Greyhound Yard Dorchester
Grimescar
Halstock
Holt
Housesteads
Inchtuthill
Lancaster
Layerthorpe
Leicester
Leucarum
Lime Street
London
Lullingstone
Maiden Castle Temple
Maiden Hatch
Mansell Street
Maryport
Medbourne
Mumrills
Narborough
Newsteads
Norfolk Street
Norton Disney
Norwich Caistor by Yarmouth
Norwich Caistor St Edmunds
Oxfordshire
Piddington
Prestatyn
Reading
Reculver
Richborough
Rivenhall
Sandwich
Shelford Quarry
Silchester
Slack
South Shields
Sparsholt
St Mary de Lode
St Oswalds Priory
Stanton Low
Tarbock
Templebrough
Thurnham
Wall
Wallsend
Wantage
Winchester
Witham
Wollaston Bathhouse D'cr
Wroxeter
Yewden
York
York Minster

# Appendix 2: Measurement of dimensions

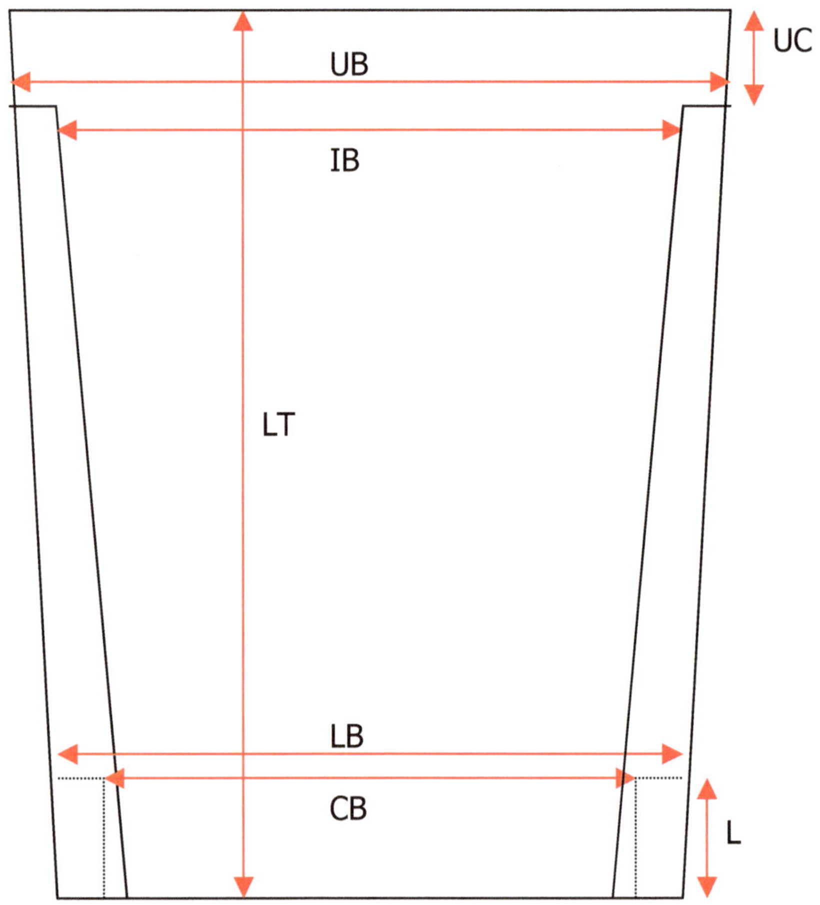


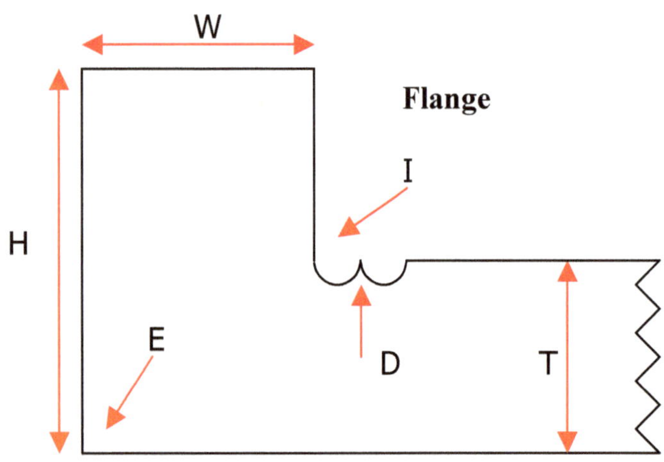

## Lower Cutaway dimensions

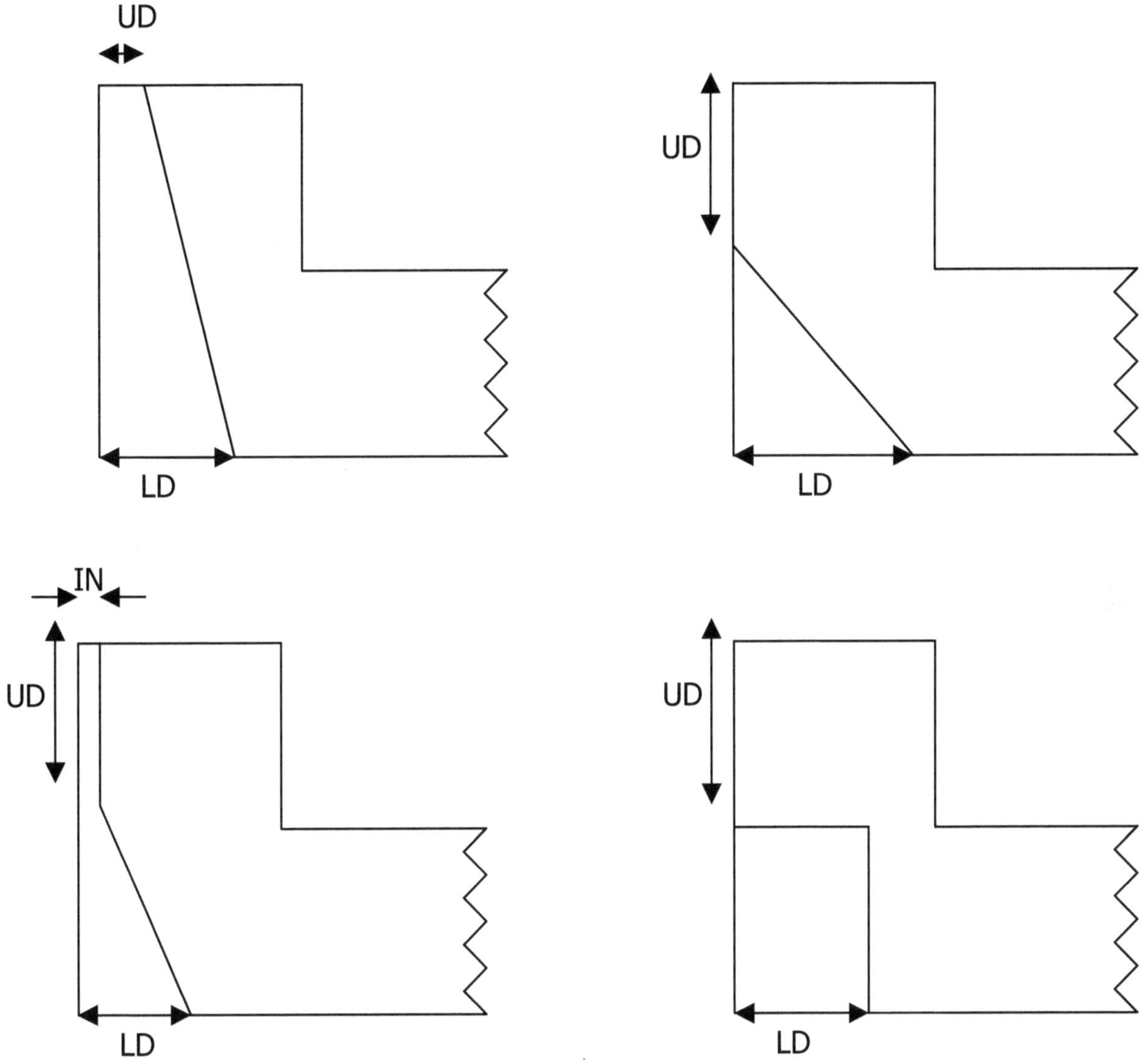

## Imbrex measuremnets

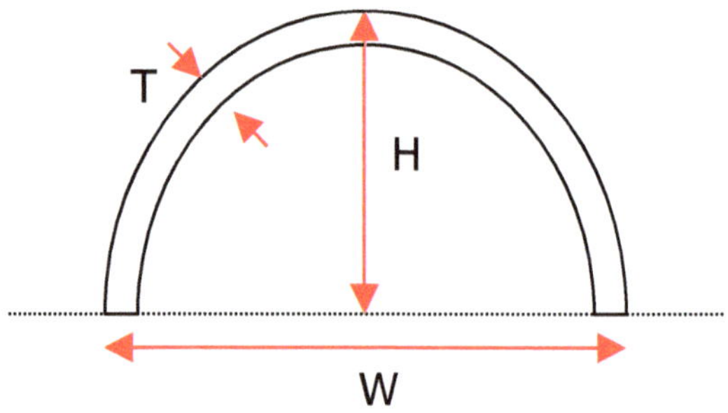


L = overall length
T = thickness

Lower (wider) end
LH = lower height
LW = lower width

Upper (narrower) end
UH = upper height
UW = upper width

# Appendix 3: Recording method and forms

1. There are three input sheets (*large tile, lower cutaway and upper cutaway*) only one of which is selected dependent on the shape of the fragment to be recorded. If the tegula is complete, or a complete length or breadth can be measured, then the *large tile sheet* should be used. If only a fragment containing a cutaway is present then the appropriate *upper or lower cutaway sheet* should be used. If no part of a cutaway is visible then the tile should not be recorded.

2. The *upper and lower cutaway sheets* each contain a subset of the information on the *large tile sheet*. Using these sub-sheets simplifies the recording and significantly speeds up the subsequent transfer to the computer database.

3. When recording large pieces where both flanges are present it is normally sufficient to base all the measurements on just one of the flanges, however if there are significant differences between the flanges then both sides should be recorded by noting those measurements that differ on a second line on the *large tile sheet*.

4. Imbrices have their own *imbrex sheet* but only imbrices with either complete lengths or complete widths are recorded.

5. Frequently other aspects of the tiles will deserve comment and these can be recorded on a comment sheet using the same tile number as on the input sheet. This can also be useful for noting photograph numbers.

6. Every tegula must have a unique tile number as must every imbrex. Separate databases are used for the tegulae and imbrices so the same numbers may be used between these two databases but there should be no duplication within the tegula or imbrex series. There is no requirement for these numbers to be sequential.

7. All measurements should be in millimetres without decimal points.

8. Most of the data will be used for statistical analysis so consistency of measurement is more important that absolute precision (which is often frustrated by the irregularities in the tiles). If measurement to + or – 2 mm can be achieved then this would be very satisfactory.

9. Where an element cannot be measured in full or a feature cannot be observed (because part of the tile is missing) then the field should be left blank. A zero indicates that the dimension actually is of zero length, or that the feature has been omitted (not just cannot be seen).

10. An explanation of each of the measurements is given in the definitions sheet and reference may also be made to Appendix 2: measurement of dimensions.

# Definitions

| LT | Length measured along the surface of tile |
|---|---|
| UB | Upper breadth measured at the point where upper cutaways begin |
| LB | Lower breadth measured at the point where lower cutaways begin |
| IB | Inner breadth measured at the upper end between the flanges (consistent with width measurements such that in principle IB+2W=UB) |
| CB | Breadth between the lower cutaways measured on the underside of the tile (such that in principle CB+2LD=LB) |
| H | Height of the flange measured at the point where the lower and upper cutaways respectively commence |
| W | Width measured from the outside of the flange to where the inside of the flange blends into the base of the tegula and taken at the point where the lower and upper cutaways respectively commence |
| T | Estimate of the average thickness of the base of the tile at the upper and lower ends respectively (this is normally a judgement because thickness tends to be very uneven) |
| I | The shape the inside of the flange forms with the base as below   |
| E | The shape the outerside of the flange makes with the base as above |
| D | The number of finger channels imprinted in the base of the tegula as the tiler formed the flanges (normally 0,1 or 2) |

| TP | Lower cutaway type – see list of descriptions |
|---|---|
| L | Lower cutaway length measured from the start of the cutaway at the side of the flange to the end of the tegula |
| UD | Distance from the top corner of the flange to where the cutaway cuts the flange measured where the cutaway starts (for Type 5 cutaways this is the vertical distance from the top of the flange to where the angular cut starts) – see Appendix 2: measurement of dimensions |
| LD | Distance from the bottom corner of the flange to where the cutaway cuts the base of the tegula measured where the cutaway starts – see Appendix 2: measurement of dimensions |
| IN | Indent – only relevant to Type 5 cutaways – the horizontal distance from the top corner of the flange to where the vertical cut starts – see Appendix 2: measurement of dimensions |
| UC | Length of the upper cutaway (in the case of sloping or diagonal cutaways it is measured at the mid-point) |
| C | Concavity measured by placing a straight-edge lengthwise on the underside of the tegula and positioning it to the line of length that shows the least concavity, then measuring the maximum distance between the straight-edge and the bottom of the tegula |
| N | Nail hole: P=preformed, B=blind, C=chipped out, 0=absent |
| S | Upper cutaway stepdown – a step change in the level of the upper surface of the tegula normally starting from the beginning of the upper cutaway and often accompanied by a change in the surface texture of the tile: S=stepdown, 0=absent |
| P | Signature (normally on lower half of the tegula) – see list of main signature descriptors |
| Context | Any identification on the tile and/or the context from which it comes |

## Tegula signature descriptors

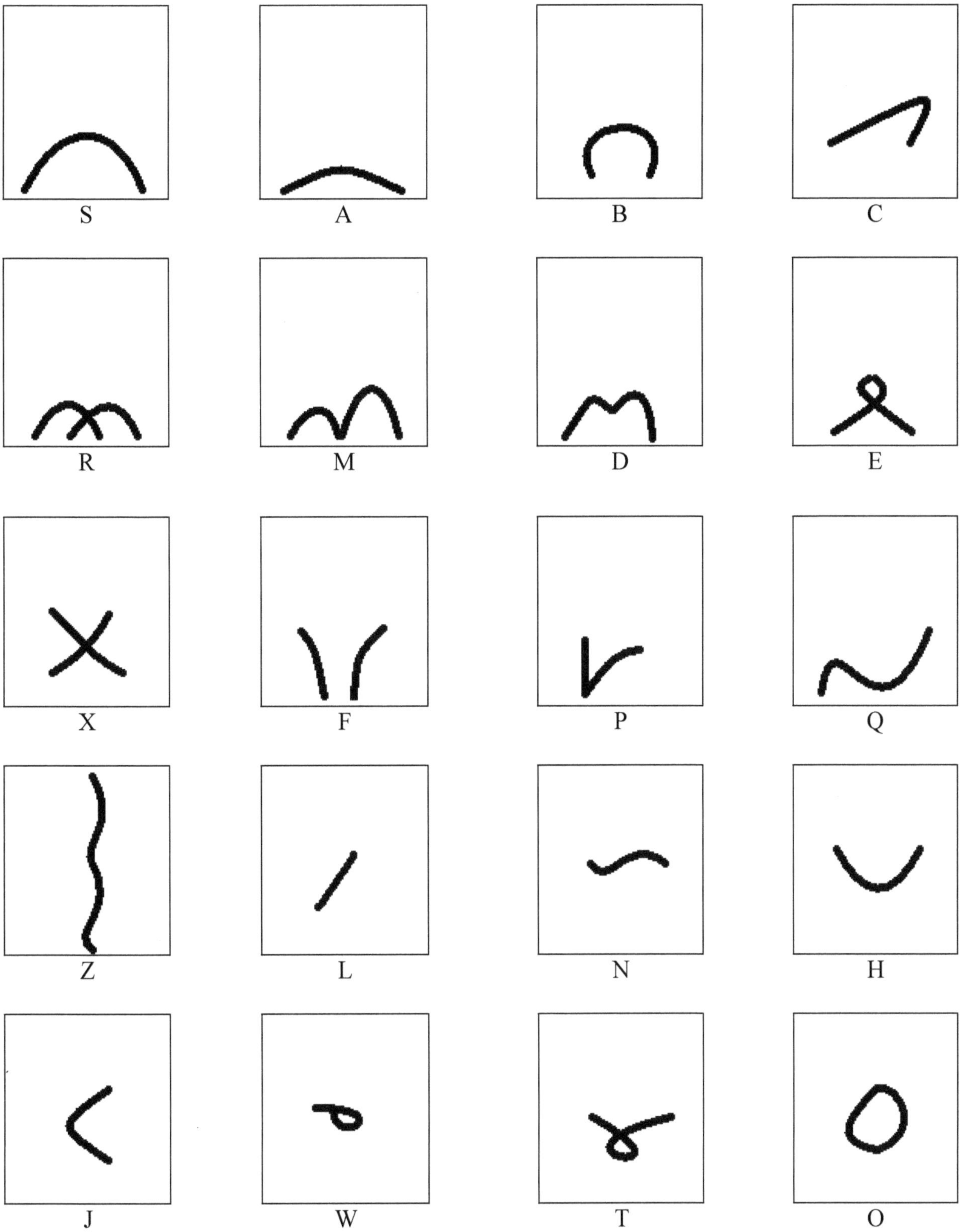


To be recorded as (Signature type) + (Size Modifier: nil, L or S) + (Number of repeats)
eg S1 would be one semi-circle of similar size to that shown above whilst SL3 would equate to three large semicircles

**Lower cutaway recording sheet**

| Site | Recorder | Date |
|---|---|---|
| Context, dating etc | | |

| Tile | FLANGE | | | | | | CUTAWAY | | | | | | |
|---|---|---|---|---|---|---|---|---|---|---|---|---|---|
| No | H | W | T | I | E | D | TP | L | UD | LD | IN | Context | P |
| | | | | | | | | | | | | | |
| | | | | | | | | | | | | | |
| | | | | | | | | | | | | | |
| | | | | | | | | | | | | | |
| | | | | | | | | | | | | | |
| | | | | | | | | | | | | | |
| | | | | | | | | | | | | | |
| | | | | | | | | | | | | | |
| | | | | | | | | | | | | | |
| | | | | | | | | | | | | | |
| | | | | | | | | | | | | | |
| | | | | | | | | | | | | | |
| | | | | | | | | | | | | | |
| | | | | | | | | | | | | | |
| | | | | | | | | | | | | | |
| | | | | | | | | | | | | | |
| | | | | | | | | | | | | | |
| | | | | | | | | | | | | | |
| | | | | | | | | | | | | | |
| | | | | | | | | | | | | | |
| | | | | | | | | | | | | | |
| | | | | | | | | | | | | | |
| | | | | | | | | | | | | | |
| | | | | | | | | | | | | | |
| | | | | | | | | | | | | | |
| | | | | | | | | | | | | | |
| | | | | | | | | | | | | | |
| | | | | | | | | | | | | | |
| | | | | | | | | | | | | | |
| | | | | | | | | | | | | | |
| | | | | | | | | | | | | | |
| | | | | | | | | | | | | | |
| | | | | | | | | | | | | | |
| | | | | | | | | | | | | | |
| | | | | | | | | | | | | | |
| | | | | | | | | | | | | | |
| | | | | | | | | | | | | | |

## Upper cutaway recording sheet

| Site | | | | Recorder | | | | Date | |
|---|---|---|---|---|---|---|---|---|---|
| Context, dating etc | | | | | | | | | |
| Tile number | FLANGE | | | | | | UC | Context | S |
| | H | W | T | I | E | D | | | |
| | | | | | | | | | |
| | | | | | | | | | |
| | | | | | | | | | |
| | | | | | | | | | |
| | | | | | | | | | |
| | | | | | | | | | |
| | | | | | | | | | |
| | | | | | | | | | |
| | | | | | | | | | |
| | | | | | | | | | |
| | | | | | | | | | |
| | | | | | | | | | |
| | | | | | | | | | |
| | | | | | | | | | |
| | | | | | | | | | |
| | | | | | | | | | |
| | | | | | | | | | |
| | | | | | | | | | |
| | | | | | | | | | |
| | | | | | | | | | |
| | | | | | | | | | |
| | | | | | | | | | |
| | | | | | | | | | |
| | | | | | | | | | |
| | | | | | | | | | |
| | | | | | | | | | |
| | | | | | | | | | |
| | | | | | | | | | |
| | | | | | | | | | |
| | | | | | | | | | |
| | | | | | | | | | |
| | | | | | | | | | |
| | | | | | | | | | |
| | | | | | | | | | |
| | | | | | | | | | |
| | | | | | | | | | |
| | | | | | | | | | |

## Imbrex recording sheet

| Site | | | | Recorder | | | Date | |
|---|---|---|---|---|---|---|---|---|
| Tile No | L | LW | LH | UW | UH | T | Context | P |
| | | | | | | | | |
| | | | | | | | | |
| | | | | | | | | |
| | | | | | | | | |
| | | | | | | | | |
| | | | | | | | | |
| | | | | | | | | |
| | | | | | | | | |
| | | | | | | | | |
| | | | | | | | | |
| | | | | | | | | |
| | | | | | | | | |
| | | | | | | | | |
| | | | | | | | | |
| | | | | | | | | |
| | | | | | | | | |
| | | | | | | | | |
| | | | | | | | | |
| | | | | | | | | |
| | | | | | | | | |
| | | | | | | | | |
| | | | | | | | | |
| | | | | | | | | |
| | | | | | | | | |
| | | | | | | | | |
| | | | | | | | | |
| | | | | | | | | |
| | | | | | | | | |
| | | | | | | | | |
| | | | | | | | | |
| | | | | | | | | |
| | | | | | | | | |
| | | | | | | | | |
| | | | | | | | | |
| | | | | | | | | |
| | | | | | | | | |
| | | | | | | | | |
| | | | | | | | | |
| | | | | | | | | |
| | | | | | | | | |

## Large tegula recording sheet

| Site | | | | | | Context, dating etc | | | | | | | | | | | | | | | | | | | |
|---|---|---|---|---|---|---|---|---|---|---|---|---|---|---|---|---|---|---|---|---|---|---|---|---|---|
| Recorder | | | | | | | | | | | | | | | | | | | | | | | | | |
| Date | | | | | | F L A N G E | | | | | | | | | LOWER CUTAWAY | | | | | | | | | | |
| Tile No | WHOLE TILE | | | | | LOWER | | | | | | UPPER | | | | | | | | | | | | | |
| | LT | UB | LB | IB | CB | H | W | T | I | E | D | H | W | T | TP | L | UD | LD | IN | UC | C | N | S | P | CONTEXT |
| | | | | | | | | | | | | | | | | | | | | | | | | | |
| | | | | | | | | | | | | | | | | | | | | | | | | | |
| | | | | | | | | | | | | | | | | | | | | | | | | | |
| | | | | | | | | | | | | | | | | | | | | | | | | | |
| | | | | | | | | | | | | | | | | | | | | | | | | | |
| | | | | | | | | | | | | | | | | | | | | | | | | | |
| | | | | | | | | | | | | | | | | | | | | | | | | | |
| | | | | | | | | | | | | | | | | | | | | | | | | | |
| | | | | | | | | | | | | | | | | | | | | | | | | | |
| | | | | | | | | | | | | | | | | | | | | | | | | | |
| | | | | | | | | | | | | | | | | | | | | | | | | | |
| | | | | | | | | | | | | | | | | | | | | | | | | | |
| | | | | | | | | | | | | | | | | | | | | | | | | | |
| | | | | | | | | | | | | | | | | | | | | | | | | | |
| | | | | | | | | | | | | | | | | | | | | | | | | | |
| | | | | | | | | | | | | | | | | | | | | | | | | | |
| | | | | | | | | | | | | | | | | | | | | | | | | | |
| | | | | | | | | | | | | | | | | | | | | | | | | | |
| | | | | | | | | | | | | | | | | | | | | | | | | | |
| | | | | | | | | | | | | | | | | | | | | | | | | | |
| | | | | | | | | | | | | | | | | | | | | | | | | | |
| | | | | | | | | | | | | | | | | | | | | | | | | | |
| | | | | | | | | | | | | | | | | | | | | | | | | | |
| | | | | | | | | | | | | | | | | | | | | | | | | | |
| | | | | | | | | | | | | | | | | | | | | | | | | | |
| | | | | | | | | | | | | | | | | | | | | | | | | | |

# Appendix 4: Catalogue of site dating evidence

This catalogue provides the dating evidence from all the sites where useful information was obtained. The conclusions are given using the system of *strong*, *probable* and *possible* dates outlined in Section 4.4. *Strong* conclusions are underlined and *possible* ones shown in square brackets thus [B=125] would signify that tiles with Group B cutaways were *possibly* in production circa AD 125; B=120-140 would indicate that production *probably* took place some time in the period AD 120-140; whilst B=120+ would show that there was *strong* evidence for Group B production sometime after AD 120. The analysis includes all the possible dating evidence acquired in this survey, even when that data are doubtful and unhelpful.[1]

## Batten Hanger, Sussex

Fourth century villa, possibly overlying an earlier structure and itself built over in the early fifth century.[2] Two tegulae from Group D with upper cutaways formed by mould inserts.

D=300

## Beauport Park, Sussex

A *Classis Britannica* bathhouse where almost all the tegulae were stamped. The excavators believed the bathhouse was constructed in the last part of the first century[3] based upon finds from the adjacent iron-working site but they provided no evidence from the bathhouse itself to support this view. Pottery found within the bathhouse would suggest a date range of AD 130-200.[4] These dates fit better with the findings from Philp's excavation of the neighbouring *Classis Britannica* forts at Dover, where no stamped tile was found in the first fort which was dated to circa AD 120, but was present in the second fort which was dated to circa AD 125.[5] Tiles from both sites would appear to come from the same source[6] and many of the dies are common to both sites. At Dover the only die that appeared stratified in any quantity was RIB 2481.7 where thirteen examples were found only in the Period III use and demolition deposits which may be dated to AD 190-210.[7] At Beauport Park this die appears with Group C and D tegulae.

One Group A, fifteen Group B, thirty-five Group C and twenty-one Group D. Almost all the tegulae are stamped with the exception of the single Group A tile which has no visible stamp and appears to be of a different manufacture and should therefore be regarded as an anomalous introduction. Group B therefore equates to the initial construction and Groups C and D can be dated by comparison with the Dover fort Phase III and placed at either end of this period by adopting the assumption that the typology sequence is correct.

B=125
[C=190]
[D=210]

## Benwell, Hadrian's Wall

One tegula stamped (incuse) by *Ala I Hispanum Asturum*[8] from the fort on Hadrian's Wall. The granary of the fort was constructed by the *Classis Britannica* in AD 122-6[9] and *Cohors I Vangionum Millaria Equitata* was present in the second century[10] but this may only have been part of the unit[11] which was at Risingham by AD 205.[12] *Ala I Hispanum Asturum* was certainly stationed at Benwell by AD 205[13] but may have been there by AD 178 under the governorship of Ulpius Marcellus.[14] If *Ala I Hispanum Asturum* had shared the fort with part of *Cohors I Vangionum Millaria Equitata* then the occupation may have been considerably earlier. One tegula from Group B.

B=170-210

## Bignor, Sussex

A three phase villa spanning the third and fourth centuries, possibly overlying an earlier Roman farm. Period I was a timber frame building constructed circa AD 190-200 and destroyed by fire circa AD 225-250. Period II was stone built, albeit possibly with a half-timbered elevation, and was in use from the middle to the end of the third century. At the start of the fourth century a new villa (Period III) was constructed on the same wall alignments but with the addition of two large wings. This villa remained in use into the fifth century.[15]

There were ten Group C and five Group D tegulae on display at the villa, unfortunately all unprovenanced.[16] There are two main references in Frere's report to tegulae that may have been lifted during excavation: the first is to tegulae reused in the Period III structural alterations outside rooms 6,[17] which must therefore have come from the original Period II roof dating to circa AD 225-250. The second reference is to tiles found in situ from the final collapse of the building and which should therefore date to the fourth century.[18] Even if the tegulae on display at the villa are not the ones that Frere excavated, it still seems likely that they came from very similar contexts.

C=225-250
D=300

## Braughing, Herts

A single period bathhouse constructed circa AD 70-90.[19] All tegulae from Group A.[20]

A=70-90

## Caerleon, Wales

1. Tegulae with stamps that have the cognomen *Antoniniana* which are datable to AD 213-222.[21] Six tegulae with die RIB 2459.54, five with die RIB 2459.55 and five with RIB 2459.56, all the tiles are from Group C.

C=213-222

2. Tiles with RIB 2459.3 stamps were found beneath the *frigidarium* drain of the Fortress Baths and dated by Boon to slightly

[1] The numbers of tegulae identified on individual sites will not always tally with those shown in Figure 3.28 because some tegulae have come from undated contexts or unstratified material. On sites with abundant material not all the tegulae will necessarily have been recorded in this survey, so totals will not always equate with those in excavation reports
[2] Pers comm. James Kenny, Chichester District Archaeologist.
[3] Brodribb & Cleere 1988, 243.
[4] de la Bedoyere 1988, 244.
[5] Williams 1981, 127.
[6] Peacock 1977, 245.
[7] Williams 1981, 126.
[8] RIB 2464.
[9] RIB 1340.
[10] RIB 1328 and 1350.
[11] Holder 1982, 123.
[12] RIB 1234.
[13] RIB 1337.
[14] Holder 1982, 110.
[15] Frere 1982, 137-143.
[16] Aldsworth & Rudling 1995, 177 refer to three tegulae with contextual provenance but unfortunately these were not available at the Bignor villa.
[17] Frere 1982, 169.
[18] Frere 1982, 170.
[19] Partridge 1977, 32.
[20] Rook 1977, 56. These tiles have not been examined by the author.
[21] RIBII.4, 125.

before AD 100.[22] Four tegulae found elsewhere within Caerleon stamped with RIB 2459.3 all from Group A.[23]

A=90-100

3. Phase IVa construction of the building on the Museum Garden site took place circa AD 130-140.[24] Four tegulae all from Group B from the Phase IVa construction or Phase V demolition.

B=130-140

4. On the same site tegulae with stamps RIB 2459.3, 5, 7 and 10 were excavated from the earliest phases dating before AD 100.[25] In total eight examples of these stamps have been identified from sites within Caerleon and all are from Group A.

A=90-100

## Caerwent, Wales

1. Evidence from the most recent excavations of the forum and basilica, places the construction as commencing circa AD 100-125.[26] Wacher[27] infers that construction was part of the Hadrianic stimulation and therefore at the end of this date range. Six tegulae with Group A cutaways found in the pre-construction background.

A=100-120

2. Tegulae were used as levelling courses in the late third century reconstruction of the basilica. Only two of the 104 tegulae examined from forum and basilica come from Group D whilst 67 are Group C, which implies that the transition to Group D only just preceded the reconstruction of the basilica

D=240-280

## Caistor-by-Norwich, Norfolk

The first forum was probably constructed in the second half of the second century[28] and destroyed by fire in the early third. It was replaced by the second forum in circa AD 270 according to Frere's reconstruction of the original excavation records.[29] Eleven tegulae have been retained from the excavation of which nine are complete. Four of the tiles are Group D and probably[30] seven are Group C. If the cutaway sequence is correct then logically the Group C tegulae should come from the first forum and the Group D from the second

[C=170]
D=270

[22] Boon 1984, 15-16.
[23] In addition to the early tiles from the Fortress Baths and the late *Antoniniana* tiles discussed above, Boon makes a number of associations with tile stamps from the Hadrianic/Antonine period all of which correlate with Group B cutaways, but these have not been included on the timeline analysis. However, he also identified the stamp RIB 2459.36 as being "perhaps c 140" (Boon 1984, 30) whereas the seven *tegulae* that could be positively identified with this stamp were all from Group C. See further discussion on Boon's dating in Section 5.3.
[24] Zienkiewicz 1993, 28.
[25] Zienkiewicz 1993, 129.
[26] Guest 2000, 27.
[27] Wacher 1995, 378.
[28] Frere 1971, 9.
[29] Frere 1971, 13.
[30] The tiles are currently a museum display of a partly reconstructed roof; as a consequence it was not possible to closely inspect all the cutaways.

## Caistor-by-Yarmouth, Norfolk

An early third century fort built on previously unoccupied ground[31] yielded three Group C and two Group D tegulae. The Group C tegulae should be assigned to the original construction and the Group D to the later phases of its occupation.

C=200

## Carlisle

1. The only incuse stamps produced by *Legio* II and XX have been found at Carlisle. The stamps are of similar design, the tegulae almost identical in size and the tiles have been found together on the same tile tomb.[32] There are two inscriptions linking detachments of *Legio* II and XX in a combined vexillation; only the one found at Netherby[33] (just a few miles from Carlisle) can arguably be dated to AD 219.[34] It seems plausible to link the incuse stamp designs with this period of joint operation; however, by AD 219 these stamps would have been expected to carry the *Antoniniana* cognomen so the dating may be slightly earlier. Non-diagnostic but stratified examples of these stamps have also come from the Annetwell Street excavations and are dated to AD 180-320[35] which supports the above dating.

Two tegulae (one from each legion) both with Group C cutaways.

C=200-210

2. Two examples of a further incuse stamp with the inscription IMP[36] have come from Group B but insecurely provenanced tegulae. However this same stamp has also been found in the Annetwell Street excavations where the context was dated as AD 170-183[37] although these tiles were too fragmentary to be diagnostic.

B=170-183

## Castleford, Yorkshire

Four phases of the fort and four (separate) phases of the associated vicus have been identified.[38] Phase I and II of the fort span the period AD 71-100 and have yielded two Group A tegulae. Phase III was a period of abandonment lasting from AD 100-250: it yielded two (presumably residual) Group A tegulae but none from Group B or C. The start of Phase IV was less certain[39] but was taken to run from AD 250-400. During this period, building material was taken from the vicus and redeposited in the fort for use in drains and wall foundations.[40] Six Group A tegulae and four Group C tegulae come from this phase but as it is all secondary material it cannot be used for dating.

Phase 1 of the vicus ran from AD 71-86 and yielded one Group A tegula. Phase 2 (AD 85-140) also yielded one Group A tegula. Phase 3 (AD 140-180) yielded two Group B and two Group C tegulae, and Phase 4 (AD 180-400) yielded one Group C tegula. Neither site produced any Group D tegulae but stone roof tiles were present which would imply that clay roof tiles gave way to stone prior to the introduction of Group D.

Fort Phase I: A=71-100
Vicus Phase 1: A=71-86
Vicus Phase 3: [B=140]
[C=180]

[31] Darling with Gurney 1993, 6.
[32] Ferguson 1895, 261. Boon 1984, 15 states that this burial rite is likely to be subsequent to Hadrian.
[33] RIB 980.
[34] Hassall & Tomlin 1989, 331.
[35] Caruana & Hird, 34-35.
[36] RIB 2483.
[37] Caruana & Hird, 34.
[38] Abramson et al 1999,19.
[39] Abramson et al 1999,20.
[40] Abramson et al 1999, 20.

## **Chelmsford Mansio, Essex** (Mildmay Road excavation)

The mansio was originally constructed circa AD 120-140 but was built in stages with the east wing being later and a second phase of the bathhouse being added in the late second or early third century. Tegulae from the east wing were used to level the ground for the second phase of the bathhouse, and again for resurfacing following a late third century fire.[41] One group C tegula comes from the construction of the second phase of the bathhouse and ten from the east wing tiles used for resurfacing and levelling.

[C=170-200]

## **Chester**

Four tegulae from Chester all from Group B and bearing *Legio* XX stamp RIB 2463.56 which ends with the lettering "VVV". Two non-diagnostic tegulae with a *Legio* XX stamp ending "VVV" (probably RIB 2463.57) were also found at Tarbock along with five consular dated stamps of AD 126 made by Viducius.[42] The two stamp types appear to be virtually contemporaneous production[43] and it seems extremely likely that the third letter "V" of the Tarbock "VVV" tiles refers to Viducius. It is therefore reasonable to assume that this should also apply to the other *Legio* XX stamps ending "VVV" which by association, should also date closely to AD 126.

B=126

## **Chignall, Essex**

A villa estate with buildings in use from the early second century to the fourth.[44] One complete Group D tegula came from a Period III.3 pit[45] which is dated as AD 245-285.[46] The source for the tile is believed to be Structure 4[47] which was first erected after AD 120 and was modified in the mid-third century.[48] On the basis that the tegula originated from the second phase of Structure 4:

[D=250]

## **Chilgrove 2, Hants**

A villa site where the first masonry built building was constructed in the late third to early fourth century.[49] Four tegulae all from Group D.

D=270-330

## **Colchester, Essex**

St Mary's Hospital: a Boudiccan destruction horizon site.[50] Four tegulae all from Group A.

A=50-60

## **Crookhorn, Hants**

Part of a villa estate incorporating an early fourth century tilery,[51] the villa itself being 600 metres to the east.[52] The earliest structure on the tilery site was an aisled building which the excavators dated as mid/late third century[53] although it was initially thought to be rather earlier.[54]

The tilery site has yielded two Group B, twelve Group C, four Group D and ten regional Type 7 tegulae. One of the Group B tegulae was stamped[55] and matches a tegula found at the villa itself which was also Group B. A further example (now lost) was found at the Langstone villa some three miles away.[56] The Crookhorn villa seems to have run from late first to the early fourth century[57] whilst the finds suggest that the main occupation of the Langstone villa was probably in the second and third centuries.[58] It is possible that the stamped tiles were made in the fourth century, but the dating of the villas and the fact that there were only two Group B tegulae found at Crookhorn compared to 26 (unstamped) later ones suggests that they were residual material from earlier production.

The Group C could reasonably be associated with the third century aisled barn which predates the kiln, whilst the Group D and Type 7 tegulae were presumably product from that kiln. The archaeomagnetic date for the kiln was AD 330+/-20[59]which is therefore the date for the regional cutaways on the basis that these followed after Group D.

[C=250]
R=330

## **Dorchester, Dorset**

1. Greyhound Yard had a number of earlier timber structures before the first masonry building was established in the third century, followed by the development of a courtyard house in the fourth century.[60] Eleven Group C and two Group D tegulae. However, as stone tiles were also present,[61] the Group C tegulae should be associated with the first masonry building.

[C=200]

2. Wadham House: the only building on the site appears to have been a substantial masonry structure probably dating to early in the third century.[62] One Group C tegula.

C=200

3. Wollaston Bathhouse was originally constructed in the second century with subsequent additions which were modified in the late third century.[63] One Group D tegula from the collapse of the final phase.

D=275

## **Dorchester-on-Thames, Oxon**

Beech House Hotel was the site of a mid-third century courtyard house that was the earliest substantive structure on the site.[64] Six tegulae all from Group D.

D=250

---

[41] Pers Comm: Patrick Allen Essex County Archaeological Team.
[42] See Section 4.7 for argument on the AD 126 date.
[43] Cowell & Philpott 2000, 114.
[44] Clarke 1998, 1.
[45] Clarke & Black 1998, 113.
[46] Clarke 1998, 129.
[47] Clarke & Black 1998, 112.
[48] Clarke 1998, 129.
[49] Down 1979, 43.
[50] Crummy pers comm.
[51] Goodburn 1978, 464 gives the archaeomagnetic dating as c AD 330.
[52] Wright et al 1976, 384 note 37.
[53] Soffe et al 1989, 108.
[54] Goodburn 1976, 366 reports the principal structures as dating to the second/third centuries.
[55] RIB 2489.42.
[56] Wright et al 1976, 384 note 37.
[57] Taylor & Collingwood 1926, 232.
[58] Taylor & Collingwood 1922, 273.
[59] McWhirr 1979, 137.
[60] Woodward et al 1993, vii.
[61] Bellamy 1993, 172.
[62] Draper & Chaplin 1982, 25.
[63] Pers Comm: John Magilton.
[64] Rowley & Brown 1981, 8.

## Exeter, Devon

1. Two Group A tegulae from the fortress baths which were constructed AD 60-65.[65]

A=60-65

2. Three Group A tegulae from a military feature in Lower Coombe Street which could be dated to circa AD 75 from associated pottery.[66]

A=75

## Fishbourne, Sussex

The proto-palace preceded the building of the main palace in AD 75-80.[67] In the second and third centuries parts of the palace were demolished but generally alterations were insubstantial.[68] The biggest change was the construction of a new bathhouse in the north wing circa AD 100 which was itself demolished about fifty years later.[69] The whole building was destroyed by fire circa AD 280-290.[70] Recent excavations have yielded 292 diagnostic tegulae, 87% of these are Group A[71] which strongly supports the proposition that the second and third century alterations were insubstantial. (It might also suggest that the tiles on the second century bathhouse were removed from the site when it was demolished).

Two Group A tegulae came from an abortive foundation trench for the palace and must therefore predate it.

A=60-70

## Gelligaer, Wales

A Welsh fort with a fragmentary inscription indicating that construction was completed between AD 103-11.[72] Three, otherwise unprovenanced, tegulae from Group A.

A=103-111

## Gestingthorpe, Essex

A villa perhaps spanning the first century to the fourth.[73] Building 1 had two phases: the first ended with a fire in the last quarter of the second century and reconstruction for the second phase appears to have started soon after with the building remaining in use into the fourth century. Building 2 seems to have followed a similar pattern to Building 1.[74] Five complete tegulae: four from Group C and one from Group D have been found in association with Buildings 1 and 2.[75]The Group C tegulae should probably be associated with construction of the second phase.

[C=200]

## Gloucester

1. St Oswald's Priory: the site of the municipal tile works which started production in the late first century and ceased production in the third.[76] Many of their products were stamped with a range of stamps all incorporating the lettering "RPG" which may be expanded as *Res Publica Glevensium.*[77] The excavators believed that stamping of the tiles started early in the second century[78] whilst McWhirr & Viner suggest a date of circa 110-125.[79] Only three stamped tegulae with diagnostic cutaways have been identified[80] all of which were Group B. A further eight unstamped Group B tegula and a single Group A tegula have been examined from St Oswald's Priory.

B=110+

2. St Mary de Lode: a large private house or public building constructed in the second century and in use into the fourth century.[81] Both stone and clay tiles were found so presumably the clay tiles represented the initial second century building phase because stone tiles tend to be a third or fourth century feature.[82] The site has also yielded a number of (undiagnostic) tegulae carrying the RPG series of stamps discussed above which tends to confirm the dating as first half of the second century. Eight Group A tegulae and fourteen from Group B.

[A=100-140]

## Gosbecks, Essex

The masonry built theatre (phase 2) probably dates to the second half of the second century and went out of use early in the third century.[83] Sixteen Group C tegulae and a single one from Group B.

[C=175]

## Great Chesters, Hadrian's Wall

A single tegula with the stamp of *Cohors II Asturum.*[84] *Cohors II Asturum* is first evidenced in Britain on a diploma of AD 105[85] when it was presumably based at Llanio in Wales.[86] This fort was abandoned in the early Hadrianic period[87] when the unit was moved north. It is subsequently attested as the third century garrison of Great Chesters under the legionary centurion Valerius Martialis dated in AD 225.[88] There is an undated inscription from a statue base at Great Chesters for the *Cohors VI Nerviorum*[89] who are also attested at Rough Castle in the Antonine period[90] then subsequently at Bainbridge from AD 205-8[91] and remain there according to the *Notitia Dignitatum.*[92] *Cohors VI Nerviorum* is therefore most likely the first garrison of Great Chesters departing in AD 139 for the Antonine Wall. The only other units associated with Great Chesters are *Cohors (?) Raetorum* in AD 166-9[93] and *Vexillatio G R[a]eto[rum]*[94] (undated) which may well be related. *Cohors II Asturum* presumably succeeded the *Raetorum* units towards the end of the second century and occupied Great Chesters until at least AD 225. As the stamp has an ansate frame this tends to suggest a mid-second century date.[95] They may therefore have taken possession as early as AD 170. One tegula from Group B.

B=170+

[65] Bidwell 1979, 57.
[66] Exeter Archaeology pers comm.
[67] Cunliffe 1971,219.
[68] Cunliffe 1971, 154.
[69] Cunliffe 1971, 220.
[70] Cunliffe 1971, 220.
[71] Turner 2003, unpublished reports FBE.95-99 and FBE02.
[72] RIB 397.
[73] Draper 1985, 19-20.
[74] Draper 1985, 20.
[75] Johnston & Williams 1985, 80.
[76] Heighway & Parker 1982, 31.
[77] RIBII.5, 41.
[78] Heighway & Parker 1982, 31.
[79] McWhirr & Viner 1978, 365.
[80] Two from Hucclecote (RIB 2486.1 and RIB 2486.15) and one from St Oswalds Priory (RIB 2488.1).
[81] Bryant 1980, 7.
[82] McWhirr & Viner 1978, 371 and Brewer 1993, 20.
[83] Dunnett 1971, 41.
[84] RIB 2467.1.
[85] De la Bedoyere 1999, 71.
[86] RIB 407, 408.
[87] Jones & Mattingly 1990, 104 Map 4.36.
[88] RIB 1738.
[89] RIB 1731.
[90] RIB 2145.
[91] RIB 722.
[92] De la Bedoyere 1999, 78.
[93] RIB 1737.
[94] RIB 1724.
[95] See discussion on the dating of ansate forms in Section 5.16.

## Grimescar, Yorks

A military kiln works supplying tile to the Yorkshire forts of Slack and Castleshaws. The tile sent to Slack appears to have been mainly stamped *Cohors IIII Breucorum* and was supplied circa AD 104,[96] whereas the tile for Castleshaws was unstamped and probably used for the rebuilding works circa AD 120.[97] Most of the residual tile found at the kiln site is likely to date to its last phase of supply to Castleshaws before being abandoned when the unit moved to Hadrian's Wall post AD 120. Four Group A tegulae and 9 Group B tegulae were found at Grimescar. Only one of these (from Group A) is stamped: as this was an antiquarian find it may have been misattributed to Grimescar rather than Slack.

B=120-125

## Halstock, Dorset

A villa which began life in the second century and continued into the fourth incorporating additional buildings and modifications.[98] Three Group D tegulae are associated with the eastern end of Building 4[99] which was initially constructed circa AD 275-325.[100]

D=275-325

## Holt, Cheshire

Grimes places the principal period of operation of the *Legio* XX kilns between the end of the first century and the start of the Antonine period.[101] This end date has been further refined by, inter alia, Ward[102] to a little beyond AD135. Eight tegulae from Group A, nineteen from Group B and one from Group C.[103] Consistent with the established sequence, the Group B tegulae may be assigned to the end of the period.

[B=130-140]

## Inchtuthil, Scotland

Agricolan legionary fortress on the Tay.[104] One tegula from Group A.

A=84

## Lancaster

Originally a Flavian fort[105] with one tegula stamped by the *Ala Gallorum Sebosiana.*[106] The unit is first attested in Britain in an Agricolan writing tablet sent by a member detached as a *singularis consularis.*[107] The only firm evidence of the unit's location at Lancaster is on an inscription recording the rebuilding of the bathhouse circa AD 262-6.[108] *Ala Gallorum Sebosiana* is not recorded elsewhere in Britain except on a commemorative inscription set up by its *praefectus* to a successful boar hunt found near Binchester[109] but which should not be taken to imply that the unit itself was stationed there.[110] The only other unit that might have occupied Lancaster is an "*Ala Augusta*" which may refer to the *Ala Gallorum Proculeiana* or the *Ala Vocontiorum,* no later than the early second century,[111] although a name on a single tombstone is inconclusive. The stamp itself has an ansate form which tends to suggest a second century date. There are also some incuse stamps from Lancaster[112] found mainly on brick which probably date to the second half of the second century.[113] On balance it would appear likely that the *Ala Gallorum Sebosiana* was present at Lancaster in the second century, perhaps from AD 100 if the "*Ala Augusta*" was indeed the original garrison. One tegula with probable Group A cutaway.[114]

A=100-200

## Leicester

1. Norfolk Street: a late second century villa that was extended in the late third or early fourth century.[115] Nine Group D and five Group C tegulae came from the extension.

[C=180]
D=300

2. Norfolk Street: underlying the villa was evidence of earlier occupation of the site, inter alia, including a probable corn dryer[116] incorporating a number of tegulae. The backfill of the corn dryer contained mid to late second century pottery which would suggest that the construction of the corn dryer was no later than AD 180 if the villa was then built on top of it. Six tegulae all from Group C came from the corn dryer and were presumably secondary material.

[C=160-]

3. Blackfriars Pavement: the construction layer for a town house dated to AD 125-135[117] contained one Group A and two Group B tegulae.

B=130-

## Leucarum, Wales

A westerly Welsh fort originally established by Frontinus, five Group B tegulae came from the third *praetorium* dated circa AD 110 and two from the construction of the fourth *praetorium* in circa AD 115.[118] There were also two Group A tegulae excavated from these contexts.

B=110-115

---

96 Richmond 1925, 47.
97 Richmond 1925, 57. He also reports that a single stamped tile was found at Castleshaws and from this he dates an initial reconstruction at Castleshaws to AD 104 to match the dates for the reconstruction in stamped tile at Slack.
98 Lucas 1993, 131.
99 Lucas 1993, 141.
100 Lucas 1993, 136.
101 Grimes 1930, 52.
102 Ward 1997, 142.
103 There is an element of doubt concerning the provenance of this Group C tegula (Tile 1097). It was one of a batch of complete tegulae held at the Museum of Wales in Cardiff that were all meant to have originated from Holt. However one of other tegulae (and the only one other than Tile 1097 with a Group C cutaway) had a *Legio* II stamp and was similar to other tiles with this stamp from Caerleon. (There was another tegula definitely from Caerleon mixed in with their Gelligaer holding). It is therefore possible that Tile 1097, which was unstamped, also came from Caerleon. Against this the morphology of Tile 1097 differed from the Caerleon tiles and it had a finds number (25-1/97) consistent with other tiles that definitely originated from Holt (although this would depend upon when the numbering was applied).
104 Frere 1999, 105.
105 Jones & Mattingley 1990, 99.
106 RIB 2465.1.
107 Tomlin 1998, 74.
108 RIB 605.
109 RIB 1041.
110 Shotter 1988, 213.
111 Shotter 1988, 214-5.
112 RIB 2465.2.
113 See Section 5.15.
114 The cutaway was damaged and could possibly have been from Group D but no other Group D cutaways were observed in a review of (undated) Lancaster tegulae whereas Group A cutaways of similar appearance were observed.
115 Lucas 1981, 103-4.
116 Mellor & Lucas 1979, 69.
117 Lucas pers comm.
118 Marvell & Owen John 1997, 210-212.

## London

1. Lime Street: fifteen tegulae mortared to a wall as a damp-proof course. According to Betts' fabric typology the majority of these tiles date between 140-200.[119] Fourteen tegulae from Group C.

C=140-200

2. Lime Street: one of the above tegulae was in Betts' fabric 2459B which post dates the late second century fabrics. This tegula was from Group D.

D=200+

3. Mansell Street: a tile cist formed of fifteen tegulae. With one exception all the tegulae were from Betts' early fabric group 2815 which he dates as first to mid-second century,[120] however some of these fabrics are dated as 140-200 in his later unpublished analysis cited above. Betts also has a late fabric group 2815 which is distinguished from the early 2815 solely by the use of a finer moulding sand but the dates are 100 years later.[121] Eleven Group C cutaways and three Group D cutaways.

[C=140-200]
[D=140-200]

4. Mansell Street: there was one tegula from the cist in fabric 3016 which Betts dates as 250/300 – 400. This was from Group D.

D=250-400

## Maiden Castle, Dorset

A temple of Romano-Celtic form lying within the Iron Age fort. The second phase of the building was roofed with tegulae[122] and is dated no earlier than AD 379 on the basis of coins sealed within the secondary floor.[123] Three complete tegulae all from Group D, two of which incorporated upper cutaway inserts.[124]

D=379

## Maryport, Cumbria

A single tegula from the fort with a stamp of *Cohors I Hispanorum* (RIB 2474). This was the first garrison of this Hadrianic fort and its presence is confirmed by the inscription to the garrison commander M Maenius Agrippa.[125] *Cohors I Hispanorum* was replaced by *Cohors I Delmatarum* under the prefect P. Postumius Acilianus during the reign of Antoninus Pius.[126] Holder[127] places the departure of *Cohors I Hispanorum* early in the reign because the unit gained the honorific title *Aelia* in the conquest of Scotland. One tegula from Group B.

B=120-140

## Mumrills, Scotland

An Antonine Wall fort.[128] One tegula from Group B.

B=140

## Narborough, Leicestershire

A mid to late-third century villa built over a corn-dryer which in turn was built over a late-second to early-third century working area which included two bowl furnaces.[129] The walls of the corn-dryer were constructed from tegulae. The corn-dryer cannot be later than the late-third or earlier than late second, and the tegulae were presumably secondary material when installed in the corn-dryer. If the corn-dryer was built in AD 225 at the mid-point of these extremes, then the tegulae may have been produced a generation earlier in AD 200 say. Twenty nine tegulae: twenty-seven from Group C and two damaged fragments originally attributed to Group B.

[C=200]

## Norton Disney, Lincs

A fortified villa with five successive periods of construction, all apparently ended by fire.[130] The first stone built structure (Period III) was erected circa AD 200[131] and was almost certainly tiled because tegulae were used to create a fireplace.[132] The Period IV villa was constructed circa AD 230 and tegulae were found in its destruction deposits.[133] The final phase of the building is dated circa AD 300[134] when the roof incorporated both stone and ceramic tiles.[135] Nine tegulae, all from Group C, have been retained from the excavation, but have no other provenance. As these tegulae were all complete, it seems most likely that they will have come from a Period V drain[136] utilising material from Period IV (although Period III is also possible).

[C=230]

## Piddington, Northants

A multi-period villa where dates have been assigned to fabric types.[137] Fabric 1 is dated as late third/early fourth century and all four tegulae identified as being made of this fabric were Group D. Fabric 2 was also initially dated as third/fourth century but has subsequently been amended to late second to early third century[138] and this yielded three Group C and one Group B tegula. Fabric 3 has been identified as early second century because it was present in deposits that pre-date an Antonine fire:[139] five tegulae identified in Fabric 3 (which did not necessarily come from the pre-Antonine deposit) were Groups C and D.

The difficulty of using fabric evidence for the dating of tegulae is that, whilst it may be possible to fix when a fabric first appears on a site, it is not often possible to be sure when its use ceases. Caruana[140] has shown that at Carlisle not only were many of the fabrics utilised over a long period of time, but that they co-existed with many other different fabrics which were also used for the manufacture of tegulae. Similar conclusions have been drawn at York.[141] For this reason only the Fabric 1 data, which starts towards the end of the Roman period and is

[119] Betts 2002.
[120] Betts 2000, 340.
[121] Betts 2000, 341.
[122] Wheeler 1943, 74.
[123] Wheeler 1943, 75.
[124] If all the tegulae were made and installed on the roof at the same time then it is surprising that two were manufactured using upper cutaway blocks and one apparently was not. Although coming from at least two different moulds, the tegulae are dimensionally sufficiently similar not to be obviously different manufacture, nevertheless the possibility that the tiles were originally installed on another building or buildings prior to coming to the temple should not be discounted.
[125] RIB 823.
[126] RIB 850.
[127] Holder 1982, 118.
[128] Frere 1999, 131.
[129] Lucas 1983, 75.
[130] Oswald 1937, 159.
[131] Oswald 1937, 159.
[132] Oswald 1937, 147.
[133] Oswald 1937, 153.
[134] Oswald 1937, 159.
[135] Oswald 1937, 149.
[136] Oswald 1937, 153.
[137] Ward 1999, 12.
[138] Ward 1999, 73.
[139] Pers comm: Cynthia Ward.
[140] Caruana & Hird, 28.
[141] Betts 1985, 247.

therefore not at risk of reuse or extended life, has been used for the dating of tegulae.

D=280-320

## Rivenhall, Essex

A first/second century villa damaged by fire circa AD 200 when it was remodelled.[142] One Group C tegula.

[C=200]

## Sandwich, Kent

A villa tentatively identified as first or second century construction which had fallen out of use by the late third or early fourth century.[143] The tegulae came from the possible collapse of the porch[144] but were also found in a gully, which post-dated the building, in association with late third or fourth century pottery.[145] The excavators found no evidence for any alteration to the structure from examination of the (very truncated) foundations.[146] The tegulae should therefore be assigned to the original construction but the possibility of some later change to the structure, such as the incorporation of the porch, should not be ignored. Two tegulae from Group D.

[D=100]

## Shelford Quarry, Kent

A rural building which yielded one Group A tegula from a pre-Flavian context.[147]

A=60-70

## Slack, Yorkshire

Originally an Agricolan timber fort in Yorkshire which was renovated in stone circa AD 104 and had further works sometime in the early AD 120s which were still incomplete at the time of abandonment circa AD 125.[148] Most of the tile that has been retained carries stamps of *Cohors IIII Breucorum.*[149] Richmond believes that it was these tiles that were used in the stone renovation of the fort circa AD 104.[150] Fifteen Group A tegulae of which all but two were stamped and two unstamped Group B tegulae. It is reasonable to surmise that all the Group A tegulae were part of the original stone renovation circa AD 104 and that the unstamped Group B tegulae probably relate to the rebuilding post AD 120.[151]

A=104
B= 120-125

## South Shields, Tyne & Weir

A fort of Flavianic or Hadrianic origin running through to the late fourth century.[152] Twenty-five Group B, C, D and regional tegulae. Five tegulae are stamped by *Cohors V Gallorum*[153] all with regional Type 8 cutaways. *Cohors V Gallorum* was the third and fourth century garrison which according to Holder was at Cramond in the reign of Severus.[154] Bidwell & Speak place *Cohors V Gallorum* at South Shields slightly earlier from 205.[155] One of the stamped tiles appears to have stamp RIB 2473.9 which reads "COHVGG": the second "G" being attributed to the cognomen *Gordiana* and thus dateable to AD 238. However, this is probably a misreading of a peg mark that appears on many of the dies.[156] Stamped tegulae were used on the roofs of buildings in Period 7[157] which ran from late third/early fourth century to the mid-fourth century. Many stamped tiles also originated from later contexts but they could have been secondary deposits. Some stamped tegulae were used as levelling for Period 7 construction with the inference that they came from Period 6 (or earlier) roofs which would have been constructed between AD 222 and the late third or early fourth century. Recognising the evidence from Period 7 and the existence of Group D tegulae which should predate the regional form:

R=300-340

## Sparsholt, Hants

A villa where the first building was demolished in the late second or early third century to be succeeded by an aisled house and a main villa, the latter of which was constructed in the early fourth century.[158] Seven drab coloured Group C tegulae and seventeen red regional Type 7 tegulae. The red tiles came from the main villa building.[159]

R=300

## Stanton Low, Bucks

A large villa complex originating in the second century with episodes of decline and expansion until in the late third/early fourth century the complex underwent a major rehabilitation, including re-roofing.[160] Two groups of tegulae have been retained from the excavation: the first was made of a red non-calcareous fabric and the second a drab buff fabric with varying amounts of calcareous inclusions. Many of the flanges of this latter group carried stamps in relief[161] and, whilst the stratigraphical evidence was poor, those that could be located all came from late or debris levels.[162] The stamp reads AVIENU[S] which is a rare name that was borne by a fourth century senator and two late Roman literary figures.[163] A late date for this group of buff coloured tiles is also supported by upper cutaways made by an insert into the mould.

There were five Group C tegulae all in the red fabric and three Group D tegulae all in the buff calcareous fabric. All of the Group D cutaways were too fragmentary to include the stamp, but it is reasonable to assume that they were part of the same production as the flanges in the same fabric that were stamped. There is no dating information for the Group C tegulae, but the buff fabric Group D tegulae must be part of the late third/early fourth century re-roofing and general rehabilitation project.

D=300

## Wallsend, Hadrian's Wall

One tegula stamped by *Cohors IIII Lingonum.*[164] This unit came to Britain with Cerialis[165] and is only attested at Wallsend. The only other

[142] Rodwell & Rodwell 1985, xii.
[143] Parfitt 1980, 242-3.
[144] Parfitt 1980, 236.
[145] Parfitt 1980, 244.
[146] Parfitt 1980, 236.
[147] Canterbury Archaeological Trust pottery dating file.
[148] Richmond 1925, 46-47.
[149] RIB 2470.1 and 2.
[150] Richmond 1925, 47.
[151] Villy 1911, 6 provides a photograph of a "typical" complete tegula found at Slack which is unstamped and from Group B. This is not in the museum and may well have been discarded along with most of the rest of the unstamped tile.
[152] Bidwell & Speak 1994, 9.
[153] RIB 2473.1-9.
[154] Holder 1982, 117 on the basis of an altar at Cramond although the dating cannot be certain.
[155] Bidwell & Speak 1994, 155.
[156] Bidwell & Speak 1994, 154.
[157] Bidwell & Speak 1994, 156.
[158] Wilson 1973, 318.
[159] Johnston 1969, 17.
[160] Woodfield & Johnson 1989, 265.
[161] RIB 2489.7.
[162] Woodfield & Johnson 1989, 252.
[163] RIB II.5 1993, 59.
[164] RIB 2476.1.
[165] Holder 1982, 118.

unit mentioned at Wallsend is *Cohors II Nerviorum*[166] but Holder is of the view that it was not necessarily stationed there.[167] As the stamp has an ansate frame it is likely to be of second century date. One tegula from Group B.

[B= 120-200]

## Wantage, Oxon

A building constructed in Mill Street probably in the early fourth century on the site of a second century stone-tiled building[168] that was demolished sometime between the middle of the third century and the early fourth century.[169] Five Group D tegulae from the deposit representing the demolition of the earlier building and the construction of the later one.

[D=300]

## Wroxeter

Following the earlier aborted bath building, construction of the new public baths started circa AD 120 but was not finished until circa AD 150.[170] The baths were re-roofed in stone around 100 years later.[171] Five Group A tegulae, 27 Group B and two Group C were recorded. Eleven of the tegulae were complete and all of these were from Group B. Given the comparative quantities, the likelihood is that the Group B tegulae represent the original roof of the baths with the Group A tegulae being residual material perhaps from the military phase of the town.

B=140-150

## York

The tile tombs in the Yorkshire Museum include 41 tegulae with legionary stamps; six of these are *Legio* IX and the remainder *Legio* VI. *Legio* VI arrived in Britain circa AD 122 and is first attested on a dedication stone from Halton Chesters[172] during the governorship of Aulus Platorius Nepos (AD 122-126). It is reasonable to assume that they immediately started stamping tiles consistent with their predecessors practice and that adopted by the other British legions of the time. *Legio* IX which was replaced by *Legio* VI is last attested on an inscription at York dating to AD 108[173] and it therefore left Britain sometime between then and AD 122 when *Legio* VI arrived. All thirty five *Legio* VI tegulae come from Group B. Two of the *Legio* IX tegulae are Group A and four are Group B, relying on the sequencing of the cutaway groups, the Group A tegulae can be assigned to the start of tile stamping (circa AD 100) and the Group B tegulae to the end of *Legio* IX's stay in Britain.

*Legio* IX
A=90-120
B=90-120

*Legio* VI
B=120+

[166] RIB 1303.
[167] Holder 1982, 120.
[168] This is an early date for the use of stone tiles in this area.
[169] Holbrook & Thomas, 9-11 and 36.
[170] Wacher 1995, 368.
[171] Barker et al 1997, 52.
[172] RIB 1427.
[173] RIB 665.

## Appendix 5: Calculation of the required breadth for self-meshing

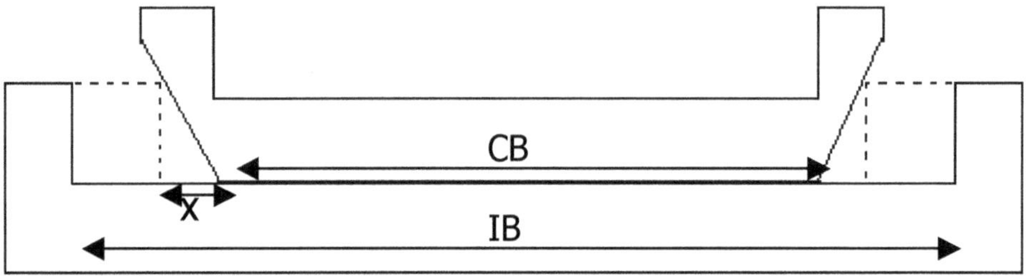


In the above stylised diagram the distance "x" is determined by where the projection of the height of the flange of the underlying tegula strikes the cutaway of the tile above. The distance "CB + 2x" represents the required breadth if the tegula is to be self-meshing. Provided that the actual breadth at the upper end "IB" is greater than or equal to the required breadth then the tegulae will mesh.

The diagram below shows the point at which the tegulae are touching and the actual breadth equals the required breadth.

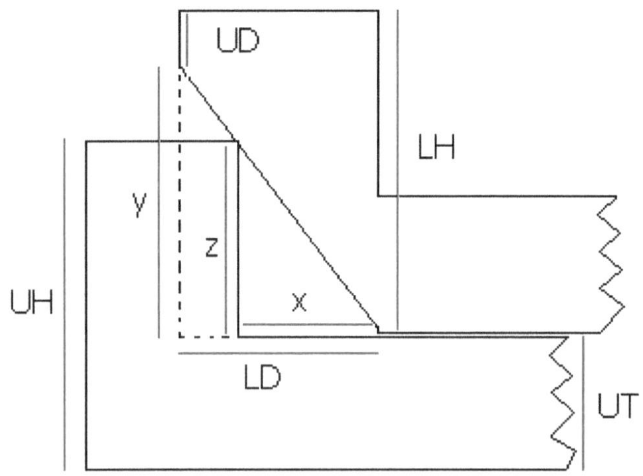


In this diagram "x" can be calculated by similar triangles: thus
x/z = LD/y or x = LD.z/y
but z = UH-UT and y = LH-UD
therefore x = LD(UH-UT)/(LH-UD)

Hence "x" can be expressed in terms of the standard measurements that were taken on each tegula in the survey, and therefore whether the tegulae were self-meshing can be directly calculated from these measurements.

The above example illustrated the calculation using a Group B lower cutaway but the same approach can be adopted for all the other cutaway forms. Formulae for each cutaway form were fed into the database such that the required breadth could be determined for all the complete tegulae that have been observed and some of these results are plotted graphically against the actual breadths in Chapter 6.

# Bibliography


Abramson, P., D.S. Berg and M.R. Fossick 1999. *Roman Castleford: Excavations 1974-85 Volume II The Structural and Environmental Evidence* (West Yorkshire Archaeology Service).

Adam, J.P. 2005. *Roman building - Materials and Techniques* (London).

Aldsworth, F. and D. Rudling 1995. "Excavations at Bignor Roman villa, West Sussex 1985-90," *Sussex Archaeological Collections* 133, 103-188.

Barker, P., R. White, K. Pretty, H. Bird and M. Corbishley 1997. *The Baths Basilica Wroxeter: Excavations 1966-90* (English Heritage Archaeological Report 8).

Bellamy, P. S. 1993. "The Clay and Stone Roof Tile," in Woodward *et al* 1993, 172-174.

Betts, I. M. 1985. *A scientific investigation of the brick and tile industry of York to the mid-eighteenth century,* (Bradford unpublished PhD thesis).

Betts, I. M. 1987. "Ceramic Building Material: Recent Work in London," *Archaeology Today* Vol 8 No 9, 26-28.

Betts, I. M. 1990. "Roman Brick and Tile," in S. Wrathmell and A. Nicholson, *Dalton Parlours Iron Age Settlement and Roman Villa* (Yorkshire Archaeology 3) 165-170.

Betts, I. M. and Foot R 1994. "A Newly Identified Late Roman Tile Group from Southern England," *Britannia* 25, 21-34.

Betts, I. M. 1995. "Procuratorial Tile Stamps From London," *Britannia* 26, 207-229.

Betts, I. M., E. W. Black and J. L. Gower 1997. *A Corpus of Relief-patterned Tiles in Roman Britain* (Journal of Roman Pottery Studies Vol. 7).

Betts, I. M. 1998. "The Roman Tile," in Cool and Philo 1998, 225-232.

Betts, I. M. 2000. "Ceramic building material and stone," in S. Barber and D. Bowsher, *The Eastern Cemetery of Roman London* (English Heritage) 340-348.

Betts, I. M. 2002. *Roman Brick and Tile Dating* (unpublished analysis).

Betts, I. M. "The building materials," in N. Bateman, C. Cowan and R. Wroe-Brown, *London's Roman amphitheatre Guildhall Yard East, City of London EC2* (MoLAS Monograph forthcoming) 219-224.

Bidwell, P. T. 1979. *The Legionary Bath-House and Basilica and Forum at Exeter* (Exeter Archaeological Reports Vol 1).

Bidwell, P. and S. Speak 1994. *Excavations at South Shields Roman Fort Vol I* (Soc of Antiq of Newcastle upon Tyne).

Birley, R. E. 1962-3. "Excavations of the Roman Fortress at Carpow, Perthshire, 1961-2," *Proc Soc Antiq Scot* 96, 184-207.

Birley, A. 2002. *Garrison life at Vindolanda: a band of brothers* (Stroud).

Bogaers, J. E. and C. B. Ruger 1974. *Der Niedergermanische Limes* (Koln).

Boon, G. C. 1974. Silchester: *The Roman Town of Calleva* (London).

Boon, G. C. 1984. *Laterarium Iscanum - the antefixes, brick and tile stamps of the Second Augustan Legion* (National Museum of Wales, Cardiff).

Bowman, A. K. 1994. *Life and letters on the Roman frontier Vindolanda and its people* (New York).

Brewer, R. J. 1993. *Caerwent Roman Town* (CADW).

Brodribb, A. C. C., A. R. Hands and D. R. Walker 1968. *Excavations at Shakenoak Farm, near Wilcote, Oxfordshire* (Oxford: A R Hands).

Brodribb, G. 1979. "A Survey of Tile from the Roman Bath House at Beauport Park, Battle, E Sussex," *Britannia* 10, 139-156.

Brodribb, G. 1980. "A further survey of stamped tiles of the Classis Britannica," *Sussex Arch Coll* 118, 183-196.

Brodribb, G. 1983. *Romano-British Tiles and Brick: An analytical survey including a corpus of surviving examples* (London, unpublished PhD thesis).

Brodribb, G. 1987. *Roman Brick and Tile* (Stroud).

Brodribb, G. and H. Cleere 1988. "The Classis Britannica Bath-house at Beauport Park East Sussex," *Britannia* 19, 217-274.

Bryant, R. 1980. "Excavations at St Mary de Lode, Gloucester 1978-1979," *Glevensis* 14, 4-12.

Cagnat, R. 1976. *Cours D'Epigraphie Latine* (Rome).

Caruana, I. and L. Hird. "Roman Tile and Brick" in I. Caruana, *The Roman Forts at Carlisle: Excavations at Annetwell Street, 1973-84* (unpublished) 27-36.

Casey, P. J. and J. L. Davies with J. Evans 1993. *Excavations at Segontium (Caernarfon) Roman fort 1975 – 1979* (London).

Chauffin, J. 1956. "Les Tuiles Gallo-Romaines du Bas-Dauphine," *Gallia* 14, 81-88.

Clarke, C. P. with E.W Black 1998. "Roman Building Materials," in Clarke 1998, 112-113.

Clarke, C. P. 1998. *Excavations to the south of Chignall Roman Villa, Essex 1977-81* (East Anglian Archaeology No 83).

Collingwood, R. G. and R. P. Wright 1965. *Roman Inscriptions of Britain Volume I Inscriptions on Stone* (Oxford).

Cool, H. E. M and C. Philo 1998. *Roman Castleford: Excavations 1974-85 Volume 1: The Small finds* (Yorkshire Archaeology 4).

Corbier, M. 2000. *L'Annee Epigraphique* (Universite de Paris).

Cowell, R. W. and R. A. Philpott 2000. *Prehistoric, Romano-British and Medieval Settlement in Lowland North West England - Archaeological excavations along the A5300 road corridor in Merseyside* (National Museums & Galleries on Merseyside).

Crummy, N. 1984. "The Roman tiles," in P. Crummy, *Excavations at Lion Walk, Balkerne Lane and Middleborough, Colchester, Essex* (Colchester Archaeological Report 3) 298-301.

Cunliffe, B. 1969. *Roman Bath* (Soc of Antiquaries of London Report No XXIV).

Cunliffe, B. 1971. *Excavations at Fishbourne 1961-1969* (Society of Antiquaries of London Report No XXVI).

# BIBLIOGRAPHY


Darling, M. J. with D. Gurney 1993. *Caister-on-Sea Excavations by Charles Green 1951-55* (East Anglian Archaeology Report No 60).

Darvill, T. 1979. "A petrological study of LHS and TPF stamped tiles from the Cotswold Region," in McWhirr 1979a, 309-350.

Darvill, T. 1980. "Some Small Groups of Stamped Roman Ceramic Tiles from the Cotswolds," *Glevensis* 14, 49-57.

Darvill, T. and A. McWhirr 1984. "Brick and Tile Production in Roman Britain: Models of Economic Organisation," *World Archaeology* Vol 15 No 3 Ceramics, 239-261.

Davies, J. L. 1993. "Stamped tile" in Casey 1993, 229-331.

De la Bedoyere, G. 1988. "The decorated samian ware," in Brodribb and Cleere 1988, 244-245.

De la Bedoyere, G. 1999. *Companion to Roman Britain* (Stroud).

De la Bedoyere, G. 2001. *The Buildings of Roman Britain* (Stroud).

DeLaine, J. 1997. *Baths of Caracalla: a study in the design, construction, and economics of large-scale building projects in imperial Rome* (JRA Suppl. 25).

Delaine, J. 2000. "Building the Eternal City: the construction industry of imperial Rome," in J. Coulston and H. Dodge (edd.), *Ancient Rome: The Archaeology of the Eternal City Building the eternal city: the construction of imperial Rome* (Oxford) 119-141.

DeLaine, J. 2001. "Bricks and Mortar - Exploring the economics of building techniques at Rome and Ostia," in D. J. Mattingly and J. Salmon (edd.), *Economies beyond agriculture in the classical world* (London) 230-268.

Dickinson, B. M. and K. F. Hartley 1971. "The evidence of potters' stamps on samian ware and on mortaria for the trading connections of Roman York," in R. M. Butler (ed.), *Soldier and Civilian in Roman Yorkshire* (Leicester) 127-142.

Dore, J. N. and J. J. Wilkes 1999. "Excavations directed by J D Leach and J J Wilkes on the site of a Roman fortress at Carpow, Perthshire, 1964-79," *Proc Soc Antiq Scot* 129, 481-575.

Dore, J. N. 1999. "Roman Pottery," in Dore and Wilkes 1999, 537-552.

Down, A. 1979. *Chichester Excavations IV* (Chichester Excavations Committee).

Draper, J. and C. Chaplin 1982. *Dorchester Excavations Volume 1: Excavations at Wadham House 1968; Dorchester Prison 1970, 1975 and 1978; and Glyde Path Road 1966* (Dorset Natural History and Arch Soc Monograph 2).

Draper, J. 1985. *Excavations by Mr H.P.Cooper on the Roman Site at Hill Farm, Gestingthorpe, Essex* (East Anglian Archaeology No 25).

Drury, P. J. 1978. *Excavations at Little Waltham 1970-71* (Chelmsford Excavation Committee).

Drury, P. J. 1988. *The mansio and other sites in the south-eastern sector of Caesaromagus* (CBA Research Report 66).

Dunnett, R. 1971. "The Excavation of the Roman Theatre at Gosbecks," *Britannia* 2, 27-47.

Faulkner, N. 2000. *The decline and fall of Roman Britain* (Stroud).

Ferguson, C. 1895. no title, *Proc Soc Antiq Series 2 xv*, 261-262.

Finley, M. I. 1965. "Technical innovation and progress in the ancient world," *Economic History Review*, 2nd series, XVIII, 29-45.

Finley, M. I. 1985. *The Ancient Economy* (London).

Frere, S. S. 1971. "The Forum and Baths at Caistor by Norwich," *Britannia* 2, 1-26.

Frere, S. S. 1982. "The Bignor Villa," *Britannia* 13, 135-195.

Frere, S. S., M. W. C. Hassall and R. S. O. Tomlin 1990. "Roman Britain in 1989," *Britannia* 21, 303-378.

Frere, S. S. and R. S. O. Tomlin 1992. *The Roman Inscriptions of Britain Volume II Instrumentum Domesticum Fascicule 4* (Stroud).

Frere, S. S. and R. S. O. Tomlin 1993. *The Roman Inscriptions of Britain Volume II Instrumentum Domesticum Fascicule 5* (Stroud).

Frere, S. S. 1999. *Britannia. A History of Roman Britain* (London).

Friendship-Taylor, R. M. and D. E. Friendship-Taylor 2002. *Iron Age and Roman Piddington: The Faunal Remains: 1 1979-1997* (Upper Nene Archaeological Society).

Fulford, M. 1989. "The Economy of Roman Britain," in M. Todd (ed.), *Research on Roman Britain: 1960-1989* (Britannia Mono. No 11) 175-201.

Fulford, M. 1994. "The Monumental and the Mundane: A Common Epigraphic Tradition," *Britannia* 25, 315-318.

Fulford, M. and J. Timby 2000. *Late Iron Age and Roman Silchester; Excavations on the site of the forum-basilica 1977, 1980-86* (Britannia Mono. No 15).

Gillam, J. P., I. M. Jobey and D. A. Welsby 1993. *The Roman Bath-House at Bewcastle, Cumbria* (Cumberland and Westmorland Antiquarian and Archaeological Society No 7).

Going, C. J. 1992. "Economic "long waves" in the Roman period? A reconnaissance of the Romano-British Ceramic evidence," *Oxford Journal of Archaeology Vol 11*, 93-117.

Goodburn, R. 1978. "Roman Britain in 1977," *Britannia* 9, 403-472.

Green, T. K. 1979. "A Tilemakers' Workshop at Itchingfield - A Reappraisal," in McWhirr 1979a, 191-200.

Greene, K. 1977. "Legionary pottery, and the significance of Holt," in J. Dore and K. Greene (edd.), *Roman Pottery Studies in Britain and Beyond* (BAR Supp. Ser. 30) 113-127.

Greene, K. 2000. "Technological innovation and economic progress in the ancient world: M. I. Finley reconsidered," *Economic History Review*, LIII, 1, 29-59.

Grimes, W. F. 1930. *Holt, Denbighshire: the works depot of the twentieth legion at Castle Lyons* (Society of Cymmrodorion, London).

Guest, P. 2000. *Caerwent Forum-Basilica Project: The Assessment Report* (Unpublished).

Hassall, M. W. C. and R. S. O. Tomlin 1978. "Roman Britain in 1977: Inscriptions," *Britannia* 9, 473-485.

Hassall, M. 1979. "Military Tile-Stamps from Britain," in McWhirr 1979a, 261-266

Hassall, M. W. C. and R. S. O. Tomlin 1984. "Roman Britain in 1983: Inscriptions," *Britannia* 15, 333-356.

Hassall, M. W. C. and R. S. O. Tomlin 1989. "Roman Britain in 1988: Inscriptions," *Britannia* 20, 327-345.

Heighway, C. M. and A. J. Parker 1982. "The Roman Tilery at St Oswald's Priory Gloucester," *Britannia* 13, 25-77.

Heil, M. 2003. "On the date of the title Britannicus Maximus of Septimus Severus and his Sons," *Britannia* 34, 268-271.

Helen, T. 1975. *Organization of Roman Brick Production in the first and second centuries AD: An interpretation of Roman Brick Stamps* (Helsinki).

Hodgson, N. and P. T. Bidwell 2004. "Auxiliary barracks in a new light: recent discoveries on Hadrian's Wall," *Britannia* 35, 121-158.

Holbrook, N. 1987. "Trial excavations at Honeyditches and the nature of the Roman occupation at Seaton," *Devon Archaeological Society Proceedings 45*, 59-74.

Holbrook, N. and A. Thomas *The Roman and Early Anglo-Saxon Settlement at Wantage, Oxfordshire - Excavations at Mill Street 1993-4* (Cotswold Archaeological Trust, unpublished).

Holder, P. A. 1982. *The Roman Army in Britain* (London).

Holmes, N. M. McQ. 1999. "Coins," in Dore and Wilkes 1999, 528-535.

Holmes, N. 2003. *Excavation of Roman sites at Cramond, Edinburgh* (Soc Ant Scotland Mono. 23).

Jarrett, M. G. 1976. *Maryport, Cumbria: A Roman Fort and its Garrison* (Kendal).

Johnston, D 1969. "Sparsholt," *Current Archaeology* No 12, 14-18.

Johnston, D. E. and D. F. Williams 1985. "Roman Tiles," in Draper 1985, 80-83.

Johnston, D. E. 1994. *Roman Villas* (Princes Risborough).

Jones, B. and D. Mattingly 1990. *An Atlas of Roman Britain* (Oxford).

Jones, L. C. 2000. "Romano-British Ceramic Tile," in Cowell and Philpott 2000, 92-94.

Keevill, G. D. 1996. "The reconstruction of the Romano-British villa at Redlands Farm, Northamptonshire," in P. Johnson with I. Haynes (edd.), *Architecture in Roman Britain* (CBA Research Report 94) 44-55.

Keppie, L. 2004. "A Roman bath-house at Duntocher on the Antonine Wall," *Britannia* 35, 179-224.

Laurence, R. 1999. *The Roads of Roman Italy: Mobility and cultural change* (London).

Ling, R. 1992. "A Collapsed Building Façade at Carsington, Derbyshire," *Britannia* 23, 233-236.

Lucas, J. 1981. "Norfolk Street, Roman Villa, Leicester (SK 575043)," *Transactions of the Leicestershire Archaeological and Historical Society* LVI, 103-104.

Lucas, J. 1983. "An excavation at Narborough (SK531978) - Interim report," *Transactions of the Leicestershire Archaeological and Historical Society* LVIII, 75-77.

Lucas, R. N. 1993. *The Romano-British villa at Halstock, Dorset. Excavations 1967-1985* (Dorset Natural History and Arch Soc Mono. 13).

MacDonald, W. L. 1965. *The architecture of the Roman empire: 1 an introductory study* (Yale University Press).

Mann, J. C. and M. G. Jarrett 1967. "The Division of Britain," *JRS* 57, 61-64.

Mann, J. C. 1988. "The history of the Antonine Wall - a reappraisal," *Proc Soc Antiq Scot* 118, 131-137.

Martins, C. 2003. "Becoming consumers: looking beyong wealth as an explanation for villa variability," in G. Carr *et al.* (edd.), *TRAC 2002* (Oxford) 84-100.

Marvell, A. G. and H. S. Owen-John 1997. *Leucarum - Excavations at the Roman auxiliary fort of Loughor, West Glamorgan 1982-84 and 1987-88* (Britannia Mono. No 12).

Mason, D. J. P. 1990. "The use of earthenware tubes in Roman Vault construction: an example from Chester," *Britannia* 21, 215-222.

May, T. 1922. *The Roman Forts of Templebrough Near Rotherham* (County Borough of Rotherham).

McWhirr, A. and D. Viner 1978. "The Production and Distribution of Tiles in Roman Britain with particular reference to the Cirencester region," *Britannia* 9, 359-377

McWhirr, A. (ed.) 1979a. *Roman brick and tile: studies in manufacture, distribution and use in the Western Empire* (BAR international series 68).

McWhirr, A. 1979b. "Roman Tile-Kilns in Britain," in McWhirr 1979a, 97-190.

Mellor, J. E. and J. Lucas 1979. "The Roman Villa at Norfolk Street, Leicester," *Transactions of the Leicestershire Archaeological and Historical Society* LIV, 68-70.

Millett, M. 1990. "The Romanization of Britain: an essay in archaeological interpretation" (Cambridge).

Millett, M. 1990. "Romanization: historical issues and archaeological interpretation," in T. Blagg T and M. Millett M (edd.), *The Early Roman Empire in the West* (Oxford) 35-44.

Minoprio, A. 1932. *A restoration of the basilica of Constantine, Rome* (British School at Rome Vol XII).

Molenkamp, C. 1989. "The stamped tiles," in Philp 1989, 57-61.

Newstead, R. and J. Droop 1936. "Excavations in the Deanery Field and Abbey Green 1935," *Annals of Archaeology and Anthropology* XXIII, 3-50.

Newstead, R. 1937. "The Roman station, Prestatyn: first interim report," *Archaeologia Cambriensis*, 208-232.

Oswald, A. 1937. "A Roman Fortified Villa at Norton Disney, Lincs," *Antiquaries Journal* 17, 138-178.

Parfitt, K. 1980. "A Probable Roman Villa on the Sandwich By-Pass," *Kent Archaeological Review*, 233-248.

Partridge, C. 1977. "Excavations and Fieldwork at Braughing, 1968-73," *Hertfordshire Archaeology* 5, 22-108.

Peacock, D. P. S. 1977. "Bricks and Tiles of the Classis Britannica: Petrology and Origin," *Britannia* 8, 235-248.

Peacock, D. P. S. 1979. "An Ethnoarchaeological Approach to the Study of Roman Bricks and Tiles," in McWhirr 1979a, 5-10.

Peacock, D. P. S. 1982. *Pottery in the Roman world: an ethnoarchaeological approach* (London).

Perring, D. 2002. *The Roman House in Britain* (London).

Philp, B. 1981. *The Excavation of the Roman Forts of the Classis Britannica at Dover 1970-77* (Kent).

Philp, B. 1989. *The Roman House with Bacchic Murals at Dover* (Kent).

Redknap, M. forthcoming. "Ceramic Building Material from the Villa Rustica 'Am Silberberg' in Bad Neuenahr-Ahrweiler," *Berichte zur Archaologie an Mittlerhein und Mosel* (Trier).

Reed, N. 1975. "The Scottish Campaigns of Septimus Severus," *Proc Soc Antiq Scot* 107, 92-102.

Reynolds, J. 1989. "The Regulations of Diocletian," in C. Roueche, *Aphrodisias in Late Antiquity* (JRA Mono. No 5) 265-318.

Richmond, I. A. 1925. *Huddersfield in Roman Times* (Tolson Memorial Museum Publications, Huddersfield).

Rodwell, W. J. and K. A. Rodwell 1985. *Rivenhall: investigations of a villa, church, and village, 1950-1977* (CBA Research Report 55).

Rook, A. 1977. "The Building Materials," in Partridge 1977, 55-58.

Rook, A. 1979. "Tiled Roofs," in McWhirr 1979a, 295-302.

Rook, A. 1992. *Roman Baths in Britain* (Princes Risborough).

Rosenthal, E. 1949. *Pottery and Ceramics - from common brick to fine china* (London).

Rowland, I. D. and T. N. Howe 1999. *Vitruvius: Ten Books on Architecture* (Cambridge).

Rowley, T. and L. Brown 1981. "Excavations at Beech House Hotel, Dorchester-on-Thames 1972," *Oxoniensia* XLVI, 1-55.

Rudling, D. R. 1986. "The Excavation of a Roman Tilery on Great Cansiron Farm, Hartfield, East Sussex," *Britannia* 17, 191-230.

Salway, P. 1981. *Roman Britain* (Oxford).

Schonberger, H. 1969. "The Roman Frontier in Germany: An Archaeological Survey," *JRS* 59, 144-197.

Shirley, E. A. M. 2000. *The Construction of the Roman Legionary Fortress at Inchtuthil* (BAR British Series 298).

Shotter, D. C. A. 1988. "The Roman Garrisons at Lancaster," in G. D. B. Jones and D. C. A. Shotter, *Roman Lancaster: Rescue Archaeology in an Historic City 1970-75* (Brigantia Monograph No 1) 212-219.

Shotter, D. C. A. 1998. "Roman coins from Holt," in Ward 1998, 68-72.

Soffe, G., J. Nicholls and G. Moore 1989. "The Roman Tilery and Aisled Building at Crookhorn, Hants, Excavations 1974-5," *Proceedings of the Hampshire Field Club and Archaeological Society* 45, 43-112.

Sokolnikoff, I. S. and E. S. Sokolnikoff 1941. *Higher Mathematics for Engineers and Physicists* (London).

St Joseph, J. K. 1973. "Air Reconnaissance in Britain, 1969-1972," JRS 63, 214-246

Stephens, G. R. 1989. "The Tile Graffito from Holt," *Bulletin Board of Celtic Studies* Vol xxxvi, 224-6.

Stephens, G. R. and M. G. Jarrett 1985. "Two Altars of Cohors IV Gallorum from Castlesteads," *Trans Cumb & West Ant Arch Soc* LXXXV, 77-80.

Strickland, T. J. 1997. "What kind of community existed at Chester during the hiatus of the 2nd c.?" in A. Goldsworthy A and I. Haynes (edd.), *The Roman army as a community* (JRA Suppl. 34) 106-109.

Swan, V. G. 1992. "Legio VI and its Men: African Legionaries in Britain," *Journal of Roman Pottery Studies* 5, 1-33.

Swan, V. G. and R. A. Philpott 2000. "Legio XX VV and Tile Production at Tarbock, Merseyside," *Britannia* 31, 55-67.

Swan, V. G. and R. M. McBride 2002. "A Rhineland Potter of the Legionary Fortress of York," in M. Aldhouse-Green and P. Webster (edd.), *Artefacts and Archaeology: Aspects of the Celtic and Roman World* (University of Wales Press, Cardiff) 190-215.

Taylor, M. V. and R. G. Collingwood 1922. "Roman Britain in 1923," *JRS* 12, 240-287.

Taylor, M. V. and R. G. Collingwood 1926. "Roman Britain in 1926," *JRS* 16, 216-244.

Timby, J. 2000. "The Ceramic Tile," in Fulford and Timby 2000, 116-122.

Tinson, B. 2002. "*'Fabric analysis of Roman tile fragments from Frilford and Silchester to explore production and distribution of architectural ceramics, and the underlying infrastructure of the Oxford region, with reference to Peacock's model of the organisation of production*" (unpublished).

Tomalin, D. J. 1987. *Roman Wight: a guide catalogue* (IOW County Council).

Tomlin, R. S. O. 1998. "Roman Manuscripts from Carlisle," *Britannia* 29, pp31-84.

Toynbee, J. M. C. 1964. *Art in Britain under the Romans* (Oxford).

Turner, D. 2003. *Brick and Tile - FBE02 and FBE.95-99 Brick and Tile* (Unpublished Fishbourne reports).

Villy, F. 1911. *The Roman Remains from Slack* (Bankfield Museum Notes, Halifax).

von Elbe, J. 1977. *Roman Germany: A Guide to Sites and Museums* (Mainz).

Wacher, J. 1995. *The Towns of Roman Britain* (London).

Ward, C. 1999. *Iron Age and Roman Piddington - The Roman Ceramic & Stone Building Materials 1979-1998* (Upper Nene Archaeological Society).

Ward, M. 1997. "A collection of samian from the legionary works-depot at Holt," in J. Bird (ed.), *Form and Fabric: Studies in Rome's Material Past in Honour of BR Hartley* (Oxbow Mono. 80) 133-143.

Ward, M. 1998. "Some finds from the Roman Works-Depot at Holt," *Studia Celtica* XXXII, 43-84.

Ward Perkins, J. B. and J. M. C. Toynbee 1949. "The Hunting Baths of Lepcis Magna," *Archaeologia* 93, 165-195.

Webster, G. 1979. "Tiles as a Structural Component in Buildings," in McWhirr 1979a, 285-294.

Wheeler, R. E. M. 1943. *Maiden Castle, Dorset* (Society of Antiquaries of London).

Williams, D. F. 1987. "A note on the petrology of the stamped tile of the Second Augustan Legion from Seaton," in Holbrook 1987, 72-73.

Williams, W. with J. Bowers, H. Davies and B. Philp 1981. "The stamped tiles," in Philp 1981, 123-142.

Wilson, D. R., R. P. Wright and M. W. C. Hassall 1973. "Roman Britain in 1972," *Britannia* 4, 270-337.

Wilson, D. R. 2004. "The North Leigh Roman villa: its plan reviewed," *Britannia* 35, 77-114.

Woodfield, C. and C. Johnson 1989. "A Roman Site at Stanton Low on the Great Ouse, Buckinghamshire. Excavated by Margaret Jones, 1957-8," *Archaeological Journal* 146, 135-278.

Woodward, P. J., S. M. Davies and A. H. Graham 1993. *Excavations at the old methodist chapel and Greyhound Yard, Dorchester, 1981-1984* (Dorset Natural History and Arch Soc Mono. 12).

Wright, R. P. 1965. "Roman Britain in 1964: Inscriptions," *JRS* 45, 220-228.

Wright, R. P. 1974. "Carpow and Caracalla," *Britannia* 5, 289-292.

Wright, R. P., M. W. C. Hassall and R. S. O. Tomlin 1976. "Roman Britain in 1975," *Britannia* 7, 378-392.

Zienkiewicz, J. D. 1993. "Excavations in the "Scamnum Tribunorum" at Caerleon: The Legionary Museum Site 1983-5," *Britannia* 24, 27-140.

www.ingramcontent.com/pod-product-compliance
Lightning Source LLC
LaVergne TN
LVHW071212210926
848482LV00015B/788